W9-CUK-465

Biological Energy Transduction: The Uroboros

From Eleazar's "Uraltes Chymisches Werk," 1760

Biological Energy Transduction: The Uroboros

RONALD F. FOX
School of Physics
Georgia Institute of Technology

A Wiley-Interscience Publication
JOHN WILEY & SONS
New York Chichester Brisbane Toronto Singapore

Library of Congress Cataloging in Publication Data:

Fox, Ronald F. (Ronald Forrest), 1943–
Biological energy transduction.

"A Wiley-Interscience publication."
Includes index.
1. Energy metabolism. 2. Energy transfer.
3. Nucleic acids–Synthesis. 4. Protein biosynthesis. I. Title.

QP171.F68 574.1'33 81-11556
ISBN 0-471-09026-3 AACR2

Printed in the United States of America

10 9 8 7 6 5 4 3 2 1

Preface

Biological Energy Transduction has a dual purpose. Physical scientists will find it useful as an introduction to the major achievements in molecular and cellular biology during the past 30 years. Biological scientists will find the emphasis on energy to be a unifying perspective that will provide them with a novel conceptualization of what they already have learned.

Although the book will be useful to researchers already active in their own discipline, it is designed for students at the advanced undergraduate or early graduate level of study. The first part of this three-part book contains material most familiar to the physical chemist. This material is reviewed primarily in order to provide a working vocabulary for the rest of the book. It is not intended as a substitute for more thorough exposure to physical chemistry, thermodynamics, and electrochemistry. The second part is an up-to-date account of what is known about the genetic code and protein biosynthesis on the one hand, and chemiosmosis on the other hand. The intimate connection between these two revolutionary advances in our understanding of molecular and cellular processes is made clear through the emphasis on energy maintained throughout the presentation. The third part of the book presents the author's speculations regarding the origins and evolution of both protein biosynthesis and energy transduction. These problems require a truly interdisciplinary approach, which, it is hoped, has been successfully developed in the first two parts of the book. It is the right combination of physical and biological reasoning that the book is designed to provide for such an approach.

The book could be the basis for a biophysics course emphasizing energy; or it could be used as a supplement in physical chemistry *and* biochemistry courses, at either undergraduate or graduate levels.

The factual content of the first two parts of the book is based on the work of thousands of researchers. The author's original contributions to this material are through the conceptual emphasis built up in the presentation. In many ways the conceptual framework is as important as the many individual factual contributions and can serve to determine future research directions. The author has worked on this conceptual framework for about 15 years, during which time he has had opportunities to test his thinking against the viewpoints of many other scientists. Significant contributions to the author's thinking have been made in this way by Max Delbrück, Fritz Lipmann, Art Weber, Joel Keizer, Larry Gold, Sidney Fox, Roger Wartell and Albert Lehninger; many others have contributed no less importantly.

The author is grateful for permission to use figures which were provided by J. Watson, A. Lehninger, J. Adler, S. Pestka and G. Stöffler. Permission to use the frontispiece was provided by the Yale University, Beinecke Rare Book library.

RONALD F. FOX

Atlanta, Georgia
June 1981

Contents

Introduction

Two great revolutions in understanding have taken place in the life sciences during the last three decades. One was in the field of gene-directed protein biosynthesis, and the other was in the field of energy metabolism. While pursued by two distinct groups of researchers, these two disciplines are deeply interconnected conceptually. Energy metabolism requires many proteins, and protein biosynthesis uses considerable metabolic energy. The virtual simultaneity of these revolutions has produced a new, comprehensive overview of the life sciences at a detailed, molecular level of description. It is the purpose of this book to report these accomplishments.

The molecular-mechanistic model of living systems, provided by biochemistry and molecular biology, is one of the outstanding achievements of the human mind. It has explained, at a detailed molecular level of description, the mechanism of heredity *and* the mechanism of biological energy transduction. To do this, much had to be learned about biological macromolecules and their properties. It is now known that all such molecules are polymers, made up from monomers. The sequence of the monomers and the three-dimensional conformation of the polymers is now known for a great many proteins and for portions of many polynucleotides.

The elucidation of the mechanism of protein biosynthesis required the determination of the structures and properties of DNA, RNA, polynucleotide polymerases, tRNAs, ribosomes, amino acyl tRNA synthetases, and many other protein components. With ribosomes, it was necessary to also learn about the self-assembly of polynucleotide–protein complexes, which require the coordinated aggregation of 50 proteins as well as 3 relatively large RNA molecules. Self-assembly, on a lesser scale, is also now understood for many enzyme complexes. The slightly more recent elucidation of the detailed mechanism of energy transduction has involved the study of the structure of membranes and their self-assembly from lipid constituents. The incorporation, again by self-assembly, of many functional proteins into membranes has also been investigated vigorously, so that now there is a good picture of how both electrons and protons are transported during energy-transducing processes in bacterial, chloroplast, and mitochondrial membranes. The molecular-mechanistic model of living systems that has been built during the last two decades has resulted in large measure from the determination of the structures of a large variety of macromolecular components and from the explication of their properties.

The accomplishments described have depended on several exploratory techniques derived from physics, including x-ray crystallography, chromatography, centrifugation, and nuclear magnetic resonance. Moreover, the explanation of the existence of molecules is one of the major achievements of the quantum theory. This is discussed in Chapter 1 according to the perspective enunciated by Erwin Schrödinger in his book *What Is Life?* Schrödinger also discussed the following question: Would the study of living systems lead to new physics? In 1945, when his book originally appeared, not even the structure of DNA was yet known, not to mention all of the other cellular components listed above. It was therefore very difficult to discuss the physics of life at a concrete detailed molecular level. Now, however, all of this has been radically changed. In this book an updated, and therefore much more detailed, account of the interface between physics and biology is presented. As it turns out, the central theme of such an endeavor is *biological energy transduction*.

Central to the study of biological energy transduction is the molecule adenosine triphosphate (ATP). It is shown in Chapter 10 that the biosynthesis of both polynucleotides and proteins requires a considerable input of Gibbs free energy, which is ultimately supplied by ATP. A detailed account of how ATP is manufactured by cellular energy transductions is presented in Chapters 8, 11, and 12. The synthesis of ATP has been studied for many years, and its production during energy transductions known as electron transport has been a mystery for almost 35 years. This mystery was finally solved with the advent of the theory of chemiosmosis, which placed the membrane at a fundamental level of importance in biological energy transductions. It was finally realized that the answer lay in maintaining an intact membrane because only in intact membranes could one observe the electrical-chemical potential that links the oxidation-reduction free energy with the phosphorylation free energy of ATP. Until then, the membrane was always disrupted and fractionated in a search for the mysterious coupling agents—which were never found. The mechanism of membrane function involves the detailed account of how electrons and protons move in and through membranes. The reduction of biology to physics could hardly go any deeper than the description of how electrons and protons carry out essential biological energy transductions. This is carried one step further in the description of photosynthetic energy transductions, in which the quantum of light, the photon, also enters the picture.

A thorough account of the importance, production, and utilization of ATP requires a careful account of thermodynamics as applied to biology. This is presented in Chapters 2, 3, and 4. There it is explained why it is necessary to distinguish between entropy and Gibbs free energy, and why it is the latter quantity that is required for a biological application of the second law of thermodynamics. Nevertheless, when modified in this way, the second law proves to be the fundamental impetus for many biological functions at the molecular level, including self-assembly (Chapter 6), energy transductions (Chapter 5), and intermediary energy metabolism (Chapter 8). The concept of "energy flow ordering" is discussed in Chapter 3, and it emphasizes further the relationship between Gibbs free energy and entropy. In Chapter 4, these concepts are applied to reaction chemistry, particularly oxidation–reduction chemistry. This application includes a discussion of membrane potentials, which is basic for an understanding of chemiosmosis, as presented in Chapter 12. Thus the account of ATP given in this book spans its thermodynamic basis as well as the detailed molecular requirements for its synthesis and utilization.

Because so many proteins are involved in all aspects of energy transduction and because their biosynthesis is one of the major energy-utilizing processes in cells, Chapter 10 is devoted to a detailed, up-to-date, account of gene-directed protein biosynthesis. Some of the most recent results in the structure and function studies of ribosomes, tRNAs, and tRNA synthetases are presented. In addition, the "mystery of the second code" is presented and discussed. This mystery involves the recognition of an amino acid and its cognate tRNA by the enzyme tRNA synthetase. Such recognition is absolutely essential if the "base-pairing" mechanism for tRNA–mRNA recognition is to have meaning. The mechanism of this recognition process is still unknown. In this book, it is suggested that the solution of this mystery may require an evolutionary perspective.

The description of energy transduction and polymer biosynthesis given in this book naturally leads to the appreciation of biological feedback processes. In Chapters 5 and 10 the concept of the "uroboros" is developed. This concept is discussed in concert with F. Jacob's concept of the "integron," which is presented in Chapters 5 and 7. The integron refers to objects that emerge from component objects of a simpler sort, which associate to form the more complicated objects. Four levels of such objects are discerned by considering the assembly of the elements hydrogen, carbon, nitrogen, oxygen, and sulfur into amino acids, the assembly of amino acids into proteins, and the assembly of proteins into membrane-bound protein complexes. At a certain level of complexity, an integron becomes an uroboros, as discussed in Chapters 10, 13, 14, and 15. The origin and evolution of an uroboros are the subjects of Part 3 of this book.

The origin and the evolution of energy metabolism and of polymer biosynthesis on the nascent earth pose difficult questions. Nevertheless, the level of conceptualization now provided by the two revolutions in understanding discussed above permits speculation that is tightly constrained by the requirements of physics and chemistry. In Part 3 of this book the origin of energy metabolism, and especially chemiosmosis, is discussed, in addition to the origin of polymer biosynthesis. The latter issue has been examined by other researchers on numerous occasions, but in this book, for the first time, there is a novel presentation of these ideas integrated together with the concomitant problem of the origin of energy metabolism. When presented together, these two problems synergistically facilitate each other's solution and, in particular, lead to a potential solution of the mystery of the second code. The reader is warned, however, that while Parts 1 and 2 of this book *report* established scientific fact, Part 3 is almost entirely speculative, without experimental foundation. Its conceptual framework, however, is entirely in harmony with that established in Parts 1 and 2.

This book is intended for biologically and physically inclined readers alike. This means that some readers will find Part 1 difficult, while others will find Part 2 difficult. Both classes of readers are urged to attempt to work through the chapters of the book in sequence, nevertheless, because its conceptual structure is built up in that way. Each chapter concludes with references in which the reader can find more relevant material for a deeper understanding. While Part 3 could be read without a proper study of Parts 1 and 2, only a very superficial understanding would be achieved.

It is the author's hope that much of what has recently been learned about biological energy transductions will provide the conceptual foundation for a more rational approach to energy management at the level of human enterprise. This issue is broached in the epilogue.

PART

1

Biochemical Thermodynamics

Part 1 of this book provides the background material necessary for an understanding of the remainder of the book. Chapter 1 begins with a paraphrase of Schrödinger's treatment of the quantum mechanical basis for the existence of molecules as presented by him in *What Is Life?* The chapter concludes with a long list of particular molecular species that figure prominently throughout the rest of the book. This list also exhibits the hierarchical structural classification of biological molecules. Chapter 2 provides an introduction to molecular statistical mechanics and its relationship to thermodynamics. Gibbs free energy and entropy are explained. Chapter 3 carries these thermodynamic ideas a step further, through the introduction of reactions and temporal behavior. It is in this chapter that crucial distinctions involving entropy and free energy are explained and, in particular, the second law of thermodynamics is discussed, with special emphasis on its biological applications.

In Chapter 4 numerous examples and special cases, each of which is of importance in later chapters, are presented. These examples include a novel treatment of oxidation–reduction (redox) reactions and an explanation of non-standard-state corrections. Chapter 5 describes a variety of biologically relevant energy transductions. Several of the examples used in this chapter acquire additional significance in later chapters. Chapter 6 is about the remarkable phenomenon of self-assembly. It contains a detailed entropy analysis of this process and several representative examples. Part 1 concludes with Chapter 7, which deals with membranes and membrane potentials. In this chapter all of the preliminaries for the subsequent developments in the book are concluded.

CHAPTER

1

Molecular Structure

Molecules are incredibly small and remarkably sturdy structures. The water molecule, H_2O, is just a few angstroms across, and hemoglobin, a tetramer with a molecular weight (MW) of 64,000, is 65 Å in diameter. At physiological temperatures molecules undergo rapid thermal motions. A water molecule has an average thermal speed of 37,000 cm/sec, or about 1300 km/hr, and makes well over 10^{12} collisions per second. The hemoglobin molecule collides just as frequently, but at a slower speed, about 22 km/hr.

Quantum mechanics has provided the explanation for the stability of molecular structures experiencing such thermal motions. The quantities of energy required to break apart the chemical bonds that hold together the constitutent atoms in a molecule must satisfy two properties. The first property is that quantum mechanics requires discrete amounts of energy for breaking a bond. Too small an amount of energy will have no effect in breaking a bond. Only if the applied energy is sufficient to span the energy gap between a bonded and an unbonded configuration will the bond have a chance to be broken. The second property is that the required amount of energy for such bond breaking is usually much larger than the average thermal energy supplied by collisions. For water molecules, the bond energy is roughly 730×10^{-14} erg per bond, whereas the average thermal energy imparted per collision is only 4×10^{-14} erg, nearly 200 times smaller. Quantum mechanics requires that the energy be imparted in a single collision if a transition is to occur; the fact that myriads of collisions are occurring, each one incapable of supplying the required energy at once, but together more than capable of doing so, does not circumvent the quantum mechanical constraints.

Occasionally, the thermal motion gives rise to a highly energetic collision with enough energy to break a bond. The probability of this occurring is governed by Boltzmann's exponential factor and can be expressed, according to Schrödinger [1],

$$t = \tau \exp\left(\frac{E}{k_B T}\right) \tag{1.1}$$

where t is the "time of expectation" for a collision involving the energy E, k_B is Boltzmann's constant ($k_B = 1.38 \times 10^{-16}$ erg/°K), and T is the temperature, usually taken to be about 300°K for physiological reactions. The quantity τ is the reciprocal of the collision frequency, a quantity that depends on the specific system being considered and is on the order of 10^{-13} to 10^{-14} sec. Looking again at a water molecule and picking $\tau = 10^{-13}$ sec, it follows that, for $T = 300°$K, the time of expectation for a collision energetic enough to supply the O—H bond energy, 730×10^{-14} erg, is 10^{65} sec, roughly 3×10^{57} years!

Not all chemical bonds of biological importance are the covalent bonds of the type just considered for water. There are also several types of much weaker interactions, such as van der Waals bonds, hydrogen bonds, and hydrophobic associations. Their energy content is in the range from a few k_BT up to about 10 times k_BT for $T \sim 300°$K. Consequently, these "bonds" are continually breaking and reforming as a result of thermal motions. Even for $E = 10\, k_BT_{300}$, t is roughly 2 nsec, in the formula above, an extraordinarily short time of expectation! It is the exponential character of Boltzmann's formula that provides this dramatic range of time scales from nanoseconds to 10^{57} years.

Quantum mechanics introduces another fundamental property of molecules that concerns their intrinsic indistinguishability when identical molecules are assembled together. Because the quantum mechanical energy levels available to a molecule are discrete, an assembly of a macroscopic amount of molecules [1 mole contains Avogadro's number = 6×10^{23} (6.02486) of molecules] will necessarily contain myriads of molecules with identical quantum numbers. With myriads of billiard balls, it is possible to number each one so that they become distinguishable, but with molecules, quantum mechanics dictates that this is impossible. The molecules, of the same type, with identical quantum numbers, are *indistinguishable*. This property will prove to have very far reaching consequences for the description of molecular events at the thermodynamic level.

The structure of biological molecules has been determined for an incredibly large number of molecules. That a rational basis exists for this great diversity of structures only became clear after many years of demanding work on structure determination. A key insight gleaned during this process was that giant biological molecules, such as proteins and polynucleotides, are polymers, built up from a relatively small number of different kinds of monomers that are themselves small, ordinary-size molecules. The monomers, in turn, can be understood as combinations of relatively few very small molecules like H_2O and CO_2. Before the pertinent molecular structures are presented, it will be useful to begin with the periodic table of the elements.

The first four periods of the table of the elements contain nearly all of the elements of biological significance. Some of the elements in the first thirty in the table only have biological significance in some particular species of organism [2]. The structures explored in this book will involve only the elements listed in the first four periods of the table, and these will prove sufficient for a discourse on fundamental principles.

In addition to these basic elements and their associated ions (Na^+, Mg^{2+}, Cl^-, K^+, Ca^{2+}, Mn^{2+}, Mn^{3+}, Mn^{4+}, Fe^{2+}, Fe^{3+}, Co^{2+}, Co^{3+}, Cu^+, Cu^{2+}, and Zn^{2+}), three of the most fundamental particles of physics will appear repeatedly throughout the book: the electron, e^-; the proton, H^+; and the photon, $h\nu$. The physical values associated with these three particles are given below.

Electron, e^-	Proton, H^+	Photon, $h\nu$
Mass = 0.911×10^{-27} g	Mass = 1.672×10^{-24} g	Energy = $h\nu$
Charge = -4.80×10^{-10} esu	Charge = $+4.8 \times 10^{-10}$ esu	ν = frequency (cycles/sec)
1 esu = $(3 \times 10^9)^{-1}$ C	*Units*	h = Planck's constant
	$1 \dfrac{\text{esu}^2}{\text{cm}} = 1$ erg	$= 6.62 \times 10^{-27}$ erg-s
	1 MW = mass of proton	

Molecular ions containing either H^+ or O^- are common.

Small molecules [3, 4], some of which are of importance in discussions of geochemistry on the early earth, are shown in Table 1.1.

Composite molecules and monomers, which can be visualized as being built up from the small molecues, are shown in Table 1.2. The last structure shown in this table is that of amino acids, which contain various residues R. The structures and molecular weights of these residues are given in Table 1.3.

For sugars, $(CH_2O)_n$, several values of n are possible and several isomers for each n. The structures and molecular weights of several sugars are shown in Table 1.4.

Purines and pyrimidines are illustrated in Table 1.5. Other important molecules are carbamyl phosphate, porphyrin, and quinones, shown in (1.2), (1.3), and (1.4), respectively.

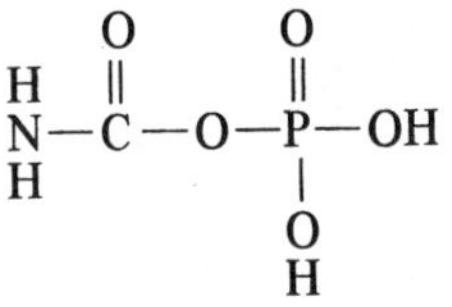

```
        O     O
H       ||    ||
N—C—O—P—OH
H             |
              O
              H
```
(1.2)

Carbamyl phosphate
MW = 141

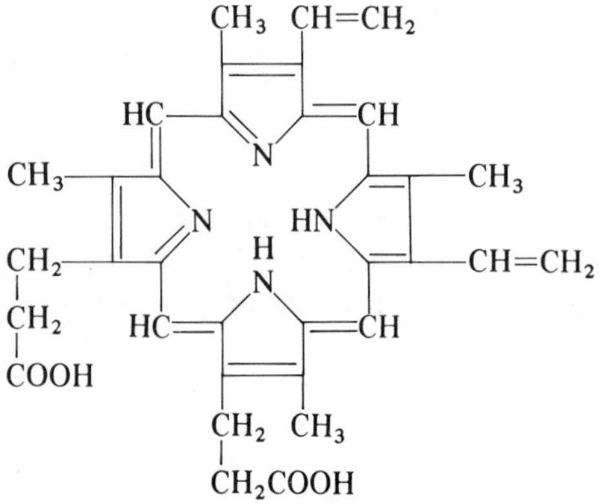

Protoporphyrin IX
MW = 538

(1.3)

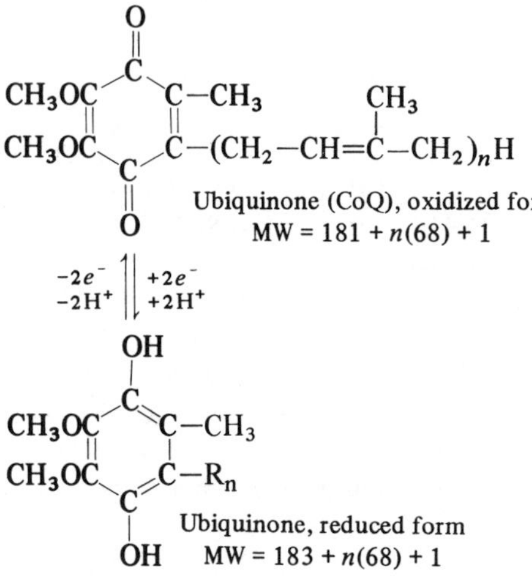

Ubiquinone (CoQ), oxidized form
MW = 181 + n(68) + 1

$$\underset{-2H^+}{-2e^-} \rightleftharpoons \overset{+2e^-}{+2H^+}$$

Ubiquinone, reduced form
MW = 183 + n(68) + 1

CoQ_6 (n = 6)
CoQ_{10} (n = 10)

(1.4)

Oligomers are small polymers made up from a few monomers by dehydration condensation linkages, the linkages that couple two molecules together while eliminating a molecule of H_2O. For example, two amino acids may combine through a peptide bond to form a dipeptide, as shown in (1.5).

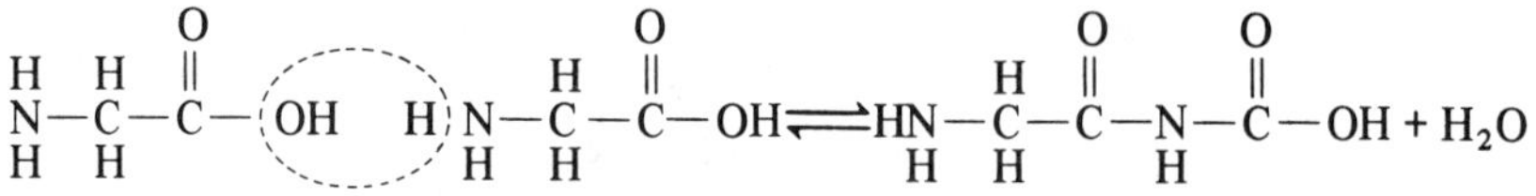

$$H_2N-CH_2-C(=O)-OH + H-NH-CH_2-C(=O)-OH \rightleftharpoons HN(H)-CH_2-C(=O)-NH-CH_2-C(=O)-OH + H_2O$$

Glycine Glycine Glycylglycine

(1.5)

Table 1.1 Structure and Molecular Weight of Small Molecules

Molecule	Structure	Molecular Weight
Molecular hydrogen, H_2	H—H	2
Molecular nitrogen, N_2	N≡N	28
Molecular oxygen, O_2	O=O	32
Carbon (graphite), $(C)_n$	$\begin{array}{c} \vert \quad \vert \quad \vert \\ —C—C—C— \\ \vert \quad \vert \quad \vert \\ —C—C— \\ \vert \quad \vert \end{array}$	$(12)_n$
Water, H_2O	H–O–H (bent; O bonded to two H)	18
Carbon dioxide, CO_2	O=C=O	44
Methane, CH_4	$\begin{array}{c} H \\ \vert \\ H—C—H \\ \vert \\ H \end{array}$	16
Ammonia, NH_3	$\begin{array}{c} H \diagdown N \diagup H \\ \vert \\ H \end{array}$	17
Hydrogen cyanide, HCN	H—C≡N	27
Sulfuric acid, H_2SO_4	$\begin{array}{c} O \\ \Vert \\ H—O—S—O—H \\ \Vert \\ O \end{array}$	98
Phosphoric acid, H_3PO_4	$\begin{array}{c} O \\ \Vert \\ H—O—P—O—H \\ \vert \\ O \\ \vert \\ H \end{array}$	98
Formaldehyde, H_2CO	$\begin{array}{c} O \\ \Vert \\ C \\ H \diagup \quad \diagdown H \end{array}$	30

Table 1.2 Structure and Molecular Weight of Composite Molecules and Monomers

Molecule	Structure	Molecular Weight
Acetic acid, CH_3COOH	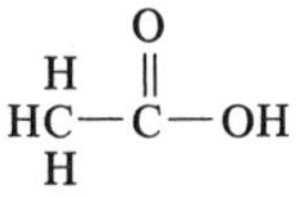 O H ‖ HC—C—OH H	60
Glycolaldehyde, $C_2H_4O_2$	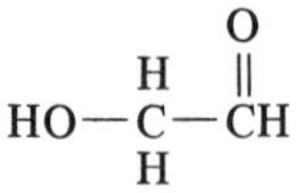 O H ‖ HO—C—CH H	60
Glycine, $C_2H_5O_2N$	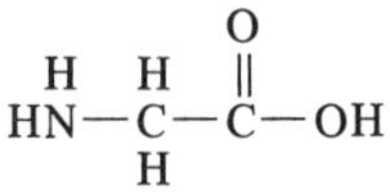 O H H ‖ HN—C—C—OH H	75
Amino acids,[a] $C_2H_4O_2NR$	R O H | ‖ HN—C—C—OH H	74 + (R)

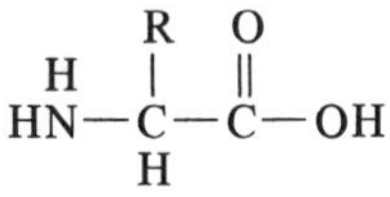

[a]See Table 1.3 for the structure and molecular weight of the residue R in amino acids.

Table 1.3 Structure and Molecular Weight of the Residue R in Amino Acids

	Residue R	
Amino Acid	Structure	Molecular Weight
Glycine	—H	1
Alanine	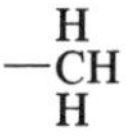 H —CH H	15
Valine	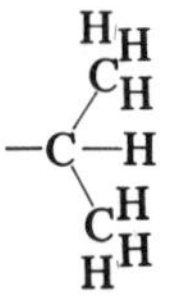 HH CH / —C—H \\ CH HH	43
Leucine	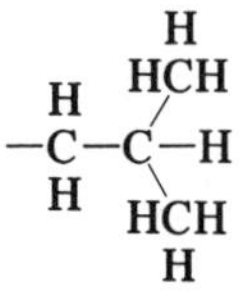 H HCH H / —C—C—H H \\ HCH H	57

Table 1.3 (*Continued*)

Amino Acid	Residue R: Structure	Residue R: Molecular Weight
Isoleucine	H H H C—CH —C H H HCH H	57
Lysine	H H H H H —C—C—C—C—NH⁺ H H H H H	73
Arginine	H H H H H —C—C—C—N N⁺ H H H C H N H H	101
Methionine	H H H —C—C—S—CH H H H	75
Cysteine	H —C—S—H H	47
Serine	H —C—OH H	31
Threonine	H O ‖ H —C—CH H H	45
Asparagine	H H —C—C—NH H ‖ O	58
Glutamine	H H H —C—C—C—NH H H ‖ O	72
Aspartic acid	O H ‖ —C—C—O⁻ H	58

Table 1.3 (*Continued*)

Amino Acid	Residue R: Structure	Residue R: Molecular Weight
Glutamic acid	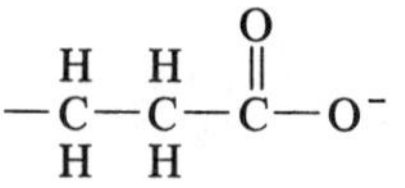 O H H ‖ —C—C—C—O^- H H	72
Tyrosine	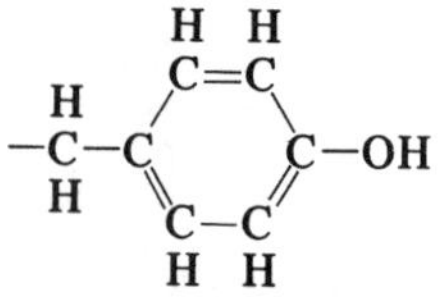 H H C=C H —C—C C—OH H C—C H H	107
Tryptophan	H H C —C—C——C C H H ‖ ‖ | C C C H N C H H H	130
Histidine	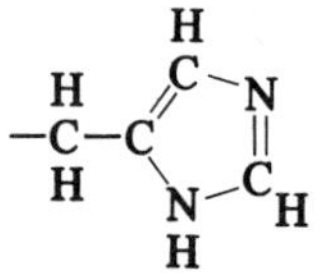 H H C N —C—C ‖ H N—C H H	81
Phenylalanine	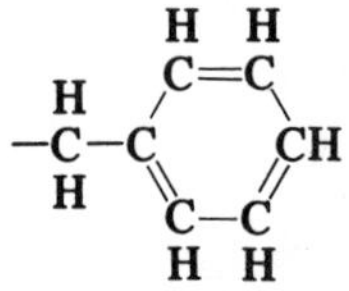 H H C=C H —C—C CH H C—C H H	91

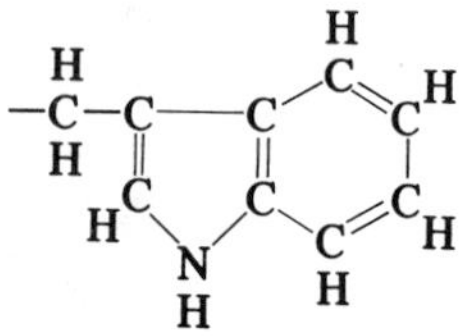

Table 1.4 Structure and Molecular Weight of Sugars, $(CH_2O)_n$

n	Sugar	Structure	Molecular Weight
3	Glyceraldehyde	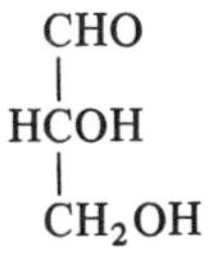 CHO–HCOH–CH_2OH	90
4	Erythrose	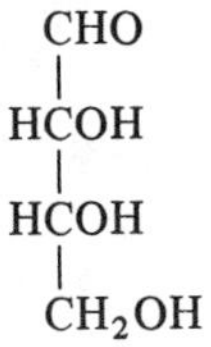 CHO–HCOH–HCOH–CH_2OH	120
5	Ribose	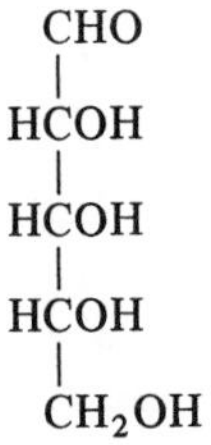 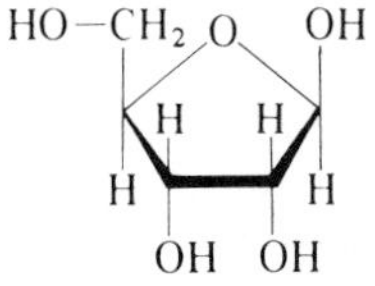 CHO–HCOH–HCOH–HCOH–CH_2OH (D-Ribose); HO–CH_2, O, OH, H, H, H, H, OH, OH (Ribofuranose)	150
5	Ribulose	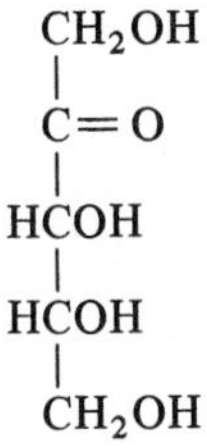 CH_2OH–C=O–HCOH–HCOH–CH_2OH	150
5	Xylulose	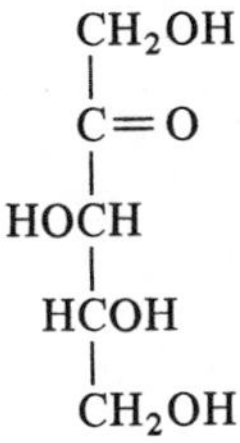 CH_2OH–C=O–HOCH–HCOH–CH_2OH	150

Table 1.4 (*Continued*)

n	Sugar	Structure	Molecular Weight
6	Glucose	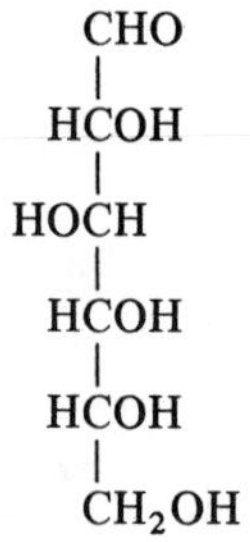 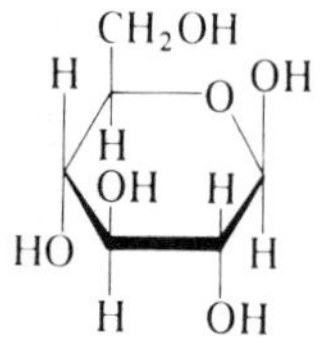 CHO–HCOH–HOCH–HCOH–HCOH–CH_2OH D-Glucose [ring structure: CH_2OH, H, O, OH, H, OH, H, HO, H, H, OH] β-D-Glucopyranose	180
6	Fructose	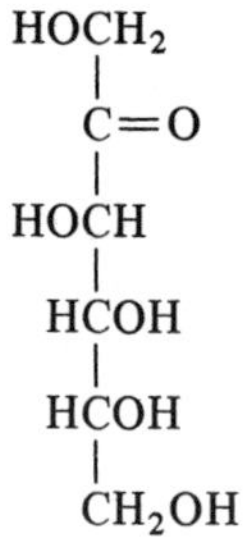 $HOCH_2$–C=O–HOCH–HCOH–HCOH–CH_2OH	180
7	Sedoheptulose	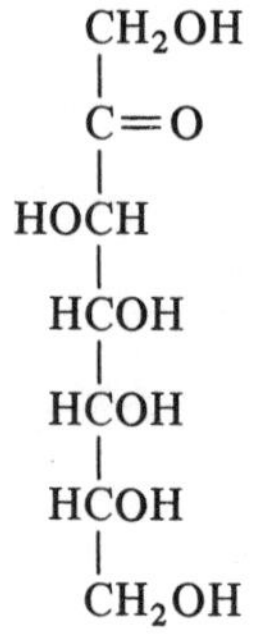 CH_2OH–C=O–HOCH–HCOH–HCOH–HCOH–CH_2OH	210

Table 1.5 Structure and Molecular Weight of Purines and Pyrimidines

Compound	Structure	Molecular Weight
Adenine (6-aminopurine)	NH_2 C N_1 6 5 C—N 7 9 8 CH HC 2 3 4 C—N N H	135
Guanine (2-amino-6-oxypurine)	O C HN_1 6 C—N CH H_2N—C 2 C—N N H	151
Uracil (2,4-dioxypyrimidine)	O C HN_3 4 CH O=C 2 CH N H	112
Cytosine (2-oxy-4-aminopyrimidine)	NH_2 C N 4 CH O=C 2 CH N H	111
Thymine (5-methyl-2,4-dioxypyrimidine)	O C HN_3 4 5 C—CH_3 O=C 2 CH N H	126

Two sugars may be combined by a glycosidic bond, as in (1.6).

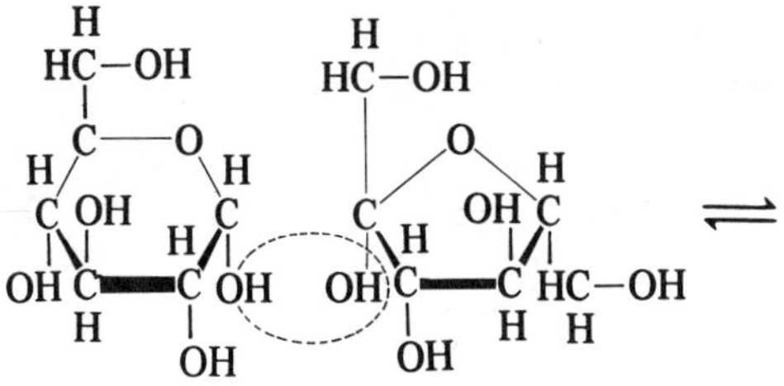

α-D-Glucopyranose β-D-Fructofuranose

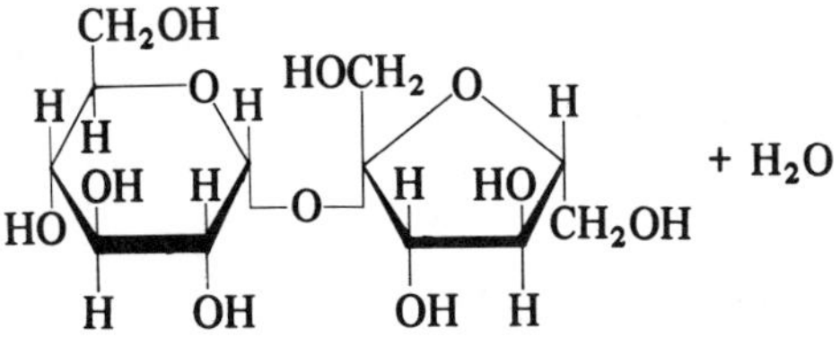

Sucrose (α-D-glucopyranosyl-β-D-fructofuranoside) + H_2O

(1.6)

Nucleotides are oligomers composed of several different types of small molecules or monomers and involve phosphodiester bonds, as shown in (1.7).

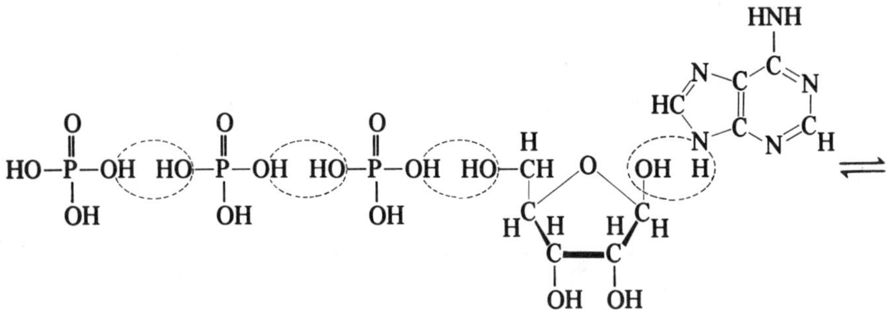

ATP + $4H_2O$

(1.7)

The structure of ATP (MW = 507) and all other nucleotides is summarized in (1.8).

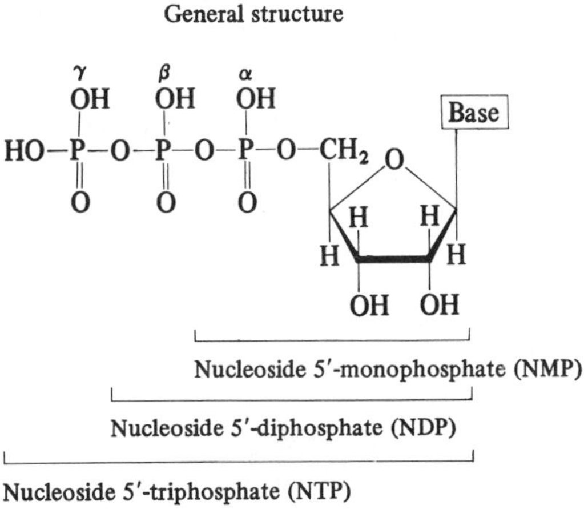

General structure

γ β α

OH OH OH

HO—P—O—P—O—P—O—CH_2 O Base

O O O

H H H H

OH OH

Nucleoside 5′-monophosphate (NMP)

Nucleoside 5′-diphosphate (NDP)

Nucleoside 5′-triphosphate (NTP)

(1.8)

Abbreviations

	Ribonucleoside 5′-mono-, di-, and triphosphates		
Base	Abbreviations		
Adenine	AMP	ADP	ATP
Guanine	GMP	GDP	GTP
Cytosine	CMP	CDP	CTP
Uracil	UMP	UDP	UTP

	Deoxyribonucleoside 5′-mono-, di-, and triphosphates		
Adenine	dAMP	dADP	dATP
Guanine	dGMP	dGDP	dGTP
Cytosine	dCMP	dCDP	dCTP
Thymine	dTMP	dTDP	dTTP

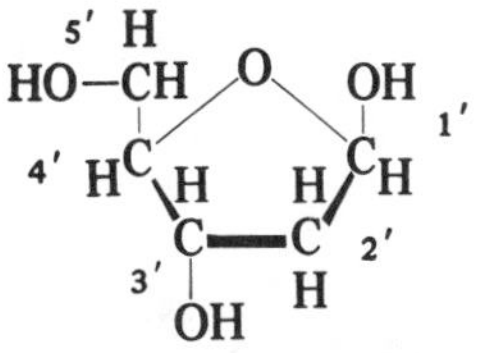

5′ H

HO—CH O OH

4′ HC H H CH 1′

C—C 2′

3′ OH H

2′-Deoxyribofuranose

Coenzymes are composed of several types of monomers and involve a variety of dehydration linkages, such as peptide bonds and ester bonds. Portions of these molecules are commonly known as vitamins. A few important examples are shown in (1.9) through (1.16).

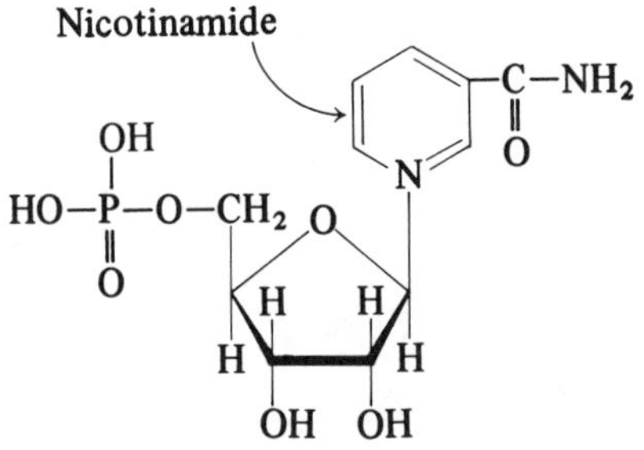

Nicotinamide

Nicotinamide mononucleotide (NMN)

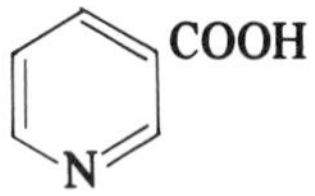

Nicotinic acid (niacin) (1.9)

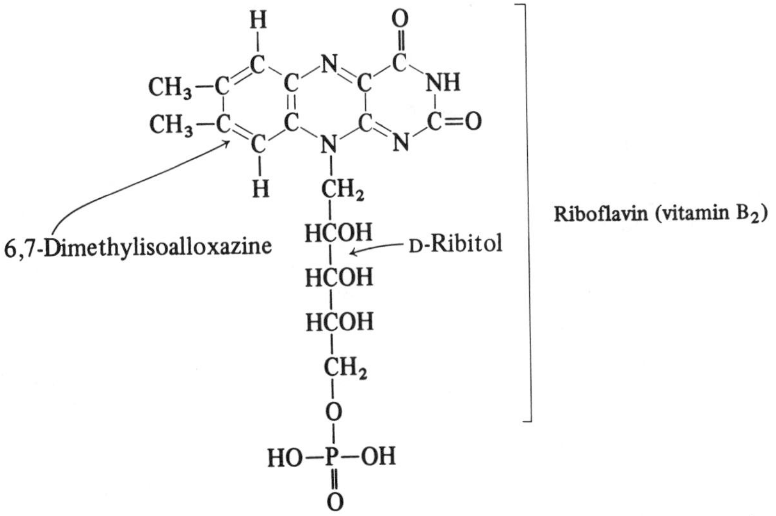

6,7-Dimethylisoalloxazine

D-Ribitol

Riboflavin (vitamin B_2)

Flavin mononucleotide (FMN) (1.10)

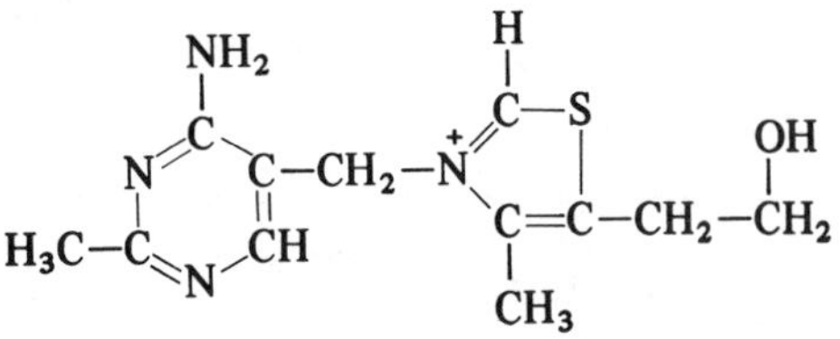

Thiamine (vitamin B_1) (1.11)

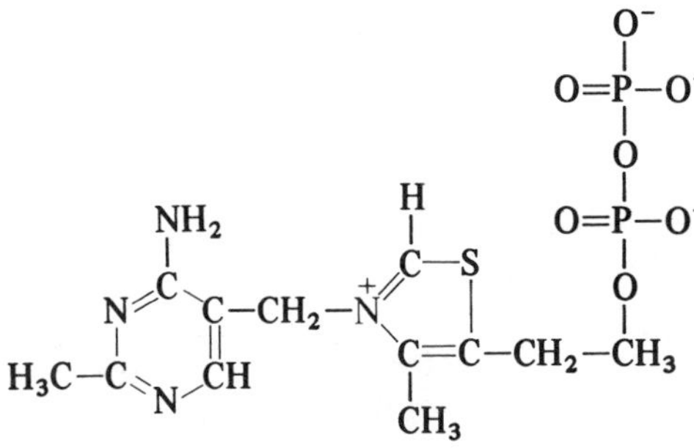

(1.12)

Thiamine pyrophosphate (TPP)

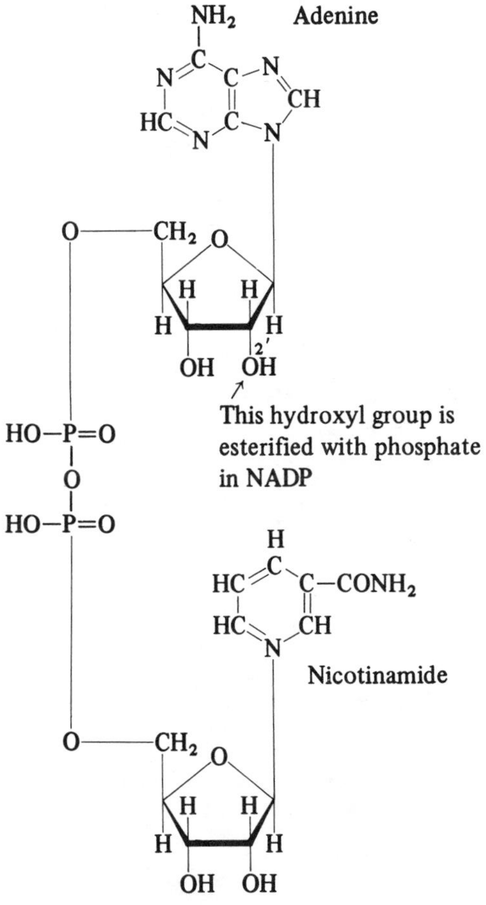

Adenine

This hydroxyl group is esterified with phosphate in NADP

(1.13)

Nicotinamide

Nicotinamide-adenine dinucleotide (NAD)

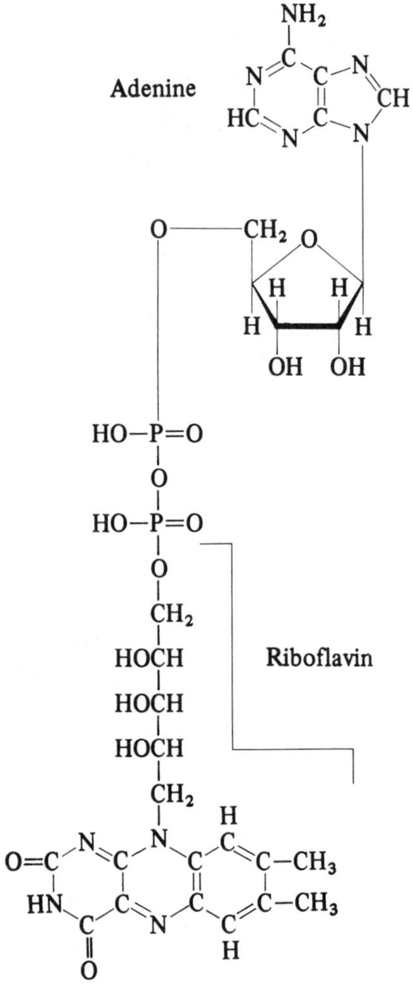

Adenine

Riboflavin

(1.14)

Flavin-adenine dinucleotide (FAD)

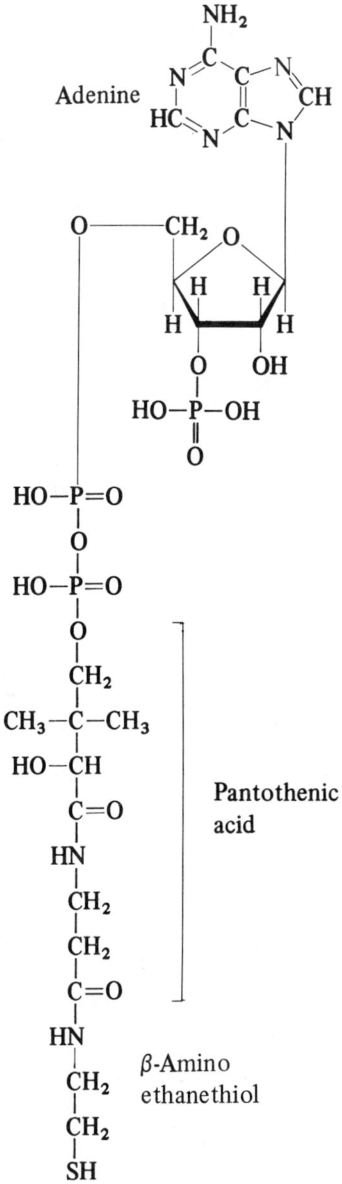

Adenine

NH_2

HO—P—OH

HO—P=O

O

HO—P=O

O

CH_2

CH_3—C—CH_3

HO—CH

C=O

HN

CH_2

CH_2

C=O

HN

CH_2

CH_2

SH

Pantothenic acid

β-Amino ethanethiol

(1.15)

Coenzyme A (CoA)

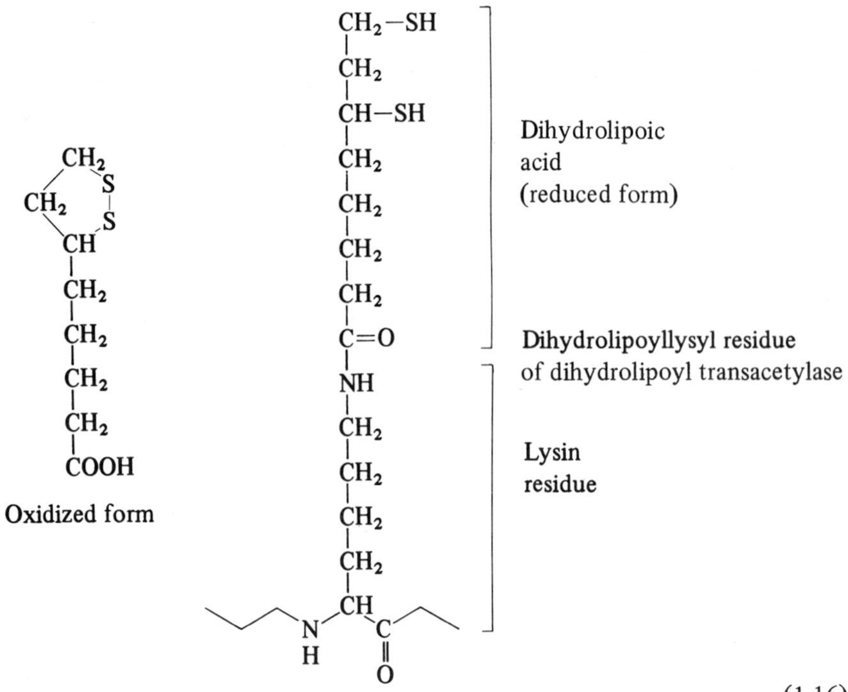

$$\begin{array}{l}
\text{Oxidized form:}\\
\text{ring } CH_2\text{–}S\text{–}S\text{–}CH\text{–}CH_2 \text{ (with } CH_2\text{)}\\
CH\text{–}CH_2\text{–}CH_2\text{–}CH_2\text{–}CH_2\text{–}COOH
\end{array}$$

Oxidized form

$$\begin{array}{l}
CH_2{-}SH\\
|\\
CH_2\\
|\\
CH{-}SH\\
|\\
CH_2\\
|\\
CH_2\\
|\\
CH_2\\
|\\
CH_2\\
|\\
C{=}O\\
|\\
NH\\
|\\
CH_2\\
|\\
CH_2\\
|\\
CH_2\\
|\\
CH_2\\
|\\
{-}N(H){-}CH{-}C({=}O){-}
\end{array}$$

Dihydrolipoic acid (reduced form)

Dihydrolipoyllysyl residue of dihydrolipoyl transacetylase

Lysin residue

Lipoic acid (1.16)

Lipids are of many types. The glycerides (mono-, di-, and tri-) are composed of glycerol and fatty acids connected by dehydration ester bonds, as shown in (1.17). The fatty acid residues can be of several types (Table 1.6).

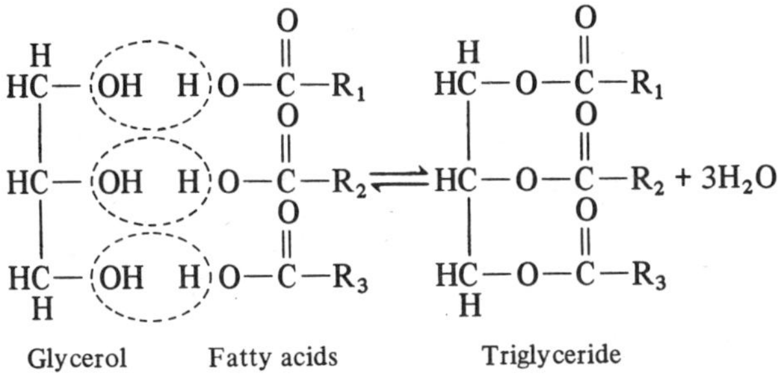

$$\begin{array}{lll}
H_2C{-}OH & HO{-}\overset{O}{\overset{\|}{C}}{-}R_1 & H_2C{-}O{-}\overset{O}{\overset{\|}{C}}{-}R_1\\
HC{-}OH \;+ & HO{-}\overset{O}{\overset{\|}{C}}{-}R_2 \;\rightleftharpoons & HC{-}O{-}\overset{O}{\overset{\|}{C}}{-}R_2 + 3H_2O\\
H_2C{-}OH & HO{-}\overset{O}{\overset{\|}{C}}{-}R_3 & H_2C{-}O{-}\overset{O}{\overset{\|}{C}}{-}R_3\\
\text{Glycerol} & \text{Fatty acids} & \text{Triglyceride}
\end{array} \qquad (1.17)$$

The diglycerides are important constitutents of membranes. These molecules contain two fatty acid side chains, and in the third position they contain the phosphate ester of an alcohol. Consequently, they are referred to as "phosphoglycerides."

Table 1.6 Fatty Acid Residues in Glycerides

Number of Carbon Atoms	Structure	Systematic Name	Common Name
		Saturated fatty acids	
12	$CH_3(CH_2)_{10}COOH$	*n*-Dodecanoic	Lauric acid
14	$CH_3(CH_2)_{12}COOH$	*n*-Tetradecanoic	Myristic
16	$CH_3(CH_2)_{14}COOH$	*n*-Hexadecanoic	Palmitic
18	$CH_3(CH_2)_{16}COOH$	*n*-Octadecanoic	Stearic
20	$CH_3(CH_2)_{18}COOH$	*n*-Eicosanoic	Arachidic
24	$CH_3(CH_2)_{22}COOH$	*n*-Tetracosanoic	Lignoceric
		Unsaturated fatty acids	
16	$CH_3(CH_2)_5CH=CH(CH_2)_7COOH$		Palmitoleic
18	$CH_3(CH_2)_7CH=CH(CH_2)_7COOH$		Oleic
18	$CH_3(CH_2)_4CH=CHCH_2CH=CH(CH_2)_7COOH$		Linoleic
18	$CH_3CH_2CH=CHCH_2CH=CHCH_2CH=CH(CH_2)_7COOH$		Linolenic
20	$CH_3(CH_2)_4CH=CHCH_2CH=CHCH_2CH=CHCH_2CH=CH(CH_2)_3COOH$		Arachidonic

The ***polymers*** [5] constitute the class of giant macromolecules, the proteins, the polynucleotides DNA and RNA, and the polysaccharides (e.g., starch, cellulose, and glycogen). These polymers are dehydration condensates containing many monomeric subunits. This is depicted in (1.18).

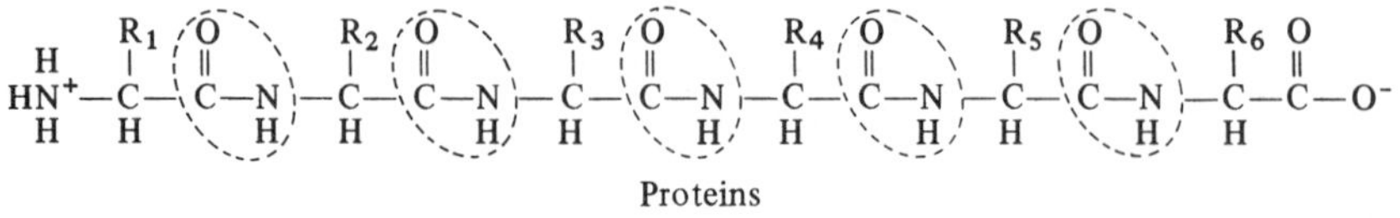

Proteins

(1.18)

The peptide bonds, circled with dots, are the dehydration linkages mentioned in the discussion of oligomers, where the structure of a dipeptide was shown in (1.5). Generally, proteins have from 50 to 350 residues, instead of the six depicted in (1.18). The zwitterion form of the amino and carboxyl termini is depicted in (1.18) because these charged forms are dominant at pH 7, as is also true for free amino acids. The molecular weights of proteins range up to more than 100,000 for a single polypeptide chain.

Polynucleotides are macromolecules composed of chains of nucleotides. The basic linkage between nucleotides in DNA and RNA is the 5′-3′ phosphodiester bond, as shown in (1.19). Long chains of such phosphodiester linkages

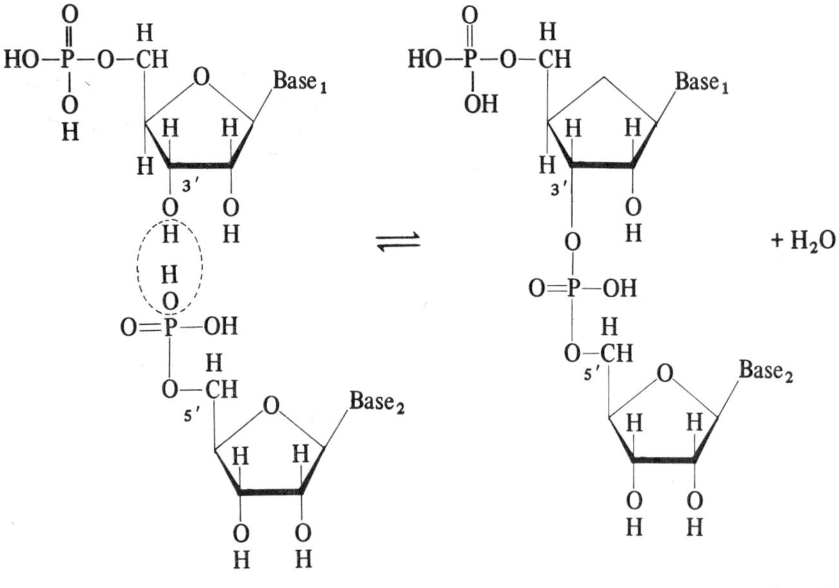

$$\rightleftharpoons \quad + H_2O$$

(1.19)

DNA

RNA

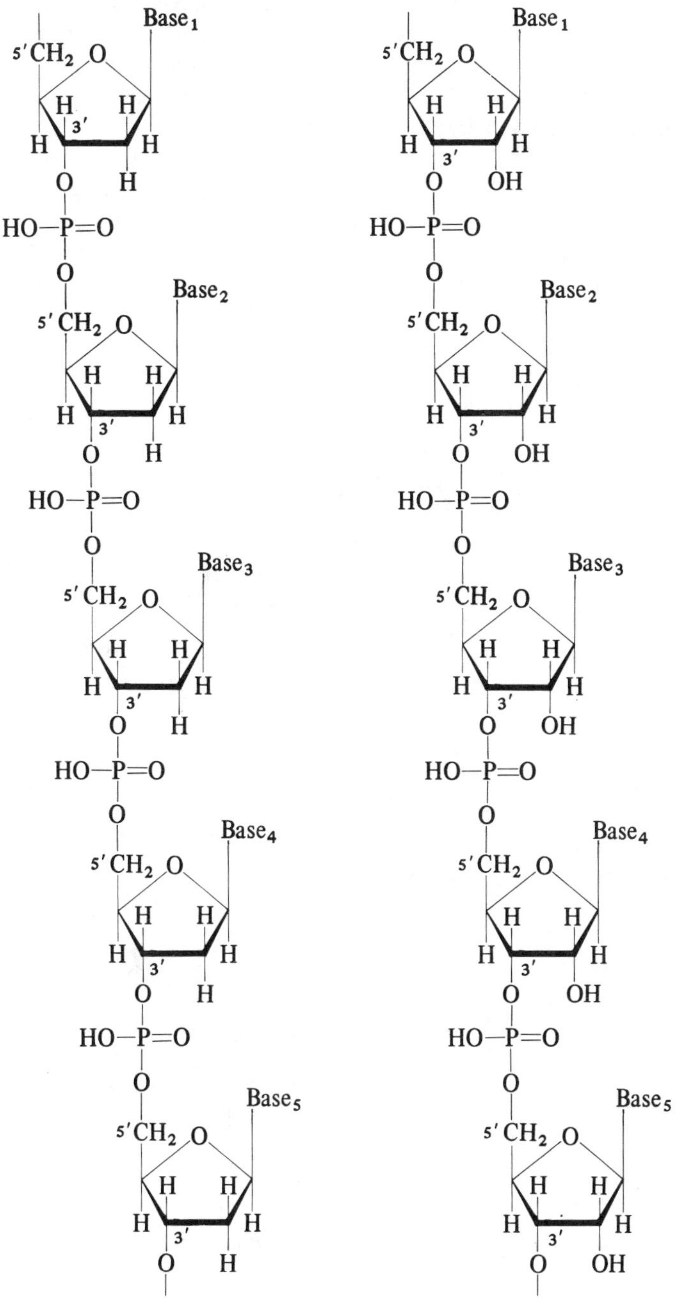

(1.20)

produce DNA and RNA molecules with molecular weights in the millions. The distinction between DNA and RNA is twofold: DNA utilizes deoxyribose, whereas RNA utilizes ribose; DNA utilizes the bases adenine, guanine, cytosine, and thymine, whereas RNA utilizes adenine, guanine, cytosine, and uracil. Single-stranded DNA and RNA are shown in (1.20). The orientation of the 5′-3′ phosphodiester bond determines the molecule's polarity.

Double-stranded DNA consists of two single strands, of opposite polarity, bound together by hydrogen bonds according to the base-pairing rules: adenine and thymine make two hydrogen bonds, and guanine and cytosine make three hydrogen bonds. These bonds are shown in (1.21), where they are designated A=T and G≡C. Since they are nearly identical in shape, a long molecule of stacked base pairs will look like a helix. A helix of the B form is depicted in (1.22) [3] [4].

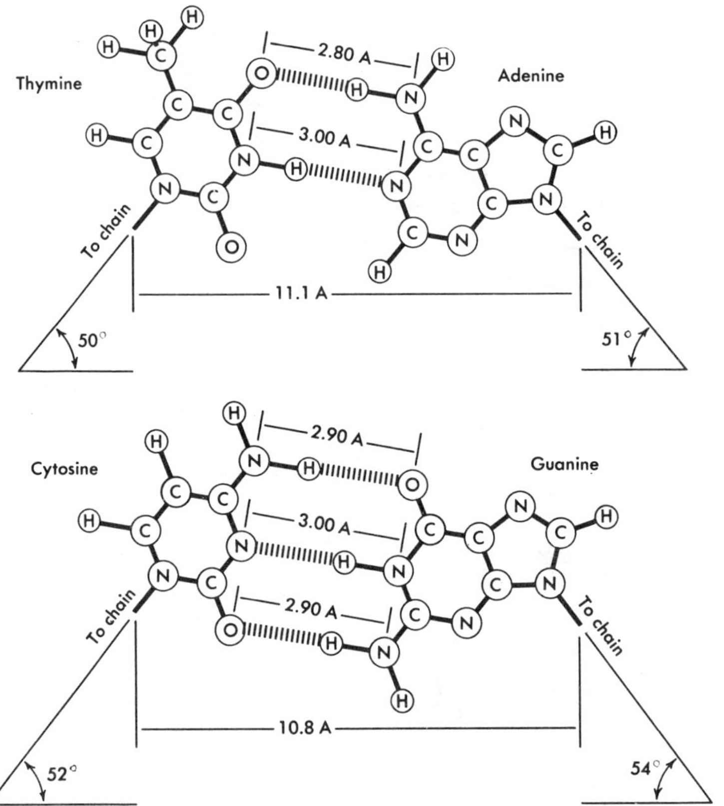

Reproduced, with permission, from J. D. Watson, *Molecular Biology of the Gene*, 2nd ed., Benjamin (1970).

(1.21)

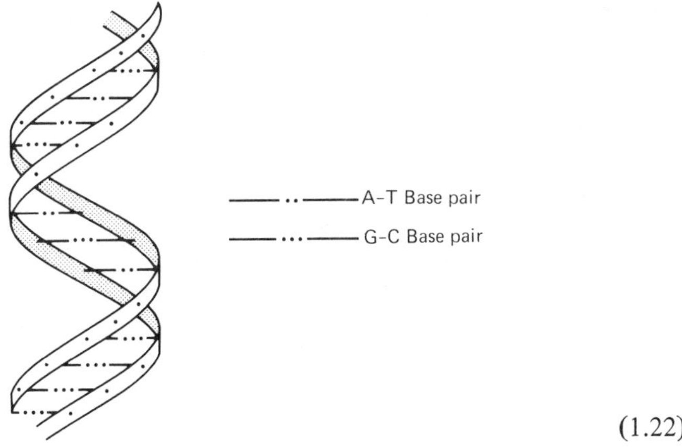

(1.22)

Aggregations of proteins, and of proteins and polynucleotides, are called ***complexes***. Explicit examples can be found in Chapter 6.

Membranes are giant complexes of lipids and of lipids and proteins. Explicit examples of these structures are diagrammed in Chapters 7 and 11.

References

1 Erwin Schrödinger, *What Is Life?* (Cambridge University Press, Cambridge, England, 1944).

This penetrating analysis, written before the molecular structure of DNA was established, is still worth reading. Although usually mentioned in the context of discussions of "negentropy," it is also eminently useful to read when molecular structure is broached, especially for biological molecules. Chapters 5 and 6 of *What Is Life?* were inspired by the work of Max Delbrück, of which Schrödinger had this to say (p. 73):

> From Delbrück's general picture of the hereditary substance it emerges that living matter, while not eluding the "laws of physics" as established up to date, is likely to involve "other laws of physics" hitherto unknown, which however, once they have been revealed, will form just as integral a part of this science as the former.

Schrödinger's Chapter 7 contains his comparison of DNA with a "clockwork" and summarizes his view of the "new principle."

2 A. E. Needham, *The Uniqueness of Biological Materials* (Pergamon, Oxford, 1965).

This book contains many examples of ordinary and exotic uses of common and rare elements in organisms.

3 A. Lehninger, *Biochemistry* (Worth, New York, 1970).
4 L. Struyer, *Biochemistry* (Freeman, San Francisco, 1976).
5 R. E. Dickerson and I. Geis, *The Structure and Action of Proteins* (Harper & Row, New York, 1969).

CHAPTER

2

Thermodynamics and Statistical Mechanics

The purpose of this chapter is to present a quantitative language for the discussion of thermal, electrical, and physical molecular processes. Physical chemists have used statistical mechanics and thermodynamics to explain a great many basic physical processes occurring in organisms. In this chapter, an elementary account of "molecular" thermodynamics is presented. Basic laws and formulas are given, and only elementary mathematical manipulations enter into the discussion of the various equations. Fuller accounts of derivations and related reasoning can be found in the references at the end of the chapter.

Statistical mechanics provides a microscopic explanation of thermodynamic relations as well as numerical values for various thermodynamic quantities. The partition function, $Z_N(T, V)$, gives all thermodynamic information as a function of the temperature T, the volume V, and the number of molecules N. The computation of Z is the primary step in statistical mechanics. The "classical limit" of the quantum mechanical partition function is [1-6]

$$Z_N(T, V) = \left(\frac{2\pi m k_B T}{h^2}\right)^{3/2} \frac{V^N}{N!} [j(T)]^N \tag{2.1}$$

where k_B is Boltzmann's constant, h is Planck's constant, and m is the mass of a single molecule. The term $j(T)$ is the molecular partition function; it is determined by the *internal state structure of a molecule* (see Appendix A), whereas the T- and V-dependent factors arise from the degrees of freedom of the molecule's center of mass. General principles require a fully quantum mechanical, statistical mechanical analysis. However, the "classical limit," exhibited in (2.1), is a valid, accurate approximation whenever

$$\frac{N}{V} \ll \left(\frac{2\pi m k_B T}{h^2}\right)^{3/2} \tag{2.2}$$

The left-hand side of this inequality is the density of molecules in units of number of molecules per unit volume, whereas the right-hand side is the "density of states per molecule," which is of quantum mechanical origin. The density of states per molecule will be of considerable importance in Chapter 6 when self-assembly is discussed. Quantum mechanics makes a second significant contribution to the classical limit formula through the appearance of $N!$ in the denominator of Z_N. The indistinguishability of identical molecules, which was emphasized in Chapter 1, leads to the quantum mechanical requirement for "symmetrized" N-molecule wave functions. The symmetrization procedure generates the $N!$ that survives in the classical limit. The importance of this $N!$ will be seen repeatedly throughout this book.

The validity of the classical limit in (2.2) can be checked for liquid H_2O. Since a liquid is far more dense than a gas at room temperature and pressure, the validity of (2.2) for liquid H_2O is a severe test of the classical limit. At nearly room temperature ($T = 300°K$), H_2O has a density

$$\frac{N}{V} = 55.55 \text{ moles/liter} = \frac{55.55(6 \times 10^{23})}{10^3} \text{ molecules/cm}^3$$
$$= 3.3 \times 10^{22} \text{ molecules/cm}^3 \qquad (2.3)$$

(1 mole = Avogadro's number of molecules = 6.025×10^{23} molecules.) The mass of H_2O is $18 \times 1.6 \times 10^{-24}$ g $= 28.8 \times 10^{-24}$ g. Therefore, the density of states per molecule is

$$\left(\frac{2\pi m k_B T_{300}}{h^2}\right)^{3/2} = 6.67 \times 10^{25} \text{ states/cm}^3$$

Clearly, $3.3 \times 10^{22} \ll 6.67 \times 10^{25}$, verifying the classical limit. Since no molecular species considered in the biophysical situations discussed in this book is significantly denser than liquid H_2O, this check justifies using the classical limit partition function generally. In this limit, quantum mechanics is exhibited in the $N!$ and in the density of states per molecule. Both of these quantum mechanical consequences are of fundamental importance in the rest of this book.

For a mixture of molecular species, a generalization of the partition function in (2.1) can be obtained:

$$Z_{N_1 N_2 \cdots N_M}(T, V) = \prod_{i=1}^{M} \left(\frac{2\pi m_i k_B T}{h^2}\right)^{(3/2)N_i} \frac{V^{N_i}}{N_i!} [j_i(T)]^{N_i} W(T, V, N_1 \cdots N_M) \qquad (2.4)$$

where $\Pi_{i=1}^{M}$ denotes the "product" over index values of i from 1 to M. The quantity m_i is the mass of a molecule of species i, and there are N_i such molecules. The factor $W(T, V, N_1 N_2 \cdots N_M)$ contains all interaction effects that result in the presence of intermolecular potentials. It is this factor that creates all the real difficulties from a computational point of view. This amazing formula for the partition function of a molecular mixture is the output of the theory of quantum statistical mechanics. With it, the laws of thermodynamics can be deduced, and explicit formulas for thermodynamic quantities like specific heats can be obtained. The two constants h and k_B have entered thermodynamics through (2.4).

The partition function gives rise to thermodynamics through the fundamental identity for the Helmholtz free energy F,

$$F = -k_B T \ln[Z_{N_1 N_2 \cdots N_M}(T, V)] \tag{2.5}$$

where the Napierian logarithm ln is used. The Helmholtz free energy for the mixture is the thermodynamic function that governs a system in contact with a fixed temperature (thermal), reservoir. Other thermodynamic functions, including the entropy S, are deducible from Z. More about the significance of these other functions will be said in the next chapter. The internal energy U and the entropy S are given by

$$U = kT^2 \left(\frac{\partial}{\partial T} \ln Z\right)_V = F - T\left(\frac{\partial F}{\partial T}\right)_V \tag{2.6}$$

and

$$S = -\left(\frac{\partial F}{\partial T}\right)_V = k \ln Z + \frac{U}{T} = \frac{U - F}{T}$$

where the subscript V means that the T derivative is performed at constant V. The pressure P and the heat capacity C_V are given by

$$P = -\left(\frac{\partial F}{\partial V}\right)_T \tag{2.7}$$

and

$$C_V = -T\left(\frac{\partial^2 F}{\partial T^2}\right)_V$$

With these quantities, both the enthalpy H and the Gibbs free energy G can be expressed as follows:

$$H = U + PV \tag{2.8}$$

and

$$G = H - TS = F + PV = U + PV - TS$$

The entropy governs the behavior of an *isolated* system. The enthalpy is appropriate in studies of systems in contact with constant pressure reservoirs, and the Gibbs free energy is used for systems in contact with both thermal and pressure reservoirs. The Gibbs free energy G proves to be the most useful thermodynamic function for most biophysical processes. The fixed temperature is about 300°K, and the pressure is "atmospheric" (1 atm). (1 atm-liter = 101.3 J.)

Returning to (2.4) and (2.5) and using the shorthand notation

$$f_i = \left(\frac{2\pi m_i k_B T}{h^2}\right)^{3/2} Vj_i(T) \tag{2.9}$$

gives

$$F = -k_B T \sum_{i=1}^{M} [N_i \ln f_i - \ln (N_i!)] - k_B T \ln W \tag{2.10}$$

For sufficiently large N, $\ln (N!)$ is very well approximated by

$$\ln (N!) \cong N \ln \left(\frac{N}{e}\right) + \frac{1}{2} \ln (2\pi N) \tag{2.11}$$

where $e = 2.718 \ldots$. This is already accurate to better than 1% for $N \geqq 10$. For $N > 10^3$, it is no longer necessary to keep track of $\frac{1}{2}\ln (2\pi N)$ for accuracy as good as 1%. This is the approximation most commonly used because molar quantities involve $N \simeq 10^{23}$. Therefore (2.10) can be written

$$F = -k_B T \sum_{i=1}^{M} \left[N_i \ln f_i - N_i \ln \left(\frac{N_i}{e}\right)\right] - k_B T \ln W \tag{2.12}$$

The quantity $\ln f_i$ can be written

$$\ln f_i = \ln \left[\left(\frac{2\pi m_i k_B T}{h^2}\right)^{3/2} V\right] + \ln [j_i(T)]$$

In Appendix B to this chapter, justification is given for associating a free energy function F_i^0 with j_i, in parallel with the relationship (2.5) for the entire system:

$$F_i^0 = -k_B T \ln [j_i(T)] \tag{2.13}$$

or

$$j_i(T) = \exp\left(-\frac{F_i^0}{k_B T}\right)$$

F_i^0 is the Helmholtz free energy per molecule of type i. It is related to an internal energy per molecule of type i, U_i^0, and to an absolute entropy per molecule of type i, S_i^0, by

$$F_i^0 = U_i^0 - TS_i^0 \tag{2.14}$$

at temperature T. This means that (2.12) can be written

$$F = \sum_{i=1}^{M} \left\{N_i U_i^0 - TN_i S_i^0 - k_B TN_i \ln \left[\left(\frac{2\pi m_i k_B T}{h^2}\right)^{3/2} V\right] + k_B TN_i \ln \left(\frac{N_i}{e}\right)\right\} - k_B T \ln W \tag{2.15}$$

The internal energy for the mixture according to (2.6) is just

$$U = \sum_{i=1}^{M} N_i U_i^0 + \frac{3}{2} k_B T \sum_{i=1}^{M} N_i + k_B T^2 \left(\frac{\partial}{\partial T} \ln W\right) \tag{2.16}$$

whereas the entropy of the mixture, S, is given by (2.6) as

$$S = \frac{3}{2} k_B \sum_{i=1}^{M} N_i + \sum_{i=1}^{M} N_i S_i^0 + k_B \sum_{i=1}^{M} N_i \ln \left[\left(\frac{2\pi m_i k_B T}{h^2}\right)^{3/2} V\right] - k_B \sum_{i=1}^{M} N_i \ln \left(\frac{N_i}{e}\right) + k_B \ln W + k_B T \left(\frac{\partial \ln W}{\partial T}\right) \tag{2.17}$$

This entropy contains the term $-k_B \sum_{i=1}^{M} N_i \ln (N_i/e)$, which is called the "entropy of mixing" and was originally introduced, before the advent of quantum mechanics, by Gibbs. Its origin here is in the $N_i!$ factors that resulted from a quantum mechanical accounting of *indistinguishable*, indentical molecules.

In some books V is replaced by the ideal gas law:

$$V = \frac{Nk_BT}{P} \qquad N = \sum_{i=1}^{M} N_i$$

and then the identity

$$\ln\left[\left(\frac{2\pi m_i k_B T}{h^2}\right)^{3/2} V\right] - \ln\left(\frac{N_i}{e}\right) = \ln\left[\left(\frac{2\pi m_i k_B T}{h^2}\right)^{3/2} \frac{k_B T}{P}\right] - \ln\left(\frac{N_i}{eN}\right) \quad (2.18)$$

is used to represent the entropy of mixing as $-k_B \sum_{i=1}^{M} N_i \ln (N_i/eN)$. Here (2.17) will be used instead.

The thermodynamics of ideal molecule mixtures is obtained from the general formulas above for the fundamental thermodynamic variables by omitting all the W-dependent terms. The quantity W contains all of the intermolecular interaction effects. Real molecular mixtures require the inclusion of W in the analysis of thermodynamic events. Physical chemists have traditionally chosen to approach the inclusion of W in a phenomenological manner because explicit computations involving W are rarely possible, except for highly idealized and simplified models. The objective (see references) has been to preserve the form of the *ideal* molecule mixture equations as much as possible while simultaneously incorporating intermolecular interaction effects through an effectively measurable phenomenological parameter. This is achieved by introducing the concepts of "activity" and activity coefficients. In the equations above these concepts must be latent in W.

Most biophysical processes occur at ambient temperature and pressure. Consequently, the Gibbs free energy is the suitable thermodynamic function to use. According to (2.8), it is given by

$$G = F + PV$$

where

$$P = -\left(\frac{\partial F}{\partial V}\right)_{T,N_1\cdots N_M} \quad (2.19)$$

Frequently, the chemical potential, defined by

$$\mu_i \equiv \left(\frac{\partial G}{\partial N_i}\right)_{T,P,N_j \neq N_i} = \left(\frac{\partial F}{\partial N_i}\right)_{T,V,N_j \neq N_i} \quad (2.20)$$

will be used in this book. The second equality in (2.20) is proved in chemical thermodynamics books and underscores the subtle significance of thermodynamic quantities like T, P, and V. The relationship between W and "activities" is most easily exhibited through its explicit influence on the chemical potential μ_i. From (2.15) and (2.20) it follows that

$$\mu_i = U_i^0 - TS_i^0 - k_BT \ln \left[\left(\frac{2\pi m_i k_B T}{h^2} \right)^{3/2} V \right] + k_BT \ln \left(\frac{N_i}{e} \right) + k_BT$$

$$- k_BT \left(\frac{\partial}{\partial N_i} \ln W \right)_{T,V,N_j \neq N_i}$$

$$= U_i^0 - TS_i^0 - k_BT \ln \left[\left(\frac{2\pi m_i k_B T}{h^2} \right)^{3/2} V \right] + k_BT \ln (\gamma_i N_i) \tag{2.21}$$

where the activity coefficient γ_i is defined by

$$\gamma_i \equiv \exp \left[- \left(\frac{\partial \ln W}{\partial N_i} \right)_{T,V,N_j \neq N_i} \right] \tag{2.22}$$

The deviations from *ideal* behavior are entirely contained in γ_i. The activity of the *i*th molecular species, A_i, is defined by

$$A_i \equiv \gamma_i N_i \tag{2.23}$$

In the following chapters, it is A_i that is directly measured in many cases, not N_i. Because γ_i cannot be computed easily (or at all in some cases), the relationship between A_i and N_i must be established empirically.

The Gibbs free energy for a molecular mixture at constant T and P is generally given by

$$G = \sum_{i=1}^{M} N_i \mu_i = \sum_{i=1}^{M} N_i [\mu_i^0 + k_BT \ln (A_i)] \tag{2.24}$$

in which

$$\mu_i^0 \equiv U_i^0 - TS_i^0 - k_BT \ln \left[\left(\frac{2\pi m_i k_B T}{h^2} \right)^{3/2} V \right]$$

Appendix 2A

Equation (3.1) shows a *translational* contribution to the partition function with the factor

$$\left[\left(\frac{2\pi m k_B T}{h^2}\right)^{3/2} V\right]^N \tag{2A.1}$$

This can be viewed as the contribution of the center of mass of the molecule to its thermal properties. The function $j(T)$ contains all internal state contributions, such as those from vibration, rotation, and electron levels. Some of these can be expressed just as explicitly as in (2A.1).

The expression for $j(T)$ illustrates the general definition of a partition function, treated quantum mechanically:

$$Z = \sum_{\substack{\text{energy}\\\text{states}}} \exp\left(-\frac{E}{k_B T}\right) \tag{2A.2}$$

Note the word "states" under the summation symbol. Energy *levels* may contain more than one *state*, so that (2A.2) can also be written

$$Z = \sum_{\substack{\text{energy}\\\text{states}}} G(E) \exp\left(-\frac{E}{k_B T}\right) \tag{2A.3}$$

where $G(E)$ is the degeneracy of the energy level E. When energy levels are closely spaced relative to $k_B T$, the summations are well approximated by integrals:

$$Z \cong \int G(E) \exp\left(-\frac{E}{k_B T}\right) dE \tag{2A.4}$$

where $G(E)$ is now a degeneracy density (per unit energy).

The different types of internal energy are approximately additive. For example, if there are just two degrees of freedom, then $E = E_1 + E_2$. The summation over energy levels can be accomplished by separately summing over E_1 and E_2 if the degeneracy $G(E)$ factors: $G(E) = G_1(E_1)\, G_2(E_2)$. This is shown as follows:

$$\begin{aligned}\sum_E G(E) \exp\left(-\frac{E}{k_B T}\right) &= \sum_{E_1}\sum_{E_2} G_1(E_1)\, G_2(E_2) \exp\left(-\frac{E_1}{k_B T}\right) \exp\left(-\frac{E_2}{k_B T}\right)\\ &= \left[\sum_{E_1} G_1(E_1) \exp\left(-\frac{E_1}{k_B T}\right)\right]\left[\sum_{E_2} G_2(E_2) \exp\left(-\frac{E_2}{k_B T}\right)\right]\end{aligned} \tag{2A.5}$$

Therefore $Z = Z_1 Z_2$, and the computation of contributions from different internal degrees of freedom is greatly simplified.

Contribution from Translation

The energy levels for a mass m in a cubic box of volume $V = L^3$ are given by

$$E_{jkl} = \frac{h^2}{8mL^2}(j^2 + k^2 + l^2) \tag{2A.6}$$

where j, k, l are integers and the degeneracies $G(E_{jkl}) = 1$. The contribution from translation is

$$Z_{\text{trans}} = \sum_j \sum_k \sum_l \exp\left[-\frac{h^2}{8mL^2 k_B T}(j^2 + k^2 + l^2)\right] \tag{2A.7}$$

Now, $k_B T_{300} = 4 \times 10^{-14}$ erg and $h^2/8mL^2 \sim 5.6 \times 10^{-34}$ erg for a molecule of molecular weight 50 in a 10^3-cm^3 box. These are very closely spaced levels! The spacing is even less for more massive molecules. Note also that this expression factors into terms corresponding to one spatial dimension each:

$$Z_{\text{trans}} = \left[\sum_k \exp\left(-\frac{h^2}{8mL^2 k_B T} k^2\right)\right]^3 \tag{2A.8}$$

This expression can be accurately estimated by the integral

$$\sum_k \exp\left(-\frac{h^2}{8mL^2 k_B T} k^2\right) \to \int_0^\infty \exp\left(-\frac{h^2}{8mL^2 k_B T} k^2\right) dk = \left(\frac{2\pi m k_B T}{h^2}\right)^{1/2} L \tag{2A.9}$$

Therefore,

$$Z_{\text{trans}} = \left(\frac{2\pi m k_B T}{h^2}\right)^{3/2} V \tag{2A.10}$$

which is precisely the quantity already exhibited in (2.1). From (2.6), U and S are obtained:

$$U_{\text{trans}} = \frac{3}{2} k_B T \tag{2A.11}$$

and

$$S_{\text{trans}} = \frac{3}{2} k_B + k_B \ln \left[\left(\frac{2\pi m k_B T}{h^2} \right)^{3/2} V \right]$$

Notice especially that the translational entropy *increases* with *increasing* mass, temperature, and volume.

Contribution from Rotation

For a "heteronuclear" diatomic molecule, which has two angular degrees of freedom (in the general case three rotational degrees of freedom are possible), the energy levels are

$$E_J = J(J+1) \frac{h^2}{8\pi^2 I} \tag{2A.12}$$

where I is the moment of inertia, J is an integer, and the degeneracies $G(E_J) = 2J + 1$. The contribution from rotation is

$$Z_{\text{rot}} = \sum_J (2J+1) \exp\left[-(J^2 + J) \frac{h^2}{8\pi^2 I k_B T} \right] \tag{2A.13}$$

Values of I are about $1 \to 50 \times 10^{-40}$ g cm^2. Therefore the energy level spacing is on the order of $(2J+2)(4 \times 10^{-15})/1 \to 50$ ergs. Here the level spacing increases with increasing J by the factor $2J + 2$. Until $J \sim 10$, the spacing is small compared with $k_B T$, so an integral approximation is good for the lower levels, while for levels higher than $J \sim 10$, the Boltzmann factor $\exp[-(J^2 + J)(h^2/8\pi^2 I k_B T)]$ is very small. Thus, an integral approximation for all J is very accurate:

$$Z_{\text{rot}} \to \int_0^{\infty} (2J+1) \exp\left[-(J^2 + J) \frac{h^2}{8\pi^2 I k_B T} \right] dJ$$

$$= \frac{8\pi^2 I k_B T}{h^2} \tag{2A.14}$$

The internal energy and entropy for rotation again follow from (2.6):

$$U_{\text{rot}} = k_B T \tag{2A.15}$$

and

$$S_{\text{rot}} = k_B + k_B \ln \left(\frac{8\pi^2 I k_B T}{h^2} \right)$$

The rotational entropy increases with increasing temperature and with increasing moment of inertia, I. One way I can increase is by an increase in the nuclear masses. Thus both the translational entropy and the rotational entropy increase with increasing mass.

Contribution from Vibration

A single vibrational mode has energy levels

$$E_n = h\nu_0 \left(n + \frac{1}{2} \right) \tag{2A.16}$$

where n is an integer, ν_0 is the natural frequency, and the degeneracies $G(E_0) = 1$. The contribution from vibration is

$$\begin{aligned} Z_{\text{vib}} &= \sum_{n=0}^{\infty} \exp\left[-\frac{h\nu_0}{k_B T} \left(n + \frac{1}{2} \right) \right] \\ &= \exp\left(-\frac{h\nu_0}{2k_B T} \right) \left[1 - \exp\left(-\frac{h\nu_0}{k_B T} \right) \right]^{-1} \end{aligned} \tag{2A.17}$$

No integration was necessary since this summation is elementary. The internal energy and entropy for vibration follow from (2.6):

$$U_{\text{vib}} = \frac{h\nu_0}{2} + \frac{h\nu_0}{\exp(h\nu_0/k_B T) - 1} \tag{2A.18}$$

and

$$S_{\text{vib}} = k_B \ln \left[1 - \exp\left(-\frac{h\nu_0}{k_B T} \right) \right]^{-1} + h\nu_0 \left[\exp\left(\frac{h\nu_0}{k_B T} \right) - 1 \right]^{-1}$$

The entropy S_{vib} decreases as ν_0 increases. If the masses of the nuclei that vibrate are increased, the frequency decreases and the entropy increases. Usually, $h\nu_0$ for molecular bonds and groups ranges from several $k_B T$ ($5k_B T$ to $10k_B T$) down to $\frac{1}{10}k_B T$ or lower for really massive groups.

Contribution from Electron Levels

The energy levels are E_l, where l is an integer. (0, 1, 2, . . .), with degeneracies $G(E_l)$. The contribution from electron levels is

$$Z_{\text{elect}} = \sum_{l=0}^{\infty} G(E_l) \exp\left(-\frac{E_l}{k_B T}\right)$$

$$= \exp\left(-\frac{E_0}{k_B T}\right) \sum_{l=0}^{\infty} G(E_l) \exp\left(-\frac{E_l - E_0}{k_B T}\right) \quad (2A.19)$$

For most electronic situations, $E_l - E_0 \gg k_B T$ for $l \geqslant 1$. Therefore

$$Z_{\text{elect}} \simeq \exp\left(-\frac{E_0}{k_B T}\right) G(E_0) \quad (2A.20)$$

Therefore

$$U_{\text{elect}} = E_0 \quad (2A.21)$$

and

$$S_{\text{elect}} = k_B \ln G(E_0)$$

Appendix 2B

Assume that in (2A.3) of Appendix 2A the discrete energy levels E_l are labeled with the integer index l so that E_l increases with l. The degeneracy factor G_l can be thought of as a function of energy: $G_l \to G(E_l)$. For many molecules, the energy levels are very closely spaced, so that (2A.3) can be replaced by an integral:

$$j(T) = \int_0^{\infty} G(E) \exp(-\beta E)\, dE \quad (2B.1)$$

whose $\beta \equiv 1/k_B T$ is a simple shorthand notation and $G(E)$ is the degeneracy per unit energy, as distinct from G, which is a number associated with energy E_l. The exponential factor in (2B.1) is rapidly decreasing. If $G(E)$ is an increasing function of E, which is a rather common situation, the product, $G(E) \exp(-\beta E)$, goes through a maximum value at some energy value E^*. This maximum is the sharper the more rapidly increasing $G(E)$ is. Writing $G(E) = \exp[\ln G(E)]$, it is possible to expand the integrand in (2B.1) around the maximum value at E^* to obtain

$$j(T) \simeq \int_0^\infty \exp[\ln G(E^*) - \beta E^*] \times \exp\left(\frac{1}{2}(E - E^*)^2 \left\{\frac{G''(E^*)}{G(E^*)} - \left[\frac{G'(E^*)}{G(E^*)}\right]^2\right\}\right) dE \quad \text{(2B.2)}$$

where $G'(E^*)$ and $G''(E^*)$ are the first and second derivatives of $G(E)$ with respect to E, evaluated at E^*. The condition for a maximum is $G'(E^*) = \beta G(E^*)$, and this was used to get (2B.2). Now, if $G(E)$ increases *less* rapidly than $\exp[\beta E]$ in the neighborhood of E^*, then

$$\frac{G''(E^*)}{G(E^*)} < \beta^2 = \left[\frac{G'(E^*)}{G(E^*)}\right]^2$$

Therefore

$$D \equiv \frac{G''(E^*)}{G(E^*)} - \left[\frac{G'(E^*)}{G(E^*)}\right]^2 < 0$$

This allows one to perform the integration in (2B.2), obtaining

$$j(T) \simeq \sqrt{\frac{2\pi}{D}}\, G(E^*) \exp(-\beta E^*) \quad \text{(2B.3)}$$

It is easily established that $\sqrt{2\pi/D}\; G(E^*)$ is dimensionless, and this quantity approximates the degeneracy of the energy E^*. From (2.13) it follows that

$$F^0 = U^0 - TS^0 = -k_B T \ln\left[\sqrt{\frac{2\pi}{D}}\, G(E^*) \exp(-\beta E^*)\right]$$
$$= E^* - k_B T \ln\left[\sqrt{\frac{2\pi}{D}}\, G(E^*)\right] \quad \text{(2B.4)}$$

Equation (2.6) can be used to justify the identifications

$$U^0 = E^*$$
$$S^0 = k_B \ln\left[\sqrt{\frac{2\pi}{D}}\, G(E^*)\right] \quad \text{(2B.5)}$$

Thus the internal energy per molecule, U^0, is the "most probable" energy value, E^*, and the absolute entropy per molecule is Boltzmann's constant multiplied by the Napierian logarithm of the degeneracy of the energy level at E^*. Such connections between entropy and ln (degeneracy) were appreciated by Boltzmann and Planck before the advent of quantum mechanics.

It is remarkable that both Gibbs and Boltzmann had profound insights regarding entropy, the entropy of mixing and (2B.5), which today are much more readily explained by means of quantum theory. More has been said about each of these results in Schrödinger's *Statistical Thermodynamics*, in Chapters 8 and 6, respectively [7].

References

1 Kerson Huang, *Statistical Mechanics* (Wiley, New York, 1963).

Only the advanced student will be able to make use of this book. However, it is one of the best places to find an account of the classical limit of the partition function (Section 10.2, p. 213). The origin of the $N!$ is explained.

2 Walter J. Moore, *Physical Chemistry*, 4th ed. (Prentice-Hall, Englewood Cliffs, N.J., 1972).

3 F. Daniels and R. Alberty, *Physical Chemistry*, 3rd ed. (Wiley, New York, 1967), Appendix, p. 747.

4 R. Fowler and E. A. Guggenheim, *Statistical Thermodynamics* (Cambridge Press, Cambridge, England, 1952).

5 R. E. Dickerson, *Molecular Thermodynamics* (Benjamin, New York, 1969).

This is a remarkably well written book of unusual clarity and utility, highly recommended for the serious student of the statistical mechanical approach to chemistry. It gives an excellent introduction to the molecular partition function. See especially Chapters 2 and 5.

6 G. N. Lewis and M. Randall, *Thermodynamics*, 2nd ed., revised by K. S. Pitzer and L. Brewer (McGraw-Hill, New York, 1961).

7 E. Schrödinger, *Statistical Thermodynamics*, 2nd ed. (Cambridge University Press, New York, 1952).

CHAPTER

3

Molecular Mixtures and Reactions

The preceding chapter considered thermodynamic functions important in determining the equilibrium properties of a mixture. Equation (2.24) expresses the Gibbs free energy as a function of the number of molecules of species i in a volume V. These amounts change as a result of reactions, until equilibrium is achieved. In some situations inputs and outputs are applied and equilibrium cannot be achieved, although steady states become a possibility. To understand steady states and all types of transient behavior, reactions must be understood.

The reactions considered here are binary: two molecules come together and react, and then two new molecules come apart. Let M species of molecules react and let N_i denote the population of the ith species. Assume that many different types of reactions are occurring simultaneously among the M species and let k denote the kth reaction. The rate laws without inputs can be written generally

$$\frac{d}{dt}N_i = \sum_{\substack{k \\ jmn}} [W^{(k)}_{(ij|mn)} A_m A_n - W^{(k)}_{(mn|ij)} A_i A_j] \tag{3.1}$$

The summation is over all reactions k and all species indices j, m, and n. The term $W^{(k)}_{(ij|mn)}$ is the reaction *rate* for the kth reaction, in which a molecule each of species m and n react to produce a molecule each of species i and j. Generally, for a particular reaction, say the kth, only a few of the M species are involved, and for the most indices $W^{(k)}_{(ij|mn)}$ is simply zero. The order of the indices in this expression is important. The reverse reaction, in which a molecule each of species i and j reacts to produce a molecule each of species m and n, is

$$W^{(k)}_{(mn|ij)}$$

This is not equal to $W^{(k)}_{(ij|mn)}$ in general. However, the order of i and j, or of m and n, is immaterial:

$$W^{(k)}_{(ij|mn)} = W^{(k)}_{(ji|mn)} = W^{(k)}_{(ij|nm)} \tag{3.2}$$

At constant temperature T and pressure P the forward and reverse reactions for each k satisfy

$$\frac{W^{(k)}_{(ij|mn)}}{W^{(k)}_{(mn|ij)}} = \exp\left[-\beta(\mu_i^0 + \mu_j^0 - \mu_m^0 - \mu_n^0)\right] \tag{3.3}$$

where $\beta = (1/k_B T)$, as before, and from (2.21)

$$\mu_i^0 = U_i^0 - T\overline{S}_i^0 \tag{3.4}$$

is the intrinsic Gibbs free energy per molecule of type i, $\overline{S}_i^0$ being defined by

$$\overline{S}_i^0 \equiv S_i^0 + k_B \ln\left[\left(\frac{2\pi m_i k_B T}{h^2}\right)^{3/2} V\right]$$

for convenience. Equation (3.3) is an important connection between purely thermodynamic, equilibrium properties like μ_i^0 and *rate* properties in $W^{(k)}$. However, only a rate *ratio* is determined, not the absolute rates. These cannot be obtained from thermodynamics alone. Quantum mechanics must be used to determine the absolute rates.

Using (3.3) in (3.1) leads to the observation that equilibrium is characterized by

$$A_i^{\text{eq}} \propto \exp(-\beta\mu_i^0) \tag{3.5}$$

This proportionality implies that $d/dt N_i^{\text{eq}} = 0$ for each i. The proportionality factor is determined by identifying the *constraints* that are operative for this mixture. From (3.1) it is easily seen that

$$\sum_{i=1}^{M} \frac{d}{dt} N_i = \frac{d}{dt} \sum_{i=1}^{M} N_i = 0 \tag{3.6}$$

which is the constraint $\Sigma_{i=1}^{M} N_i = N$, a constant. This constraint together with (3.5) implies that

$$A_i^{\text{eq}} = \frac{N}{Q_G} \exp(-\beta\mu_i^0) \tag{3.7}$$

where $Q_G \equiv \Sigma_{i=1}^{M} \exp(-\beta\mu_i^0)$. This, then, tells how N, T, and the μ_i^0 terms determine the equilibrium amounts of each molecular species.

Thermodynamics also states that equilibrium for a system governed by the Gibbs free energy is determined by *minimizing* the Gibbs free energy subject to any constraints, in this case the constraint $\Sigma_{i=1}^{M} N_i = N$. The variation of G is, from (2.19), (2.20), and (2.24),

$$\delta G = \sum_{i=1}^{M} \delta N_i \, [U_i^0 - T\bar{S}_i^0 + k_B T \ln (A_i)] = 0 \tag{3.8}$$

and the variation of the constraint is

$$\delta N = \sum_{i=1}^{M} \delta N_i = 0 \tag{3.9}$$

Together these equations are satisfied by

$$A_i^{\text{eq}} \propto \exp(-\beta\mu_i^0) \tag{3.10}$$

just as in (3.5). Thus this thermodynamic criterion is compatible with the results of the rate analysis, (3.5).

It should be observed that the present discussion has been given in terms of the Gibbs free energy because the system of interest is assumed to be in contact with thermal and pressure reservoirs. These conditions are typical of biochemical situations. In the absence of such reservoirs, it is the entropy that governs equilibrium by being *maximized* at equilibrium. Thus the distinction between using the Gibbs free energy or the entropy will be emphasized throughout.

Several situations must be described and distinguished to avoid confusion and shallow reasoning. Some systems will be in contact with "reservoirs," which maintain a particular physical property, such as temperature, at a constant value. Other systems will also be subject to inputs and outputs that are generally time-dependent quantities. The distinction between closed and open systems is also required. In fact, a slightly finer distinction is required [1-6].

Three situations must be distinguished. An *isolated* system is a *closed* system not in contact with any reservoirs. A *closed* system is a system that may also be in contact with reservoirs for temperature, pressure, or both. An *open* system is a system that may or may not be in contact with reservoirs but is definitely coupled to inputs and outputs. Closed systems are governed by autonomous equations in N_i, as in (3.1), whereas open systems also depend on inputs that may not depend on N_i exclusively, such as in

$$\frac{d}{dt}N_i = \sum_k \sum_{j,m,n} [W^{(k)}_{(ij|mn)} N_m N_n - W^{(k)}_{(mn|ij)} N_i N_j] + I_i(N, N_2 \cdots N_M, \vec{\alpha}) \quad (3.11)$$

where I_i is a "generalized" input-output term that may depend on the N_i terms as well as on new parameters, $\vec{\alpha}$, which are independent of the N_i terms.

An isolated system's behavior is governed by the entropy, which *monotonically* approaches a maximum at equilibrium. A closed system is governed by the thermodynamic function appropriate for the kinds of reservoirs that are coupled to the system: the Helmholtz free energy F, for a T-reservoir; enthalpy H, for a P-reservoir; and the Gibbs free energy G for both T- and P-reservoirs. The appropriate function monotonically approaches its minimum at equilibrium. An open system may approach a steady state, but no thermodynamic function changes monotonically, and at the steady state no thermodynamic function becomes a maximum or a minimum.

When an open system is shut off, or closed, the appropriate thermodynamic function, determined by the reservoirs present, will monotonically approach its equilibrium value. If T- and P-reservoirs are present, it is the Gibbs free energy G that monotonically decreases to its minimum at equilibrium, but the entropy need not monotonically change and, in particular, need not even increase, as would be required for an *isolated* system. In fact, there are instances in which the entropy also decreases as G decreases, at least for a portion of the transient approach to equilibrium.

These considerations help to eliminate confusion as to whether or not living systems violate the second law of thermodynamics. The confusion results from a twofold error: ignoring the distinction between G and S that reservoirs require and failing to recognize that the second law does not simply apply to open systems. The dissipative processes in open systems nevertheless reflect a definite thermal origin.

Even though the second law does not simply apply to open systems, the dynamics of such systems is largely determined by thermal processes, which in a closed system would be the basis for the validity of the second law. Specifically, the "relaxation," or "dissipative," terms in (3.11) are precisely those in (3.1). Only input-output terms have been added to (3.1) to get (3.11). This reflects the fact that (3.11) describes an open system that is isothermal. The temperature is maintained at a fixed value by a thermal reservoir, so that the effect of inputs and outputs is modulated by isothermal *relaxation* processes identical with those in a closed system. The isothermal nature of biophysical processes is primarily a result of the considerable "thermal buffering" capacity of the solvent (H_2O). Water possesses both a high heat capacity and a high thermal conductivity, properties that maintain an isothermal system. Consequently, the thermal properties of a closed system are of great significance even for an open system. Other important properties of H_2O will be encountered in Chapters 6 and 10. How far a steady state of an open system is from equilibrium will be determined by G, the Gibbs free energy. When an open system is "shut off" and becomes a closed system, its G value will monotonically decrease until equilibrium is achieved. The entropy, on the other hand, will not necessarily show monotone behavior, and there are many special cases where it does not.

The system of equations given in (3.11) can be generalized to include spatial dependence in all variables. Coupled reaction-diffusion equations, with or without inputs, are used in place of (3.11). Preliminary work by many researchers has shown how richly varied are the solutions to reaction-diffusion networks of equations. Although much of the material covered in this book is more deeply appreciated from the reaction-diffusion point of view, it is not the purpose of this book to elaborate this more technical side of the description.

There are special cases in which the decrease in G, say after an open system has been closed, is attended, during a transient period of time, by a regular decrease in entropy S. If one thinks of the second law always in terms of entropy and the single context of an isolated, closed system, then this type of phenomenon looks very strange. A closed container of nutrient broth that is maintained at constant temperature and pressure and contains a single bacterium will develop a stage of proliferation with an exponentially increasing population that possesses the property of a monotonic decrease in G and S. This is paid for by a large decrease in U as the nutrient is converted into cells. Eventually the nutrient will be used up and death will overtake the culture; G will finally be minimized, and much of the early decrease in S will be undone.

The relationship between entropy and degeneracy has been exhibited in (2B.5) of Appendix 2B. A high degeneracy corresponds with a high "probability" and with a high "disorder"; a low degeneracy corresponds with a low probability and a low disorder. Low disorder is simply order. Consequently, it is often said that an entropy increase is an increase in disorder, and an entropy decrease is an increase in order (a decrease in disorder). The situation depicted by the system, in contact with T- and P-reservoirs, that is originally open and then closed is one in which G decreases while S may also, for a while, decrease. Such a situation is referred to in this book as "energy flow ordering," and is a basic phenomenon in cells.

In Appendixes 3A and 3B, technical remarks are added regarding the second law of thermodynamics and the concept of activity coefficients.

Appendix 3A

It is shown here that (3.1) and (3.3) can be used to "prove" the second law of thermodynamics, in parallel with Boltzmann's proof of his celebrated H-theorem for dilute gases [1].

From (2.24), the time rate of change of the Gibbs free energy for a molecular mixture at constant T and P is given by

$$\frac{d}{dt} G = \sum_{i=1}^{M} \left[\frac{dN_i}{dt} (\mu_i^0 + k_B T \ln A_i) + \frac{N_i}{A_i} \frac{dA_i}{dt} \right] \tag{3A.1}$$

subject to the constraint given in (3.6). Following the index-algebra method used by Boltzmann, it is possible to use (3.1) to show that (3A.1) can be equivalently written

$$\frac{d}{dt} G = \frac{1}{4} \sum_{k} \sum_{ijmn} [W_{(ij|mn)}^{(k)} A_n A_m - W_{(mn|ij)}^{(k)} A_i A_j] \times \left[\mu_i^0 + \mu_j^0 - \mu_m^0 - \mu_n^0 + k_B T \ln \left(\frac{A_i A_j}{A_m A_n} \right) \right] + \sum_{i=1}^{M} \frac{N_i}{A_i} \frac{dA_i}{dt} \tag{3A.2}$$

With (3.3), this becomes

$$\frac{d}{dt} G = \frac{1}{4} \sum_{k} \sum_{ijmn} [W_{(ij|mn)}^{(k)} A_n A_m - W_{(mn|ij)}^{(k)} A_i A_j] \times k_B T \ln \left[\frac{W_{(mn|ij)}^{(k)} A_i A_j}{W_{(ij|mn)}^{(k)} A_m A_n} \right] + \sum_{i=1}^{M} \frac{N_i}{A_i} \frac{dA_i}{dt} \tag{3A.3}$$

As in Boltzmann's original argument for the dilute gas, the logarithm above occurs in the form $(x - y) \ln (y/x)$, which always satisfies the inequality $(x - y) \ln (y/x) \leqq 0$ with strict equality only when $x = y$. Therefore

$$\frac{d}{dt} G - \sum_{i=1}^{M} \frac{N_i}{A_i} \frac{dA_i}{dt} \leqq 0 \tag{3A.4}$$

for all t. Notice especially that this inequality is satisfied for each value of k (i.e., for each separate reaction, as well as for the entire k summation). Equality occurs only when the values of the A_i terms are their equilibrium values, given in (3.7).

It would be desirable to prove that

$$\sum_{i=1}^{M} \frac{N_i}{A_i} \frac{dA_i}{dt} \leqq 0$$

also, so that we could conclude

$$\frac{d}{dt} G \leqq 0 \tag{3A.5}$$

which is the second law of thermodynamics, as it was quoted in the text of this chapter. It has not been possible to prove this in general, just as it has not been possible to extend Boltzmann's H-theorem to dense gases and liquids. This is not because these more general inequalities are not true, as a phenomenological law, but because the needed proofs are too difficult to construct by contemporary mathematical methods. Throughout this book, the validity of the second law is based largely on empirical evidence rather than on rigorous proofs.

Examination of graphs of activity coefficients as functions of molality, or molarity, leads to several conclusions. Activity coefficients may increase, decrease, or go through extremes as the molality is varied from 0 to 1. At typical biophysical molalities, many substances possess virtually constant γ values over the range of molalities, or molarities, spanned by the reactions that occur. In these cases

$$\frac{\partial \gamma_i}{\partial N_j} \simeq 0 \tag{3A.6}$$

This implies

$$\sum_{i=1}^{M} \frac{N_i}{A_i} \frac{dA_i}{dt} = \sum_{i=1}^{M} \frac{1}{\gamma_i} \left(\gamma_i \frac{dN_i}{dt} + N_i \sum_j \frac{\partial \gamma_i}{\partial N_j} \frac{dN_j}{dt} \right) \tag{3A.7}$$

The first summation on the right-hand side of (3A.7) is just (3.6), and the second contains (3A.6). In this sense, (3A.5) has been "proved" for the context appropriate for this book.

Appendix 3B

Activity coefficients were introduced by (2.21). Their "ideal" value is $\gamma = 1$. The noble gases are very nearly "ideal." Salts, on the other hand, are far from ideal.

For HCl, with molalities between 0.1 and 1, $\gamma = 0.78 \pm 0.02$. Purine has $\gamma = 0.84$ at 0.85 molal and $\gamma = 0.37$ at 0.5 molal. At very low molality, or molarity, $\gamma \simeq 1$ for most substances. It is often a good approximation to take $\gamma \simeq 1$ for biophysical concentrations $\leqq$ millimolar. When this is valid, so is (3A.6) [6].

It is tempting to suppose that $\gamma < 1$ always holds because the activity coefficient measures the "effectiveness" of N molecules in a reaction, and intermolecular interactions (which give rise to γ through W) would tend to make it more difficult for molecules to react. For salts, this is observed to be the case. An extreme example is given by $NiSO_4$, for which $\gamma < 0.09$ for all molalities $\geqq 0.2$. The simple amino acid alanine, however, has $\gamma \geqq 1.005$ for all molalities $\geqq 0.2$.

A considerable amount of study has been devoted to the theoretical analysis of the transition from W to γ in partition function calculations. Much progress has been made toward understanding in this way the diverse behavior observed in empirical activity coefficient measurements. The reader must consult the references for more details.

References

1 G. E. Uhlenbeck and G. W. Ford, *Lectures in Statistical Mechanics* (American Mathematical Society, Providence, R.I., 1963).

Chapter 1 of this book is a beautiful account of the connection between statistical mechanics and thermodynamics. It emphasizes the work of both Boltzmann and Gibbs, and explains both versions of the second law.

2 E. Broda, *The Evolution of the Bioenergetic Processes* (Pergamon, New York, 1975).

Chapter 1 contains an informative account of the emergence of an understanding of the need for open systems when discussing life. That Boltzmann already perceived this in 1886 is proved by a direct quotation.

3 Harold J. Morowitz, *Energy Flow in Biology* (Academic, New York, 1968).

Some ideas regarding energy flow ordering are expounded in this book.

4 Fritz Lipmann, *Wanderings of a Biochemist* (Wiley-Interscience, New York, 1971).

This is a remarkable book in many ways. In the first of a series of essays presented in the back of the book, "Projecting backward from the present stage of evolution of biosynthesis," Lipmann reviews his early ideas on open systems viewed from a quite detailed biochemical level, with particular emphasis on ATP. Figures 3 and 4 of this essay are especially instructive.

5 R. E. Dickerson, *Molecular Thermodynamics* (Benjamin, New York, 1969).

6 D. Eisenberg and D. Crothers, *Physical Chemistry* (Benjamin/Cummings, Reading, Mass., 1979).

CHAPTER

4

Examples and Special Cases

The general formalism presented up to now will gain in clarity through the exhibition of a variety of specific examples that will also prove useful for other subsequent considerations. This chapter will be used to list these examples [1-3]. Note especially how the ln A_i terms appear over and over again, as was anticipated in Chapter 2.

Ionization of H_2O

This phenomenon is of considerable importance in many situations. The basic reaction step is

$$H_2O + H_2O \underset{k_2}{\overset{k_1}{\rightleftharpoons}} H_3^+O + O^-H \tag{4.1}$$

in which three species are involved

$$\begin{aligned} N_1 &\equiv N_{H_2O} \\ N_2 &\equiv N_{H_3^+O} \\ N_3 &\equiv N_{O^-H} \end{aligned} \tag{4.2}$$

and there is only one reaction, $k = 1$:

$$W^{(1)}_{(23|11)} = k_1 \qquad \text{and} \qquad W^{(1)}_{(11|23)} = k_2 \tag{4.3}$$

The constraint is simply

$$N_1 + N_2 + N_3 = N, \text{ a constant} \tag{4.4}$$

According to (3.1), the rate equations are

$$\frac{d}{dt}N_1 = 2k_2A_2A_3 - 2k_1A_1^2$$
$$\frac{d}{dt}N_2 = \frac{d}{dt}N_3 = -\frac{1}{2}\frac{d}{dt}N_1 \tag{4.5}$$

In the expression for $(d/dt)N$, there is an overall factor of 2 not present in (3.1). This is because (3.1) was expressed for a reaction in which all four species i, j, m, and n, are distinct, whereas in the present case the species N_1 (N_{H_2O}) occurs twice on the left-hand side of (4.1). Such a multiple occurrence requires the additional "stoichiometric" coefficient 2 for the $(d/dt)N_1$ equation; no such factor appears in the equation for either $(d/dt)N_2$ or $(d/dt)N_3$, because the stoichiometrics for indices 2 and 3 are both 1.

The ionization constant K_i is defined by

$$K_i \equiv \frac{k_1}{k_2} = \exp(-\beta\Delta\mu^0) = \frac{A_2^{eq}A_3^{eq}}{(A_1^{eq})^2} \tag{4.6}$$

where

$$\Delta\mu^0 \equiv \mu_2^0 + \mu_3^0 - 2\mu_1^0$$

It is simply the ratio of the product of the activities for reaction product molecules to the product of the activities for reaction reactant molecules. Experimentally, liquid H_2O yields $\Delta\mu^0 = 23.8$ kcal/mole; equivalently [4], $\mu_{H_2O}^0 = -56.69$ kcal/mole $= \mu_{H_3^+O}^0$ and $\mu_{O^-H}^0 = -32.89$ kcal/mole:

$$K_i = 3.24 \times 10^{-18} \quad \text{at} \quad T = 298°\text{K} \tag{4.7}$$

The density of H_2O is

$$\frac{N_1^{eq}}{V} = \frac{55.6(6 \times 10^{23})}{10^3 \text{ cm}^3} = 3.33 \times 10^{22} \text{ molecules/cm}^3 \tag{4.8}$$

where 55.6 is molarity, 6×10^{23} is Avogadro's number (1 mole) and 10^3 cm^3 = 1 liter. Since the molecular weight of H_2O is 18, 3.33×10^{22} molecules/cm^3 = 1 g/cm^3. The γ values for this process are all nearly unity to better than 1% accuracy.

The symbol [i] will be used to denote N_i/V in units of moles per liter (i.e., molarity), $\gamma[i]$ will be used to denote activity density by $\gamma[i] \equiv \gamma_i(N_i/V)$. For $H_2O, \gamma[H_2O] \cong [H_2O]$, and

$$[H_2O] = 55.5 \text{ molar} \tag{4.9}$$

At equilibrium, because all activity coefficients are nearly unity,

$$\begin{aligned}[H_3^+O]\,[O^-H] &= 3.24 \times 10^{-18}\ (55.55 \text{ molar})^2 \\ &= 10^{-14} \text{ molar}^2\end{aligned} \tag{4.10}$$

Therefore, in equilibrium,

$$[H_3^+O] = [O^-H] = 10^{-7} \text{ molar} \tag{4.11}$$

if N_2 and N_3 are equal initially. If initially $N_2 \neq N_3$, then (4.4), (4.6), and (4.7) must be used to obtain the equilibrium values of $[H_3^+O]$ and $[O^-H]$.

Two other constants are frequently used instead of K_i, although all three are related. The dissociation constant K_d is defined by

$$K_d \equiv \frac{[H_3^+O]\,[O^-H]}{[H_2O]} = K_i\ [H_2O] = 1.8 \times 10^{-16} \text{ molar} \tag{4.12}$$

and the water constant K_w is defined by

$$K_w \equiv [H_3^+O]\,[O^-H] = K_i\ [H_2O]^2 = 10^{-14} \text{ molar}^2 \tag{4.13}$$

pH

In the aqueous phase protons, H^+, are almost always found in the form H_3^+O, the hydronium ion. The pH is defined to be the negative of the base 10 logarithm of the *numerical value* of the hydronium activity in units of moles per liter:

$$\text{pH} \equiv -\log_{10} \frac{\gamma[H_3^+O]}{6 \times 10^{23}/10^3 \text{ cm}^3} \tag{4.14}$$

with

$$[H_3^+O] = \frac{N_{H_3^+O}}{V} \quad \text{and} \quad \gamma_{H_3^+O} \simeq 1$$

Even though this is the proper definition of pH, one often finds the expression

$$\mathrm{pH} = -\log_{10}[\mathrm{H}^+] \tag{4.15}$$

or

$$\mathrm{pH} = -\log_{10}[\mathrm{H}_3^+\mathrm{O}]$$

neither of which is technically correct because the logarithm of a quantity with units (moles per liter) is not well defined. Equation (4.14) is literally what is intended by the expressions in (4.15).

Strong Acid

A strong acid is one that in equilibrium is virtually completely ionized (e.g., HCl). Consider a 10^{-3} molar HCl solution in H_2O. The basic reaction is

$$\mathrm{HCl} + \mathrm{H_2O} \underset{k_2}{\overset{k_1}{\rightleftharpoons}} \mathrm{H_3^+O} + \mathrm{Cl^-} \tag{4.16}$$

in which four species are involved:

$$\begin{aligned} N_1 &\equiv N_{\mathrm{H_2O}} \\ N_2 &\equiv N_{\mathrm{H_3^+O}} \\ N_3 &\equiv N_{\mathrm{HCl}} \\ N_4 &\equiv N_{\mathrm{Cl^-}} \end{aligned} \tag{4.17}$$

and there is only one reaction, $k = 1$:

$$W^{(1)}_{(24|13)} = k_1 \qquad \text{and} \qquad W^{(1)}_{(13|24)} = k_2 \tag{4.18}$$

The constraint is

$$N_1 + N_2 + N_3 + N_4 = N, \text{ a constant} \tag{4.19}$$

The rate equations, according to (3.1), are

$$\frac{d}{dt}N_1 = k_1 A_1 A_3 - k_2 A_2 A_4 \tag{4.20}$$

and

$$\frac{d}{dt}N_2 = -\frac{d}{dt}N_1 = -\frac{d}{dt}N_3 = \frac{d}{dt}N_4$$

Because HCl is a strong acid, in equilibrium it is found that

$$\begin{aligned} [H_2O] &\cong 55.5 \text{ molar} \\ [Cl^-] &= 10^{-3} \text{ molar} \\ [H_3^+O] &= 10^{-3} \text{ molar} \\ [HCl] &\simeq 0 \end{aligned} \qquad (4.21)$$

It is also found that $\gamma_{HCl} \sim 0.78$ for $[HCl] \geqq 0.2$ molality. However, here $\gamma_{HCl} \cong 1$. Obviously, $[H_2O]$ should be a little less, but 55 is much bigger than 10^{-3}, and $[H_3^+O]$ also has a contribution from the H_2O ionization of 10^{-7}, which is very small in comparison with 10^{-3} and is neglected. The ionization constant K_i is

$$K_i = \frac{[H_3^+O]\,[Cl^-]}{[H_2O]\,[HCl]} \sim \infty \qquad (4.22)$$

Since $K_i = \exp(-\beta\Delta\mu^0)$, where

$$\Delta\mu^0 \equiv \mu_2^0 + \mu_4^0 - \mu_1^0 - \mu_3^0 \qquad (4.23)$$

it must be that $\Delta\mu^0 < 0$ and $|\Delta\mu^0| \gg k_B T$. The contribution of ΔH^0 and ΔS^0 to $\Delta\mu^0$ can be determined from temperature dependence because

$$\Delta\mu^0 = \Delta H^0 - T\Delta S^0 \qquad (4.24)$$

The Henderson-Hasselbalch Equation for Weak Acids

A weak acid does not almost completely ionize in equilibrium. Denote such an acid by HA and denote the anion by A^-. The reaction is

$$HA + H_2O \underset{k_2}{\overset{k_1}{\rightleftharpoons}} H_3^+O + A^- \qquad (4.25)$$

in which four species are involved:

$$
\begin{aligned}
N_1 &= N_{H_2O} \\
N_2 &= N_{H_3^+O} \\
N_3 &= N_{HA} \\
N_4 &= N_{A^-}
\end{aligned}
\tag{4.26}
$$

and there is only one reaction, $k = 1$:

$$W^{(1)}_{(24|13)} = k_1 \quad \text{and} \quad W^{(1)}_{(13|24)} = k_2 \tag{4.27}$$

The constraint is $N_1 + N_2 + N_3 + N_4 = N$, a constant. At equilibrium, again assuming all activity coefficients are essentially unity,

$$K_i = \frac{[H_3^+O]\,[A^-]}{[H_2O]\,[HA]} = \frac{k_1}{k_2} \tag{4.28}$$

The "acid" constant K_a is often used:

$$K_a = K_i[H_2O] = 55.6 \times K_i \text{ moles per liter} \tag{4.29}$$

Therefore

$$[H_3^+O] = \frac{[HA]}{[A^-]} K_a \tag{4.30}$$

The pK_a is defined as the negative base 10 logarithm of the *numerical value* of K_a in units of moles per liter:

$$\text{p}K_a \equiv -\log_{10} \left(\frac{K_a}{6 \times 10^{23}/10^3 \text{ cm}^3} \right) \tag{4.31}$$

Using this and (4.14) in (4.30) yields, including the γ terms,

$$\text{pH} = \text{p}K_a + \log_{10} \left(\frac{\gamma[A^-]}{\gamma[HA]} \right) \tag{4.32}$$

which is the *Henderson–Hasselbalch equation*.

Standard-State Relations

Consider a reaction in which two substrates, or reactants, S_1 and S_2, are converted into two products P_1 and P_2:

$$S_1 + S_2 \rightleftharpoons P_1 + P_2 \tag{4.33}$$

At any instant, the Gibbs free energy is, according to (2.24),

$$G = \sum_{i=1,2} [N_{S_i}(\mu^0_{S_i} + k_B T \ln A_{S_i}) + N_{P_i}(\mu^0_{P_i} + k_B T \ln A_{P_i})] \tag{4.34}$$

If the reaction proceeds to the right by a single step (i.e., one molecule each of substrate is consumed and one molecule each of product is produced), the free energy changes from the expression above to one in which N_{S_i} is replaced by $N_{S_i} - 1$ and N_{P_i} is replaced by $N_{P_i} + 1$. Denoting the final value of G minus the initial value by ΔG gives, neglecting terms of order $1/N$,

$$\Delta G = \sum_{i=1,2} \left[\mu^0_{P_i} - \mu^0_{S_i} + k_B T \ln \left(\frac{\gamma_{P_i} N_{P_i}}{\gamma_{S_i} N_{S_i}}\right)\right]$$

$$= \Delta\mu^0 + k_B T \ln \left(\frac{\gamma_{P_1}\gamma_{P_2} N_{P_1} N_{P_2}}{\gamma_{S_1}\gamma_{S_2} N_{S_1} N_{S_2}}\right) \tag{4.35}$$

in which $\Delta\mu^0 \equiv \mu^0_{P_1} + \mu^0_{P_2} - \mu^0_{S_1} - \mu^0_{S_2}$.

If concentration units of moles per liter and energy units of ergs per mole are used,

$$\Delta G = \Delta\mu^0 + 2.3RT \log_{10} \left(\frac{[P_1]\,[P_2]}{[S_1]\,[S_2]}\right) + 2.3RT \log_{10} \left(\frac{\gamma_{P_1}\gamma_{P_2}}{\gamma_{S_1}\gamma_{S_2}}\right) \tag{4.36}$$

where R is the gas constant, defined by $R = (6 \times 10^{23})k_B$. This is just Avogadro's number times Boltzmann's constant and gives the value $R = 8.3 \times 10^7$ ergs/mole°K ($\equiv$ 8.3 J/mole-°K). The factor of 2.3 arises from the conversion from Napierian logarithms to base 10 logarithms. Both ΔG and $\Delta\mu^0$ in (4.36) must be given in units of ergs per mole (or joules per mole), whereas in (4.35) they are given in ergs per molecule. It is common to find energy units given in calories instead of in ergs or joules, and the conversion is achieved by 1 cal = 4.186 J = 4.186×10^7 ergs.

If the values of A_{P_i} and A_{S_i} for i = 1, 2 are taken to be the equilibrium values obtained in (3.7), then (4.35) implies that $\Delta G = 0$ in this case. In general, whether ΔG is positive or negative is determined by the values of A_{P_i} and A_{S_i} for $i = 1, 2$ and in particular by which side of the equilibrium values they happen to be. When the amount of products is greater than the equilibrium values, $\Delta G > 0$; when the amount of products is less than the equilibrium values, $\Delta G < 0$.

The "standard state" is a special situation defined by the requirement that each activity is 1 molar, and T and P are specified. From (4.36) it follows that, at standard state,

$$\Delta G = \Delta\mu^0 \qquad \text{at standard state} \tag{4.37}$$

Obviously, equilibrium and standard state are distinct situations.

Nonstandard pH

The standard-state activity density of H_3^+O is 1 molar, by definition. This corresponds to a pH of 0. Most biochemical processes occur at a pH value far from 0, usually about 7, and this value remains quite stable as a result of an intrinsic pH buffering capacity.

Suppose that H_3^+O is a product in a reaction that is maintained at a fixed pH by other processes:

$$S + H_2O \rightleftharpoons H_3^+O + P^- \tag{4.38}$$

in which electric neutrality is explicitly indicated by the negatively charged product species P^-. From (4.36), it follows that

$$\Delta G = \Delta G^0 + 2.3RT \log_{10}\left(\frac{[P^-]}{[S]}\right) + 2.3RT \log_{10}\left(\frac{[H_3^+O]}{[H_2O]}\right) \tag{4.39}$$

In conformity with textbook usage, $\Delta\mu^0$ is written ΔG^0 here and later. The quantity ΔG^0 includes contributions from all molecular species, including H_3^+O and H_2O. Since $G^0_{H_3^+O} = G^0_{H_2O}$ or, equivalently, $G^0_{H^+} = 0$, ΔG^0 is *unchanged* as a result of recognizing that H^+ is really H_3^+O. All species are going to be taken at standard state except H_3^+O, which must be given an activity density value determined by the fixed pH. This means that instead of (4.37) for standard-state pH, the nonstandard pH equation is

$$\Delta G = \Delta G^0 - 2.3RT\,\text{pH} \equiv \Delta G^{0\prime} \tag{4.40}$$

as follows from (4.14). The higher the pH, the more negative, relative to ΔG^0, ΔG becomes. For pH 7, this can be a sizable effect, with the value 4×10^4 J/mole, or about 9.6 kcal/mole. Suppose that $\Delta G^0 > 0$ for some specific reaction that produces H_3^+O as one of its products. At standard state, including pH 0, the reaction will not proceed to the product side according to the second law. If $\Delta G^0 < 4 \times 10^4$ J/mole, then at nonstandard pH = 7 with all other species at standard state, the reaction will satisfy $\Delta G < 0$ and can proceed to the product side. Clearly, the pH can strongly affect the tendency of this process. It is also possible to imagine that H_3^+O occurs on the reactant, or substrate, side of a reaction, and a similar analysis also yields a strong pH effect. For reactions that do not involve H_3^+O on either side of the equation, the pH value of the medium is irrelevant and has no direct effect on ΔG.

Nonstandard $\gamma[H_2O]$

In biology, $\gamma[H_2O]$ is *always* nonstandard. Liquid H_2O has a density of about 1 g/cm^3, which corresponds to a molarity of 55.55. Even when various species of molecule are in H_2O, they reduce this value a little bit, but never to anything close to the standard-state value of 1 molar. The activity coefficient for H_2O is nearly unity: $\gamma[H_2O] \cong [H_2O]$.

The hydrolysis of ethyl acetate to ethanol and acetic acid provides a useful example:

$$\text{EtOAc} + H_2O \rightleftharpoons \text{EtOH} + \text{HOAc} \tag{4.41}$$

where EtOAc is ethyl acetate, EtOH is ethanol, and HOAc is acetic acid. At "normal" temperature ($T \simeq 300^\circ$K) and pressure (P = 1 atm), the equilibrium constant K_{eq} for this reaction is measured to be

$$K_{eq} = 0.33 \tag{4.42}$$

and the equilibrium concentrations are found to be [EtOH] = 2×10^{-1} molar, [HOAc] = 2×10^{-1} molar, [EtOAc] = 2.18×10^{-3} molar, and $[H_2O] = 55.6$ molar. Neglecting the γ terms, we obtain

$$K_{eq} = \frac{[\text{EtOH}]\,[\text{HOAc}]}{[\text{EtOAc}]\,[H_2O]} \tag{4.43}$$

From (3.3) and (3.7), and the preceding examples, such as (4.6), it follows that generally

$$\Delta G^0 = -2.3RT \log_{10} K_{eq} \tag{4.44}$$

At standard state, (4.37) holds, so that

$$\Delta G = \Delta G^0 = -2.3RT \log_{10} K_{eq} = +2760 \text{ J/mole} \tag{4.45}$$

This means that, under standard-state conditions, the reaction will proceed not in the direction of hydrolysis but in the reverse direction instead. However, with $[H_2O] = 55.6$ molar, that is, *nonstandard* $[H_2O]$, (4.36) leads to

$$\begin{aligned} \Delta G &= \Delta G^0 - 2.3RT \log_{10} (55.6) \\ &= \Delta G^0 - 10^4 \text{ J/mole} \end{aligned} \tag{4.46}$$

This means that, for the hydrolysis reaction (4.41),

$$\Delta G = (2760 - 10{,}000) \text{ J/mole} = -7240 \text{ J/mole} \tag{4.47}$$

Therefore, under standard-state conditions, *except* for H_2O, the reaction does proceed in the direction of hydrolysis, as should be expected because $[H_2O] = 55.6$ molar means that there is an abundance of H_2O.

Oxidation-Reduction Reactions

Oxidation-reduction (redox) processes always involve *pairs* of molecules, one the oxidized form and one the reduced form of some underlying molecular structure. For example, quinone has the two forms

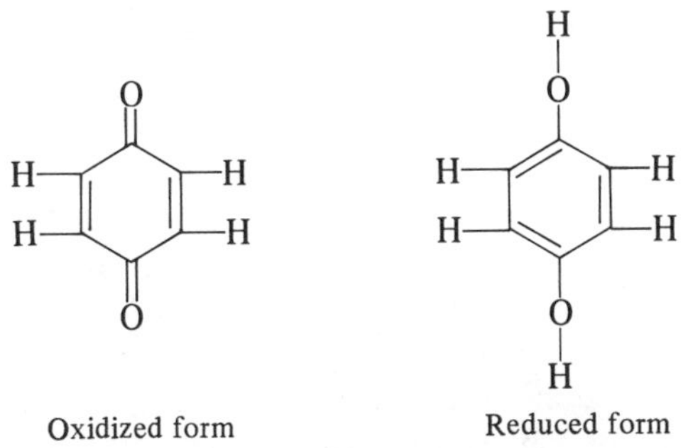

(4.48)

Oxidized form

Reduced form

This redox pair of molecules will be denoted by (Q/QH_2), where Q is quinone and QH_2 is dihydroquinone. Other types of pairs are exhibited by the ferric-ferro iron states (Fe^{3+}/Fe^{2+}) and by the redox states of the coenzyme NAD (nicotinamide-adenine dinucleotide), $(NAD^+/NADH)$. The general event is the transfer of electrons to the oxidized form, thereby producing the reduced form.

For the (Fe^{3+}/Fe^{2+}) pair, this is all that happens. In the (Q/QH_2) pair a proton accompanies each electron during the reduction, and two protons and two electrons are required for complete reduction. The (NAD^+/NADH) pair actually requires two electrons and one proton for a reduction, exhibiting a third type of reduction.

As a consequence of the variety of involvement of protons in redox reactions, pH effects are also variable.

Iron reactions are independent of pH, as exhibited by the redox reaction between two cytochromes, each of which contains an iron atom in its active site:

$$\text{Cyt}_b\ \text{Fe}^{2+} + \text{Cyt}_c\ \text{Fe}^{3+} \rightleftharpoons \text{Cyt}_b\ \text{Fe}^{3+} + \text{Cyt}_c\ \text{Fe}^{2+} \tag{4.49}$$

Using (4.36) for this case gives

$$\Delta G = \Delta G^0 + 2.3RT \log_{10} \frac{\gamma[\text{Cyt}_b\ \text{Fe}^{3+}]\gamma[\text{Cyt}_c\ \text{Fe}^{2+}]}{\gamma[\text{Cyt}_b\ \text{Fe}^{2+}]\gamma[\text{Cyt}_c\ \text{Fe}^{3+}]} \tag{4.50}$$

This is a one-electron exchange reaction. One mole of electrons has a charge of $-6 \times 10^{23} \times 1.6 \times 10^{-19}$ C = -9.6×10^4 C. This quantity is 1 faraday. (1 F). Equivalent expressions are

$$\begin{aligned} 1\,F &= 9.6 \times 10^4 \text{ C} = 9.6 \times 10^4 \text{ J/V} \\ &= \frac{9.6 \times 10^4}{4.186} \text{ cal/V} \cong 23 \text{ kcal/V} \end{aligned} \tag{4.51}$$

It is common practice in the discussion of redox reactions to convert from the energy units in (4.50) to units of volts, by dividing (4.50) by the charge per mole, using the notation

$$\Delta E = -\frac{\Delta G}{F} \quad \text{and} \quad \Delta E^0 = -\frac{\Delta G^0}{F} \tag{4.52}$$

for the voltage differences between initial and final states. The minus sign records the fact that electrons are negative. (Note that volts have units of energy per charge. There is also an energy unit of electron volts, defined by 1 eV = 1.6×10^{-12} erg. Do not confuse volts and electron volts.) Equation (4.50) can now be written

$$\Delta E = \Delta E^0 - 2.3 \frac{RT}{F} \log_{10} \frac{\gamma[\text{Cyt}_b\ \text{Fe}^{3+}]\gamma[\text{Cyt}_c\ \text{Fe}^{2+}]}{\gamma[\text{Cyt}_b\ \text{Fe}^{2+}]\gamma[\text{Cyt}_c\ \text{Fe}^{3+}]} \tag{4.53}$$

A useful number is $2.3RT/F$ at about $T = 300°K$.

$$2.3\,\frac{RT_{300}}{F} \simeq 0.06\text{ V (60 mV)} \tag{4.54}$$

The quantity ΔE^0 in (4.53) is obtained from a table of redox potentials (see Table 4.1), which show that

$$(\text{Cyt}_c\ \text{Fe}^{3+}/\text{Cyt}_c\ \text{Fe}^{2+}) \qquad E^0 = 0.22\text{ V}$$

$$(\text{Cyt}_b\ \text{Fe}^{3+}/\text{Cyt}_b\ \text{Fe}^{2+}) \qquad E^0 = 0.12\text{ V}$$

Note that ΔE^0 is $E^0_{\text{final}} - E^0_{\text{initial}}$ and final and initial are determined by the *reduced* member of each pair. Thus, in (4.49), $\text{Cyt}_c\ \text{Fe}^{2+}$ is the reduced product, so that E^0_{final} is 0.22, and therefore

$$\Delta E^0 = 0.22 - 0.12 = 0.10\text{ V}$$

in this example. Note again that, even though this number is positive, it is not an energy but a potential. When multiplied by $-F$, an energy is obtained

$$\Delta G^0 = -9600\text{ J}$$

which means that, at standard state, Cyt_b gives electrons to Cyt_c.

A pH-dependent reaction is

$$2H_2O + UQH_2 + 2Fe^{3+}\text{ cyanide} \rightleftharpoons UQ + 2Fe^{2+}\text{ cyanide} + 2H_3^+O \tag{4.55}$$

where UQ denotes ubiquinone and UQH_2 is its reduced form. As written, this looks like a pentamolecular reaction. It can be broken down into a succession of binary steps, but this would create too great a diversion at present. The analog of (4.53) is

$$\Delta E = \Delta E^0 - 2.3\,\frac{RT}{2F}\log_{10}\frac{\gamma[UQ]\,\gamma[Fe^{2+}\text{cyanide}]^2\,[H_3^+O]^2}{\gamma[UQH_2]\,\gamma[Fe^{3+}\text{cyanide}]^2\,[H_2O]^2} \tag{4.56}$$

in which division by $2F$ is used since *two* electrons are exchanged in this redox reaction. At fixed pH and with $[H_2O] = 55.6$ molar, this equation can be written

$$\Delta E = \Delta E^0 + 2.3\,\frac{RT}{F}\,\text{pH} + 2.3\,\frac{RT}{F}\log_{10}(55.6)$$

$$- 2.3\,\frac{RT}{2F}\log_{10}\left(\frac{\gamma[UQ]\,\gamma[Fe^{2+}\text{cyanide}]^2}{\gamma[UQH_2]\,\gamma[Fe^{3+}\text{cyanide}]^2}\right) \tag{4.57}$$

Table 4.1

Redox Pair	$\Delta E^{0\prime}$ at pH 7 (V)
Oxygen/water	0.815
Ferric/ferrous	0.77
Nitrate/nitrite	0.42
Ferricyanide/ferrocyanide	0.36
Oxygen/hydrogen peroxide	0.30
Cytochrome *a*; ferric/ferrous	0.29
Cytochrome *c*; ferric/ferrous	0.22
Crotonyl-SCoA/butyryl-SCoA	0.19
Methemoglobin/hemoglobin	0.17
Cytochrome b_2; ferric/ferrous	0.12
Ubiquinone; oxidized/reduced	0.10
Dehydroascorbic acid/ascorbic acid	0.08
Metmyoglobin/myoglobin	0.046
Fumaric acid/succinic acid	0.03
Methylene blue, oxidized/reduced	0.01
Yellow enzyme, FMN/$FMNH_2$	-0.122
Pyruvate + ammonium/alanine	-0.13
α-Ketoglutarate + ammonium/glutamic acid	-0.14
Oxaloacetate/malate	-0.17
Pyruvate/lactate	-0.19
Acetaldehyde/ethanol	-0.20
Riboflavin, oxidized/reduced	-0.21
Glutathione, oxidized/reduced	-0.23
Acetoacetate/β-hydroxybutyrate	-0.27
Lipoic acid, oxidized/reduced	-0.29
NAD^+/NADH	-0.32
Pyruvate/malate	-0.33
Uric acid/xanthine	-0.36
Carbon dioxide/formate	-0.42
H^+/H_2	-0.42
3-P-Glycerate + $2H^+$ + $2e^-$/glyceraldehyde-3-phosphate + H_2O	-0.55
Acetate/acetaldehyde	-0.60
Succinate/α-ketoglutarate	-0.67
Acetate + carbon dioxide/pyruvate	-0.70

The combination $\Delta E^0 + 2.3RT/F$ pH is denoted by $\Delta E^{0\prime}$, in Table 4.1, which is compiled for pH 7. The quantity $2.3RT/F \log_{10}(55.6) \simeq 0.10$ V at $T = 300°$K can be significant. A bigger effect is caused by the pH term, which for pH 7 gives

$$2.3 \frac{RT_{300}}{F} 7 = 0.42 \text{ V} \tag{4.58}$$

Thus, if protons are released during the reaction, then at pH 7 ΔE is larger by 0.42 V per mole of protons. If protons are consumed–that is, they are found on the reactant side of the equation–then ΔE is decreased by -0.42 V. The G^0 for H_2, molecular hydrogen, is zero by convention. In the redox pair (H^+/H_2), H_2 is on the product side and H^+ is on the reactant side. Therefore, at pH 7, this corresponds with an $E^{0\prime}$ value for (H^+/H_2) of -0.42 V per mole of protons (see Table 4.1 for confirmation).

The reaction

$$\text{NADH} + \text{H}_3^+\text{O} + \text{pyruvate} \rightleftharpoons \text{NAD}^+ + \text{H}_2\text{O} + \text{lactate} \tag{4.59}$$

provides an example for the case in which two electrons and one proton are given up by NADH. Therefore, neglecting the γ terms, we obtain

$$\begin{aligned}\Delta E &= \Delta E^0 - 2.3\frac{RT}{2F}\log_{10}\left(\frac{[\text{NAD}^+]\,[\text{H}_2\text{O}]\,[\text{lactate}]}{[\text{NADH}]\,[\text{H}_3^+\text{O}]\,[\text{pyruvate}]}\right)\\ &= \Delta E^0 - 2.3\frac{RT}{2F}\text{pH} + 2.3\frac{RT}{2F}\log_{10}(55.6)\\ &\quad - 2.3\frac{RT}{2F}\log_{10}\left(\frac{[\text{NAD}^+]\,[\text{lactate}]}{[\text{NADH}]\,[\text{pyruvate}]}\right)\end{aligned} \tag{4.60}$$

in which the second equality explicitly shows the pH effect and the $[H_2O]$ effect. Unlike the situation in (4.57), all denominators here involve $2F$. From Table 4.1 it is seen that for $[H_2O] = 1$ molar

$$\Delta E^{0\prime} = (-0.19) - (-0.32) = 0.13 \text{ V} \tag{4.61}$$

A representative set of redox potentials at pH 7 is given in Table 4.1. By the definition of the standard state, $\gamma[H_2O] = 1$ molar is assumed in this table.

The Nernst Equation

It has already been seen that, in a redox reaction at pH 7,

$$S_1 + S_2 + \cdots \rightleftharpoons P_1 + P_2 + \cdots \tag{4.62}$$

where n electrons are transferred,

$$\Delta G = \Delta G^0 + 2.3RT\log_{10}\left(\frac{\gamma[P_1]\gamma[P_2]\cdots}{\gamma[S_1]\gamma[S_2]\cdots}\right) \tag{4.63}$$

Table 4.2

Half-Reaction[a]	$E^{0\prime}$ at pH 7 (V)
$\frac{1}{2}O_2 + 2H^+ + 2e^- \rightarrow H_2O$	0.816
$Fe^{3+} + e^- \rightarrow Fe^{2+}$	0.771
$SO_4^{2-} + 2H^+ + 2e^- \rightarrow SO_3^{2-} + H_2O$	0.48
$NO_3^- + 2H^+ + 2e^- \rightarrow NO_2^- + H_2O$	0.42
Cytochrome a_3-Fe^{3+} + e^- → cytochrome a_3-Fe^{2+}	0.55
$\frac{1}{2}O_2 + H_2O + 2e^- \rightarrow H_2O_2$	0.30
Cytochrome a–Fe^{3+} + e^- → cytochrome a–Fe^{2+}	0.29
Cytochrome c-Fe^{3+} + e^- → cytochrome c-Fe^{2+}	0.25
2,6-Dichlorophenolindophenol* + $2H^+$ + $2e^-$ → 2,6-DCPP†	0.22
Crotonyl-S-CoA + $2H^+$ + $2e^-$ → butyryl-S-CoA	0.19
$Cu^{2+} + e^- \rightarrow Cu^+$	0.15
Methemoglobin-Fe^{3+} + e^- → hemoglobin-Fe^{2+}	0.139
Ubiquinone + $2H^+$ + $2e^-$ → ubiquinone-H_2	0.10
Dehydroascorbate + $2H^+$ + $2e^-$ → ascorbate	0.06
Metmyoglobin–Fe^{3+} + e^- → myoglobin–Fe^{2+}	0.046
Fumarate + $2H^+$ + $2e^-$ → succinate	0.030
Methylene blue* + $2H^+$ + $2e^-$ → methylene blue†	0.011
Pyruvate + NH_3 + $2H^+$ + $2e^-$ → alanine	−0.13
α-Ketoglutarate + NH_3 + $2H^+$ + $2e^-$ → glutamate + H_2O	−0.14
Acetaldehyde + $2H^+$ + $2e^-$ → ethanol	−0.163
Oxalacetate + $2H^+$ + $2e^-$ → malate	−0.175
FAD + $2H^+$ + $2e^-$ → $FADH_2$	−0.18
Pyruvate + $2H^+$ + $2e^-$ → lactate	−0.190
Riboflavin + $2H^+$ + $2e^-$ → riboflavin–H_2	−0.200
Cystine + $2H^+$ + $2e^-$ → 2cysteine	−0.22
$S^0 + 2H^+ + 2e^- \rightarrow H_2S$	−0.23
1,3-Diphosphoglyceric acid + $2H^+$ + $2e^-$ → GAP + P_i	−0.29
Acetoacetate + $2H^+$ + $2e^-$ → β-hydroxybutyrate	−0.290
Lipoate* + $2H^+$ + $2e^-$ → lipoate†	−0.29
$NAD^+ + 2H^+ + 2e^- \rightarrow NADH + H^+$	−0.320
$NADP^+ + 2H^+ + 2e^- \rightarrow NADPH + H^+$	−0.320
Pyruvate + CO_2 + $2H^+$ + $2e^-$ → malate	−0.33
Uric acid + $2H^+$ + $2e^-$ → xanthine	−0.36
Acetyl-S–CoA + $2H^+$ + $2e^-$ → acetaldehyde + CoA	−0.41
CO_2 + $2H^+$ + $2e^-$ → formate	−0.420
$H^+ + e^- \rightarrow \frac{1}{2}H_2$	−0.420
Ferredoxin–Fe^{3+} + e^- → ferredoxin–Fe^{2+}	−0.420
Gluconate + $2H^+$ + $2e^-$ → glucose + H_2O	−0.45
3-Phosphoglycerate + $2H^+$ + $2e^-$ → glyceraldehyde-3-phosphate + H_2O	−0.55
Methylviologen* + $2H^+$ + $2e^-$ → methylviologen†	−0.55

Table 4.2 (*Continued*)

Half-Reaction[a]	$E^{0'}$ at pH 7 (V)
Acetate + $2H^+ + 2e^- \rightarrow$ acetaldehyde	-0.60
Succinate + $CO_2 + 2H^+ + 2e^- \rightarrow$ α-ketoglutarate - H_2O	-0.67
Acetate + $CO_2 + 2H^+ + 2e^- \rightarrow$ pyruvate	-0.70

[a]Written as a reduction. An asterisk (*) denotes an oxidized form; a dagger (†), a reduced form.

or equivalently

$$\Delta E = \Delta E^0 - 2.3 \frac{RT}{nF} \log_{10} \left(\frac{\gamma[P_1]\gamma[P_2] \cdots}{\gamma[S_1]\gamma[S_2] \cdots} \right) \tag{4.64}$$

if n electrons are transferred. If either reactants or products include protons, it is always possible to account for this in terms of H_3^+O. Doing so automatically forces consideration of $[H_2O]$ effects in biological situations. Equation (4.64) is referred to as Nernst's equation in some books.

Nernst's equation is also used to refer to "half-reactions" and redox potentials related by

$$E = E^0 + 2.3 \frac{RT}{nF} \log_{10} \left(\frac{\gamma[\text{oxidized form}]}{\gamma[\text{reduced form}]} \right) \tag{4.65}$$

for the redox pair (oxidized/reduced). If E_1^0 corresponds with (S_1/P_1) and E_2^0 corresponds with (P_2/S_2) in (4.65), then

$$\Delta E^0 = E_1^0 - E_2^0 \tag{4.66}$$

in (4.64). In applying half-reaction analysis, it is useful to explicitly consider the redox reaction in terms of electrons and protons. Table 4.1 can be replaced by Table 4.2 when this has been done. For every H^+ depicted, think of H_3^+O, and for every e^- depicted think of an electron carrier C^-. Thus the reduction half-reaction of "cupric" copper, depicted by $Cu^{2+} + e^- \rightarrow Cu^+$, is really the full reaction

$$Cu^{2+} + C^- \rightleftharpoons Cu^+ + C \tag{4.67}$$

For this full reaction (4.64) yields

$$\Delta E = \Delta E^0 - 2.3 \frac{RT}{nF} \log_{10} \left(\frac{\gamma[Cu^+]\gamma[C]}{\gamma[Cu^{2+}]\gamma[C^-]} \right) \tag{4.68}$$

At standard-state values for all reactants and products,

$$\Delta E = \Delta E^0 \tag{4.69}$$

This ΔE^0 is comprised of a term for (Cu^{2+}/Cu^+), which is $E^0 = 0.15$ V and a term for (C/C^-) that is unspecified. In any other half-reaction, such as in the reduction of ferric iron $(Fe^{3+} + e^- \rightarrow Fe^{2+})$, it is also really the full reaction that should be considered:

$$Fe^{3+} + C^- \rightleftharpoons Fe^{2+} + C \tag{4.70}$$

and

$$\Delta E = \Delta E^0 - 2.3\frac{RT}{nF}\log_{10}\left(\frac{\gamma[Fe^{2+}]\gamma[C]}{\gamma[Fe^{3+}]\gamma[C^-]}\right) \qquad (n = 1) \tag{4.71}$$

At standard state, $\Delta E = \Delta E^0$ with $E^0 = 0.77$ V for (Fe^{3+}/Fe^{2+}) and a term for (C/C^-) that is still unspecified. However, in the coupled transfer, a full reaction comprised of two half-reactions,

$$Fe^{3+} + Cu^+ \rightleftharpoons Fe^{2+} + Cu^{2+} \tag{4.72}$$

The (C/C^-) terms exactly cancel out to give

$$\Delta E = \Delta E^0 - 2.3\frac{RT}{nF}\log_{10}\left(\frac{\gamma[Fe^{2+}]\gamma[Cu^{2+}]}{\gamma[Fe^{3+}]\gamma[Cu^+]}\right) \qquad (n = 1) \tag{4.73}$$

with $E^0 = (0.77) - (0.15)$ V $= 0.62$ V. The positivity of this result imples that ferric iron will spontaneously oxidize cuprous copper. In neither of these half-reactions is H^+ or H_2O of importance. This means that the value of E^0 for (Fe^{3+}/Fe^{2+}) and for (Cu^{2+}/Cu^+) are identically given by their $E^{0\prime}$ values for pH 7 or for any other pH and for $[H_2O]$ of any value.

Now consider the reaction depicted by $NAD^+ + 2H^+ + 2e^- \rightarrow NADH + H^+$. This is really

$$NAD^+ + H_3^+O + 2C^- \rightleftharpoons NADH + 2C + H_2O \tag{4.74}$$

and

$$\Delta E = \Delta E^0 - 2.3\frac{RT}{nF}\log_{10}\left(\frac{\gamma[NADH]\gamma[C]^2[H_2O]}{\gamma[NAD^+]\gamma[C^-]^2[H_3^+O]}\right) \qquad (n = 2) \tag{4.75}$$

If $[H_2O] = 1$ molar, then

$$\Delta E = \Delta E^{0'} - 2.3 \frac{RT}{2F} \log_{10} \left(\frac{\gamma[\mathrm{NADH}]\gamma[C]^2}{\gamma[\mathrm{NAD}^+]\gamma[C^-]^2} \right) \tag{4.76}$$

where $\Delta E^{0'} = \Delta E^0 - 2.3RT/2F$ pH. At pH = 7, $\Delta E^{0'} = \Delta E^0 - 0.21$ V. Because all contributions for (C/C^-) effects will cancel in any coupled reaction, the value of $\Delta E^{0'}$ is taken to be just the contribution from $(\mathrm{NAD}^+/\mathrm{NADH})$, which is $E^{0'} = -0.32$. This corresponds with a value of $E^0 = -0.11$ for $(\mathrm{NAD}^+/\mathrm{NADH})$ at standard-state pH = 0. If $[H_2O] = 1$ molar is *not* assumed, but instead the biological value $[H_2O] = 55.55$ molar is used, then there is a correction of $-2.3RT/2F \log_{10} (55.55) = -0.05$ V. This means that, in place of $E^{0'} = -0.32$ for $(\mathrm{NAD}^+/\mathrm{NADH})$ at pH = 7 and $[H_2O] = 1$ molar, there should be $E^{0'} = -0.37$ V for $[H_2O] = 55.55$ molar.

The ubiquinone reduction is depicted by $\mathrm{UQ} + 2H^+ + 2e^- \rightarrow \mathrm{UQH}_2$. This is really the full reaction

$$\mathrm{UQ} + 2H_3^+O + 2C^- \rightleftharpoons \mathrm{UQH}_2 + 2H_2O + 2C \tag{4.77}$$

and

$$\Delta E = \Delta E^0 - 2.3 \frac{RT}{2F} \log_{10} \left(\frac{\gamma[\mathrm{UQH}_2]\gamma[C]^2 [H_2O]^2}{\gamma[\mathrm{UQ}]\gamma[C^-]^2 [H_3^+O]^2} \right) \tag{4.78}$$

For ΔE^0, the (C/C^-) terms are ignored for the reasons already given above, and the pH and $[H_2O]$ effects are

$$\Delta E^0 - 2.3 \frac{RT}{2F} 2\mathrm{pH} - 2.3 \frac{RT}{2F} 2 \log_{10} (55.55) \tag{4.79}$$

Note the extra factors of 2. The value $E^{0'} = 0.1$ V in Table 4.2 includes the pH effect, but *not* the $[H_2O]$ term. The E^0 value at pH 0 would be 0.42 V more positive! The $[H_2O]$ correction is more *negative* by -0.1 V. It will prove essential, later, to always incorporate the additional correction for nonstandard $[H_2O]$ when using $E^{0'}$ values in biological contexts.

Membrane Potentials

Nernst's name figures again in the discussion of the effects of transmembrane electrical potentials on charged ions with respect to which the membrane is semipermeable. The internal energy of a charged molecule, U_i^0, will include an electrical contribution whenever the molecular ion is in an electrical field. If the ion carries the charge $z|e|$ in which z is a positive or negative integer, and if the ion experiences the electrical potential ψ, the energy contribution to U_i^0 is $z|e|\psi$.

Consider two chambers separated by a semipermeable membrane, permeable to an ionic species M of charge $z|e|$, as shown in (4.80). An externally applied electrical field produces a transmembrane potential such that

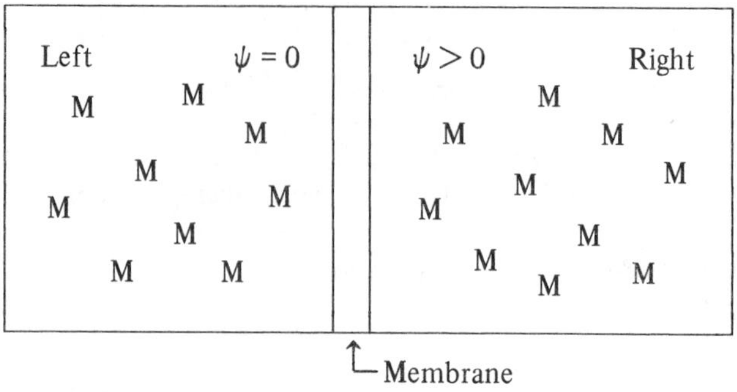

Left $\psi = 0$		$\psi > 0$ Right
M M M M M M M M M		M M M M M M M M M M

Membrane (4.80)

$$\Delta\psi = \psi_{\text{right}} - \psi_{\text{left}} > 0 \tag{4.81}$$

The Gibbs free energy on each side of the membrane is

$$\begin{aligned} G_{\text{left}} &= N_M^{\text{left}} U_M^0 + N_M^{\text{left}} P V_M^0 - N_M^{\text{left}} T \overline{S}_M^{0\text{left}} + N_M^{\text{left}} k_B T \ln (A_M^{\text{left}}) \\ G_{\text{right}} &= N_M^{\text{right}} (U_M^0 + z|e|\Delta\psi) + N_M^{\text{right}} P V_M^0 - N_M^{\text{right}} T \overline{S}_M^{0\text{right}} \\ &\quad + N_M^{\text{right}} k_B T \ln (A_M^{\text{right}}) \end{aligned} \tag{4.82}$$

It is assumed that P and T are identical on the left and right sides; $\overline{S}_M^{0\text{left}}$ and $\overline{S}_M^{0\text{right}}$ are not necessarily equal, because the volumes may be different on left and right, and from (3.4)

$$\overline{S}_M^0 = S_M^0 + k_B \ln \left[\left(\frac{2\pi M_M k_B T}{h^2} \right)^{3/2} V \right] \tag{4.83}$$

Equilibrium is achieved when the Gibbs free energy per molecule, the "chemical potential" $\mu \equiv G/N$, is equal across the membrane. Including the volume effect in this requirement implies

$$-z|e|\Delta\psi = k_B T \ln \frac{A_M^{\text{right}}/V^{\text{right}}}{A_M^{\text{left}}/V^{\text{left}}} \tag{4.84}$$

or

$$-z|e|\Delta\psi = k_B T \ln \frac{\gamma[M]^{\text{right}}}{\gamma[M]^{\text{left}}}$$

or

$$\frac{\gamma[M]^{\text{right}}}{\gamma[M]^{\text{left}}} = \exp\left(-\frac{z|e|\Delta\psi}{k_B T}\right) = \exp\left(-z\frac{F\Delta\psi}{RT}\right) \tag{4.85}$$

Thus the distribution of ions across the membrane is determined by the $\Delta\psi$. This equation is also called Nernst's equation.

It is important not to confuse this discussion of the effect of an externally applied electrical potential on an ion distribution with the *generation* of an electrical potential by ion transport. The generation of an electrical potential across a membrane will be discussed in Chapter 7.

References

1 H. Mahler and E. Cordes, *Biological Chemistry* (Harper and Row, New York, 1966), Chapter 5.

2 W. J. Moore, *Physical Chemistry*, 4th ed. (Prentice-Hall, Englewood Cliffs, N.J., 1972), Chapters 1 through 10.

3 Irwin H. Segel, *Biochemical Calculations*, 2nd ed. (Wiley, New York, 1976).

This book is full of worked examples.

4 R. E. Dickerson, *Molecular Thermodynamics* (W. A. Benjamin, New York, 1969).

An appendix provides G^0 values for many substances. The ΔG^0_{298} value for O—H (aqueous) quoted in Dickerson's appendix is -37.595 kcal/mole instead of the value -32.89 kcal/mole used here. The value chosen by Dickerson is also quoted by Mahler and Cordes in their Table 5-5 for the aqueous hydroxide ion. The error stems from the inconsistent use of K_i and K_w. If the value -37.595 kcal/mole for $\Delta G^0_{\text{O-H}}$ is used in (4.6), then $K_i = 10^{-14}$ instead of the correct value 3.24×10^{-18}. Since 10^{-14} is precisely the value of K_w, this is the source of the error. The correct value of $\Delta G^0_{\text{O-H}}$ at $T = 298°\text{K}$ is -32.89 kcal/mole.

CHAPTER

5

Free Energy Transductions

Many biologically important processes can be analyzed by preparing a closed system initially in a nonequilibrium state. This can be achieved by simply "closing" a previously open system. Because the system is now closed, the second law of thermodynamics (for constant T and P) requires that the Gibbs free energy, for each individual reaction step, monotonically decrease. This requirement is really quite stringent, and can be checked for a great many processes. When checked, it has always been found to be true and explains the tendency for processes to proceed in the directions they do.

The simplest energy transductions are elementary reaction steps. For example, the hydrolysis of the dipeptide glycylglycine,

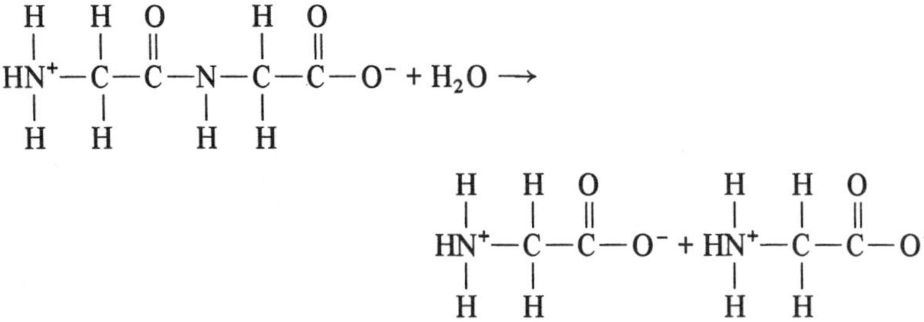

$$\begin{array}{ccccccccccc} \mathrm{H} & & \mathrm{H} & & \mathrm{O} & & & & \mathrm{H} & & \mathrm{O} \\ | & & | & & \| & & & & | & & \| \\ \mathrm{HN^+} & - & \mathrm{C} & - & \mathrm{C} & - & \mathrm{N} & - & \mathrm{C} & - & \mathrm{C}-\mathrm{O^-} + \mathrm{H_2O} \longrightarrow \\ | & & | & & & & | & & | & & \\ \mathrm{H} & & \mathrm{H} & & & & \mathrm{H} & & \mathrm{H} & & \end{array}$$

$$\begin{array}{ccccccccccc} \mathrm{H} & & \mathrm{H} & & \mathrm{O} & & \mathrm{H} & & \mathrm{H} & & \mathrm{O} \\ | & & | & & \| & & | & & | & & \| \\ \mathrm{HN^+} & - & \mathrm{C} & - & \mathrm{C}-\mathrm{O^-} & + & \mathrm{HN^+} & - & \mathrm{C} & - & \mathrm{C}-\mathrm{O^-} \\ | & & | & & & & | & & | & & \\ \mathrm{H} & & \mathrm{H} & & & & \mathrm{H} & & \mathrm{H} & & \end{array} \qquad (5.1)$$

involves a $\Delta G^{0\prime}$ of about -15×10^3 J. This means that at standard state the hydrolysis proceeds to the right. At $[H_2O] = 55.6$ molar, the reaction has a ΔG containing the term $-2.3RT\log_{10}(55.6) = -9.6 \times 10^3$ J, which implies that the hydrolysis is even more likely.

A hydrolysis reaction satisfies $\Delta G < 0$ whenever the initial state is a nonequilibrium state. As the Gibbs free energy decreases, $\Delta S > 0$ (i.e., heat is released). Thus hydrolysis converts free energy into heat.

A more useful reaction is one in which some of the free energy of the reactants is maintained in the products, such as in (5.2). For this process $\Delta G^{0\prime} < 0$, but ATP is a compound that is rich in free energy and maintains some of the free

1-3-Diphosphoglycerate (3-Phosphoglyceroyl phosphate)

$$\Delta G^{0'} = -4.5\ \text{kcal/mole} \qquad \text{ADP} \rightarrow \text{ATP} \qquad \bullet\ \text{enzyme (Phosphoglycerate kinase)} \tag{5.2}$$

3-Phosphoglycerate

energy of the energy-rich 1,3-diphosphoglycerate. This reaction constitutes an elementary free energy transduction that is simply a phosphate transfer. 1,3-Diphosphoglycerate is an intermediate of energy metabolism. On the other hand, ATP is the source of chemical free energy for an incredible variety of processes that consume free energy. The transduction from intermediary metabolism phosphate into usable phosphate currency is a very fundamental transduction.

This type of transduction is also called the "common intermediate principle" of energy transfer by *coupled reactions*. In (5.2) 1,3-diphosphoglycerate carries phosphate directly to ADP to make ATP. For the *uncoupled* hydrolysis reactions

$$\text{1,3-diphosphoglycerate} + H_2O \rightleftharpoons \text{3-phosphoglycerate} + H_3PO_4 \tag{5.3}$$

and

$$ATP + H_2O \rightleftharpoons ADP + H_3PO_4$$

the $\Delta G^{0'}$ values for hydrolysis are -11.8 and -7.3 kcal/mole, respectively. When coupled together, as in (5.2), the net $\Delta G^{0'}$ is given by subtracting -7.3 kcal/mole from -11.8 kcal/mole, which equals the value given in (5.2). This is just one example of a variety of phosphate carriers that occur in metabolism and may couple to ATP, either to make ATP from ADP or to use ATP to get a phosphate. Table 5.1 lists representative $\Delta G_0'$ values for the hydrolysis of some of these

Table 5.1

Compound	$\Delta G^{0'}$ (kcal/mole)
Phosphoenolpyruvate	-14.80
3-Phosphoglyceroyl phosphate	-11.80
Phosphocreatine	-10.30
Acetyl phosphate	-10.10
Phosphoarginine	-7.70
ATP ($\rightarrow$ ADP + P_i)	-7.30
Glucose 1-phosphate	-5.00
Fructose 6-phosphate	-3.80
Glucose 6-phosphate	-3.30
Glycerol 1-phosphate	-2.20

molecules. Note, however, that these $\Delta G^{0\prime}$ values should not be confused with ΔG values. For example, the coupled process

$$\text{ATP} + \text{arginine} \rightleftharpoons \text{phosphoarginine} + \text{ADP} \tag{5.4}$$

has a $\Delta G_0' = (+7.70 - 7.30)$ kcal/mole = +0.40 kcal/mole. Nevertheless, this transduction may well proceed to the right, with $\Delta G < 0$, if there is an excess of ATP and arginine over ADP and phosphoarginine.

When sequences of energy transductions are studied, two distinct categories appear. In the first category, which contains considerable variety, sequences are found to rationalize seemingly more complex events. The second category, although containing fewer examples, is far more intriguing and includes processes usually referred to as autocatalytic.

Simple Sequences

Coenzymes are functional chemical groups that are directly involved in biological reactions and are usually associated with a protein, the apoenzyme, in enzymes. The coenzyme is sometimes called the "prosthetic group" and often contains a molecular fragment identical with a known vitamin. Coenzymes function in catalytic cycles, in which they are used over and over again while substrate species are consumed and product species accumulate. The oxidative decarboxylation of pyruvate to acetyl-coenzyme A (CoA) is a good example:

$$\begin{gathered}\text{Pyruvate} + \text{NAD}^+ + \text{CoA-SH} \rightleftharpoons \text{acetyl-CoA} + \text{NADH} + \text{H}^+ + \text{CO}_2 \\ \Delta G^{0\prime} = -8.0 \text{ kcal/mole}\end{gathered} \tag{5.5}$$

The components of (5.6) are either substrate level or catalytic components. The amount of acetyl-CoA and NADH obtained is proportional to the amounts of pyruvate, NAD^+, and CoA supplied. This makes these components substrate level components. The rate of the transformation is proportional to the amounts of the catalytic components: TPP, LSS, and FAD. The equilibrium amount of acetyl-CoA is *not* dependent on the amounts of TPP, LSS, and FAD. In the

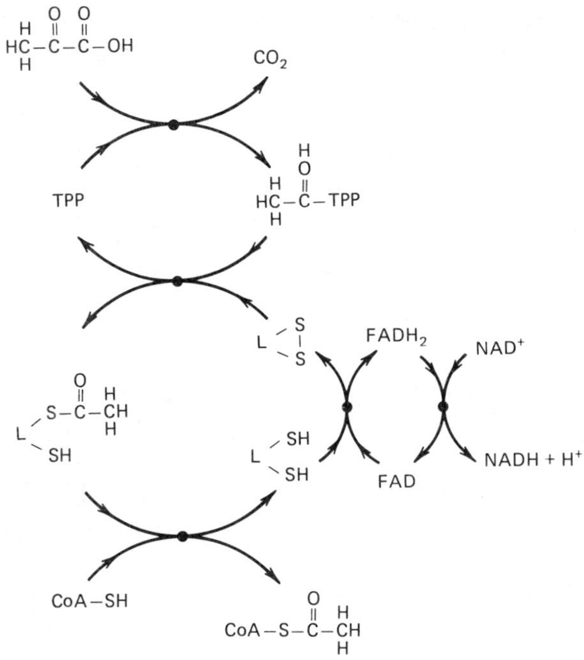

Pyruvate ($H_3C-CO-COOH$) → CO_2; TPP → $H_3C-C(OH)H-TPP$ (hydroxyethyl-TPP) → TPP;

Oxidized lipoic acid (L with S–S) → $L(SH)(S-CO-CH_3)$ (acetyl lipoamide);

$L(SH)(S-CO-CH_3)$ + CoA–SH → $CoA-S-CO-CH_3$ + $L(SH)_2$;

$L(SH)_2$ + FAD → L(S–S) + $FADH_2$;

$FADH_2$ + NAD^+ → FAD + NADH + H^+

TPP is thiamine pyrophosphate

L(S–S) is oxidized lipoic acid

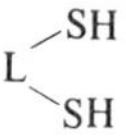

$L(SH)_2$ is reduced lipoic acid

Thiamine:
(TPP)

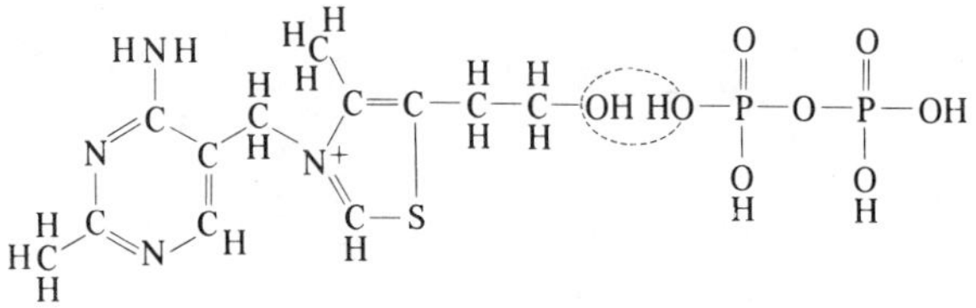

Lipoic acid:
(LSS)

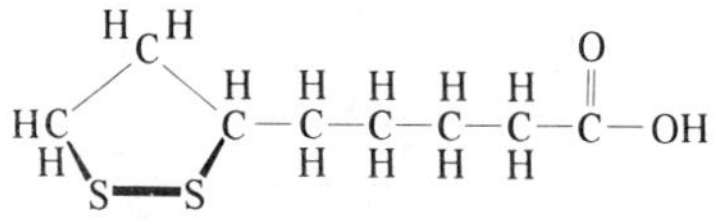

(5.6)

broader perspective of total energy metabolism, CoA and NAD^+ will be seen to be catalytic also. All of this is illustrated by the pathway fragment in (5.7). Let

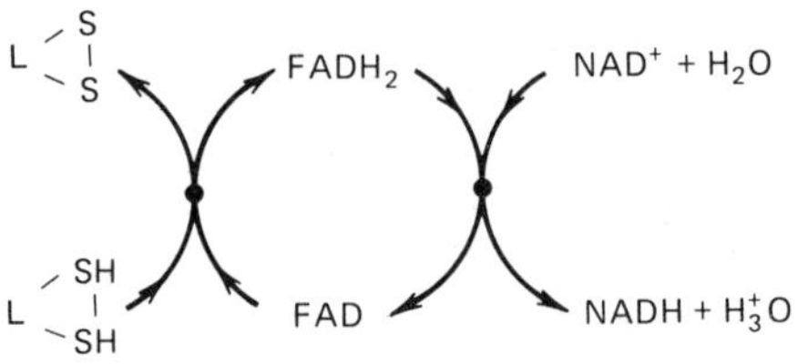

$$\begin{matrix} \mathrm{L}\begin{matrix}\diagup\mathrm{S}\\ \mid \\ \diagdown\mathrm{S}\end{matrix} & & \mathrm{FADH_2} & & \mathrm{NAD^+ + H_2O} \\ & \times & & \times & \\ \mathrm{L}\begin{matrix}\diagup\mathrm{SH}\\ \mid \\ \diagdown\mathrm{SH}\end{matrix} & & \mathrm{FAD} & & \mathrm{NADH + H_3^+O} \end{matrix} \tag{5.7}$$

N_1 = number of $\mathrm{L}\begin{matrix}\diagup\mathrm{SH}\\ \diagdown\mathrm{SH}\end{matrix}$ molecules

N_2 = number of $\mathrm{L}\begin{matrix}\diagup\mathrm{S}\\ \mid \\ \diagdown\mathrm{S}\end{matrix}$ molecules

N_3 = number of NAD^+ molecules
N_4 = number of NADH molecules
N_5 = number of FAD molecules
N_6 = number of $FADH_2$ molecules
N_7 = number of H_2O molecules
N_8 = number of H_3^+O molecules

The reaction is assumed to occur at pH = 7 with $[H_2O]$ = 55.55 molar. The rates for these reactions are

$$\begin{aligned} \mathrm{L}\begin{matrix}\diagup\mathrm{SH}\\ \diagdown\mathrm{SH}\end{matrix} + \mathrm{FAD} &\underset{k_2}{\overset{k_1}{\rightleftharpoons}} \mathrm{L}\begin{matrix}\diagup\mathrm{S}\\ \mid \\ \diagdown\mathrm{S}\end{matrix} + \mathrm{FADH_2} \\ \mathrm{FADH_2 + NAD^+ + H_2O} &\underset{k_4}{\overset{k_3}{\rightleftharpoons}} \mathrm{FAD + NADH + H_3^+O} \end{aligned} \tag{5.8}$$

and the rate equations are, ignoring the γ terms for simplicity,

$$\begin{aligned} \frac{d}{dt}N_1 &= -k_1N_1N_5 + k_2N_2N_6 = -\frac{d}{dt}N_2 \\ \frac{d}{dt}N_4 &= k_3N_6N_3N_7 - k_4N_4N_5N_8 = -\frac{d}{dt}N_3 \end{aligned} \tag{5.9}$$

[In contrast with earlier conventions in this book, the second equation in (5.9) involves a three-molecule event rather than a binary event. However, because H_2O is the third molecule and $[H_2O]$ is always so large in biology, this is not an unlikely event. Moreover, it can be viewed as a succession of two purely binary events.] Clearly, the rates of formation of $\mathrm{L}\begin{matrix}\diagup\mathrm{S}\\ \mid \\ \diagdown\mathrm{S}\end{matrix}$ and NADH (i.e., N_2 and N_4, respectively) are proportional to the amounts of FAD and $FADH_2$ (N_5 and N_6, respectively). However, the equilibrium values are determined by $(d/dt)N_2 = 0 = (d/dt)N_4$, which imply

$$\frac{N_2^{\text{eq}}}{N_1^{\text{eq}}} = \frac{k_1}{k_2}\frac{N_5^{\text{eq}}}{N_6^{\text{eq}}}, \quad \frac{N_4^{\text{eq}}}{N_3^{\text{eq}}} = \frac{k_3}{k_4}\frac{N_6^{\text{eq}}N_7^{\text{eq}}}{N_5^{\text{eq}}N_8^{\text{eq}}}, \quad \frac{N_7^{\text{eq}}}{N_8^{\text{eq}}} = 55.55 \times 10^7 \tag{5.10}$$

The ratio $(N_5^{\text{eq}}/N_6^{\text{eq}})$ can be eliminated to yield an identity for the substrate level components only [note that LSS is a substrate level component in (5.7) and is catalytic in (5.6)]:

$$N_2^{\text{eq}}N_4^{\text{eq}} = N_1^{\text{eq}}N_3^{\text{eq}}\frac{k_1k_3}{k_2k_4}(55.55 \times 10^7) \tag{5.11}$$

and

$$\frac{k_1k_3}{k_2k_4} = \exp\left[-\frac{1}{k_BT}(G_2^0 + G_4^0 - G_1^0 - G_3^0)\right]$$

Note that even the rate ratio is independent of G_5^0 and G_6^0, the contributions from the catalytic components.

By starting with values for N_1, N_2, N_3, and N_4 that do not equal the equilibrium values, the pathway fragment (5.7) can be run forward or backward. Beginning with excesses of N_1 and N_3 will drive it in the direction indicated by the arrow arcs in (5.7), but loading up with an excess of N_2 and N_4 will drive it in the reverse direction. In each case, for each of the two steps, $\Delta G < 0$ will be observed until equilibrium is reached and $\Delta G = 0$.

Throughout metabolism the interplay between substrate level components and catalytic components can be discerned. Much of evolution has been devoted to the fine tuning and regulation of the catalytic components, and these objectives have been achieved through the structures of the apoenzyme proteins.

Autocatalytic Sequences [1, 3]

The biological character of reaction sequences lies not so much in the individual steps as in the whole sequence. This begins to become clear when *autocatalytic* sequences are studied. There are, of course, many nonbiological autocatalytic processes as well. What distinguishes biological sequences from nonbiological ones will be explained at the end of Chapter 10.

Glycolysis provides an instructive example of an autocatalytic sequence. In this sequence, the free energy content of glucose is transduced into free energy currency in ATP molecules. The production of ATP during the con-

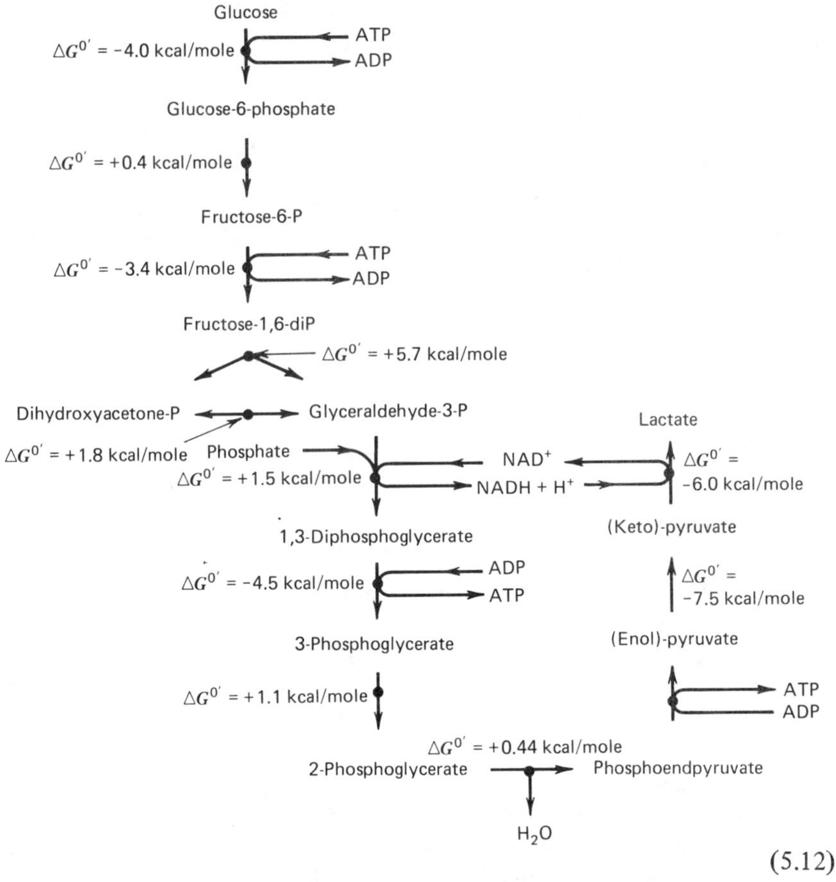

(5.12)

versions of this pathway is autocatalytic [see (5.12)]. Several aspects of this pathway are of interest in addition to its autocatalytic production of ATP. The system can be prepared in the laboratory, and was so prepared many years ago. Every transformation indicated requires a specific protein enzyme for its catalysis, *except* the (enol)-pyruvate → (keto)-pyruvate isomerization, which is one of the rare spontaneous, uncatalyzed reactions in all of biochemistry. The preparation of the system requires all of the enzymes, appropriate amounts of ions like Mg^{2+}, appropriate pH and temperature, and substrate level amounts of glucose, ADP, and phosphate. Each step obeys the second law requirement $\Delta G < 0$. [Compare the $\Delta G'_0 > 0$ steps with (5.5).] Curiously, two molecules of ATP are required at the beginning of the pathway to get it going. Fructose-1,6-diphosphate breaks up into two three-carbon moieties, dihydroxyacetone and glyceraldehyde-3-phosphate, which are interconvertible. Because it is glyceraldehyde-3-phosphate that actually connects up with the rest of the pathway, it is correct to view this part of the pathway as converting one six-carbon moiety, fructose-1,6-diphosphate, into two three-carbon moieties, that is, into two molecules of glyceraldehyde-3-phosphate. Each of these three-carbon moieties goes through the rest of the pathway and generates two molecules of ATP for a total of four ATP molecules for the original fructose. In summary, two ATP molecules "prime" the reaction sequence that produces four ATP molecules for a net gain of two ATP molecules. In this way more reactions starting with glucose can be primed to produce still more ATP, which doubles in amount during each complete passage along the pathway. This is the autocatalytic production of ATP. The redox pair (NAD^+/NADH) acts catalytically in this pathway.

This example also serves to exemplify the concepts developed in Chapter 3. Once the system gets going, it starts out making ATP. All of this obeys $\Delta G < 0$ overall and at each step. On a longer time scale the ATP will eventually hydrolyze to ADP and phosphate again, so that full equilibrium for this system is obtained when all the glucose has been degraded to lactate and all of the ATP that was concomitantly produced has been hydrolyzed. It is the intermediate stage, when there is still a great deal of ATP, that is of biological relevance, and it is this stage that integrates into the *entire* pathway system of metabolism.

Another good example of an autocatalytic sequence is the carbon cycle, during which the free energy content of ATP and NADPH, both of which are produced during the primary steps of photosynthesis, is transduced into carbohydrate (sugar) free energy. The production of carbohydrate from its precursors, H_2O and CO_2, is the "anabolic" pathway of carbohydrate metabolism, whereas the preceding example of glycolysis represents the "catabolic" pathway in carbohydrate metabolism. During the anabolic portion of carbohydrate metabolism, the free energy content of sunlight, through the molecular intermediates ATP and NADPH, is transduced into free energy in carbohydrate. The catabolic portion of this metabolism transduces this carbohydrate energy back into a more utilizable free energy currency in ATP. Plants use both portions, the former during daylight hours and the latter during nighttime hours. Animals exclusively make use of the *catabolic* portion and are consequently absolutely dependent on plants in their biochemistry. As will be seen later, this dependency is also exhibited with regard to O_2, which plants liberate during photosynthesis and which animals depend on for the complete "oxidation" of carbohydrate back into its simple precursors, H_2O and CO_2 [see (5.13)].

The step from ribulose-1,5-diphosphate up to two 3-phosphoglyceric acid molecules also involves the incorporation of CO_2 and H_2O. Consequently, the five-carbon ribulose-1,5-diphosphate can break up into two three-carbon molecules of 3-phosphoglyceric acid. Further activation by ATP and reduction by NADPH yields glyceraldehyde-3-phosphate, the three-carbon moiety from which all of the other branches of this pathway flow. The stoichiometry of this branched

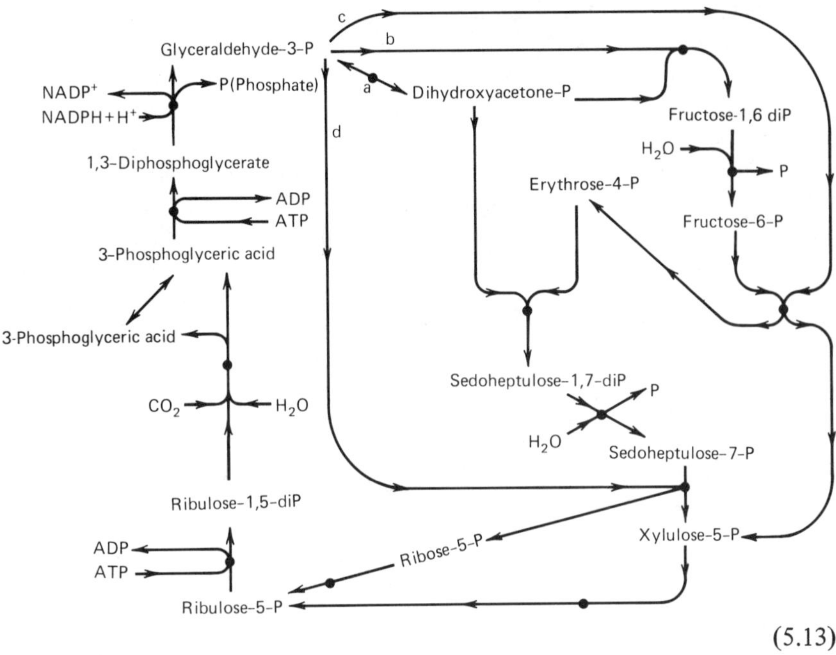

(5.13)

pathway is amusing to study. If initially there are five molecules of glyceraldehyde-3-phosphate, a total of 15 carbon atoms, then eventually three molecules of ribulose-5-phosphate will accumulate, accounting precisely for all 15 carbon atoms. The ATP- and NADPH-dependent steps will convert all of this into six molecules of glyceraldehyde-3-phosphate. This is not the factor of 2 as in glycolysis, but even a factor of 6/5 yields an exponential growth that is unmistakenly autocatalytic. If, instead, initially there are only three molecules of glyceraldehyde-3-phosphate, for a total of nine carbon atoms, then the branches leading to ribulose-5-phosphate cannot all be used. However, if one molecule of glyceraldehyde-3-phosphate is used on each of the pathways marked *a*, *b*, and *c* in (5.13), then one molecule of erythrose-4-phosphate and one molecule of xylulose-5-phosphate are produced, accounting for the nine carbons. The xylulose-5-phosphate becomes ribulose-5-phosphate, and the ATP- and NADPH-dependent steps of the cycle incorporate a molecule each of CO_2 and H_2O to yield two molecules of glyceraldehyde-3-phosphate. One of these follows pathway *a* to combine with erythrose-4-phosphate and ultimately appears in sedoheptulose-7-phosphate; the latter combines with the remaining glyceraldehyde-3-phosphate, which has taken pathway *d*, yielding ultimately two molecules of ribulose-5-phosphate. In this a total of 10 carbon atoms are involved. As both ribulose-5-phosphate molecules are processed along the ATP- and NADPH-dependent portion of the cycle, two molecules each of H_2O and CO_2 are incorporated, yielding four molecules of glyceraldehyde-3-phosphate. Now the pathway is up to 12 carbon atoms and will rapidly accumulate an exponentially increasing number of carbon atoms in the form of carbohydrate.

Every step of this pathway is catalyzed by enzymes. NADPH is an analog of NADH that contains an additional phosphate group at the $2'$ position of the ribose to which adenine is attached [see (1.13)].

Because DNA localizes in a single macromolecule the entire genetic heritage of an organism, many scientists have viewed it as an almost "living" molecule. It is surely a quintessential molecule. With regard to autocatalytic function, DNA is nonpareil as a molecule. In Chapter 1 the structure of duplex or double-helix DNA was given. It is a polymer made up of just four types of nucleotide, AMP, GMP, CMP, and TMP. Because the linkages are dehydration condensations, activated precursors are required for biosynthesis, and there are the triphosphates, ATP, GTP, CTP, and TTP. The two strands of DNA in a double helix are antiparallel and are connected together exclusively by the base-pairing hydrogen bonds that selectively link adenine with thymine and guanine with cytosine. The replication, or duplication, of DNA occurs by separating the two strands and lining up NTP's according to the base-pairing rules on each strand. The phosphodiester linkages between adjacent nucleotides form spontaneously from the triphosphate precursors and release pyrophosphate as a byproduct [see (5.14)]. This can be depicted as in (5.15). The molecule DNA polymerase is required for DNA duplication to occur. It is a catalytic component and appears to be absolutely essential for the replication of long-stranded DNA. Complete equilibrium for this system is achieved only after all the DNA has hydrolyzed back into nucleotides, which themselves will hydrolyze into bases, ribose, and phosphate. In addition, the polymerase enzyme, a protein, will also eventually hydrolyze into its monomeric constituents, amino acids. The time required for complete equilibrium is very long in comparison with the replication of DNA. The polymerase catalyzes a replication that is rapid enough for DNA to accumulate autocatalytically during a transient, initial phase, long before any significant hydrolysis of DNA or polymerase occurs. During all phases of development each step in this system obeys the second-law requirement: $\Delta G < 0$.

As a final example of autocatalysis, the proliferation of a cell population has been chosen. Consider a closed container of nutrient broth inoculated with a single bacterium. It is supposed that the nutrient broth is sufficiently diverse to support all the needs of cellular multiplication. An exponentially increasing cell population develops, until crowding and exhaustion of the nutrient take place, after which cell death and degradation relentlessly bring the entire system to equilibrium. In *complete* equilibrium all cellular membranes, proteins, and polynucleotides will have hydrolyzed into their monomeric constitutents. Nevertheless, the initial phase exhibits exponential population growth and $\Delta G < 0$ overall

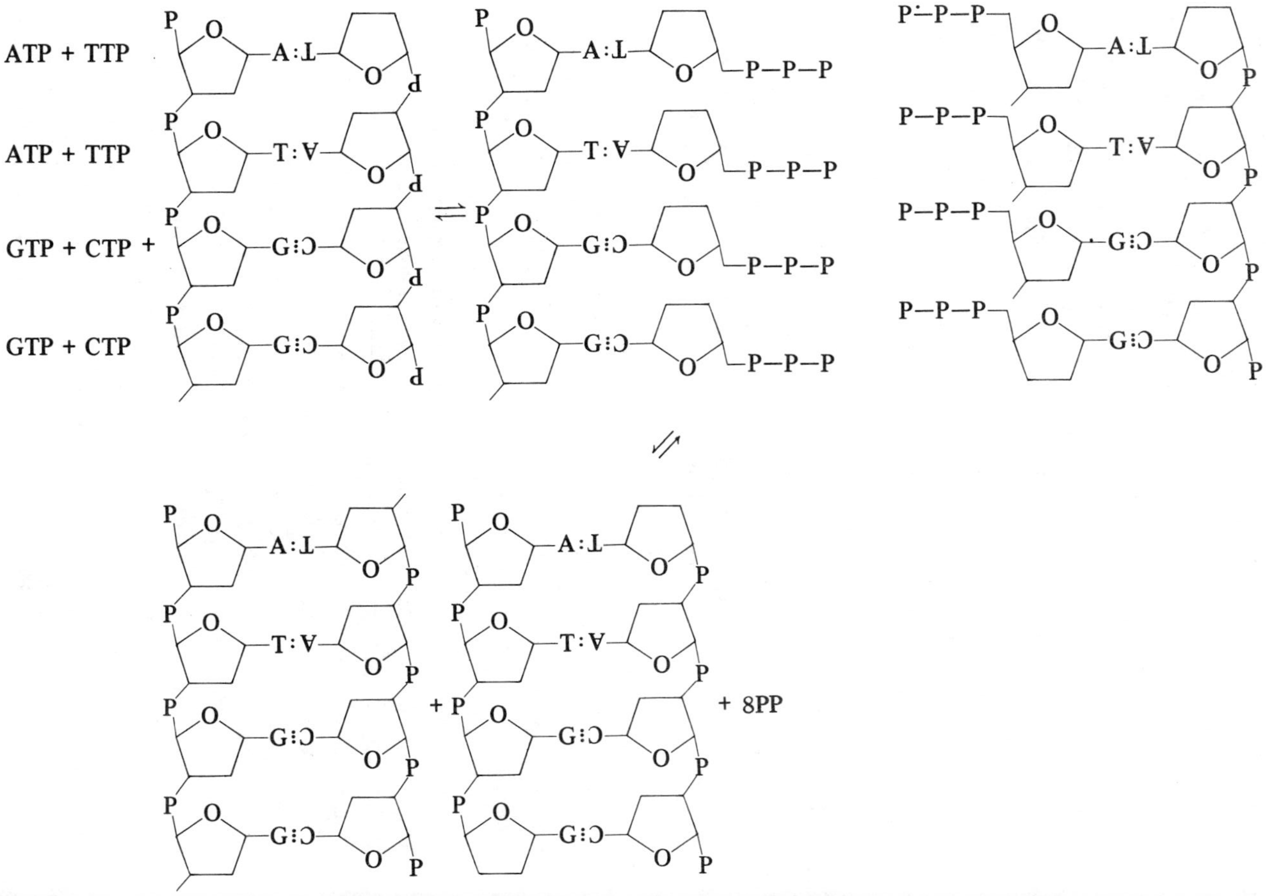

ATP + TTP

ATP + TTP

GTP + CTP +

GTP + CTP

+ 8PP

(5.14)

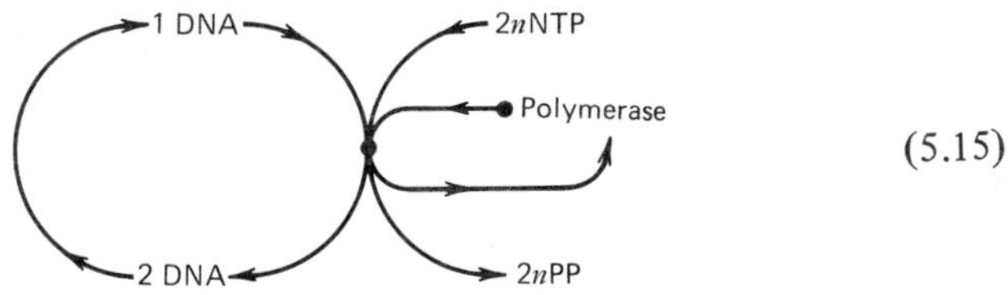

(5.15)

as well as step by step. From the source of free energy and materials in the nutrient broth, a single cell is able to create myriads of replicas. In this case all components are replicated as entire cells–rather the opposite extreme from the example of the DNA, in which case the polymerase is supplied as an external reagent. This limitation of the DNA example is shared by the examples of glycolysis and the carbon cycle as well because in each numerous enzymes had to be supplied as external reagents to support autocatalysis. The cell serves as the only example of an entirely macromolecularly self-sufficient autocatalytic object. A fundamental question that is pursued later in this book is, What is the minimal macromolecular structure capable of complete self-replication? The tentative suggestion is that it is the *cell* [2]. More will be said about this point in Chapter 7.

References

1 M. Calvin, *Chemical Evolution* (Oxford, New York, 1969).

2 F. Jacob, *The Logic of Life* (Pantheon, New York, 1973).

A historical treatment of the emergence of a molecular view of biology. The emphasis is on the entire cell and the concept of the "integron."

3 A. Lehninger, *Biochemistry*, 2nd ed. (Worth, New York, 1975).

CHAPTER

6

Self-Assembly

In the preceding chapters, the examples have primarily involved *covalent* bonds—the strong, thermally stable bonds that provide the structure of molecules. The weak, thermally labile bonds, such as hydrogen bonds, hydrophobic bonds, and van der Waals bonds, provide the structure in aggregations, or "complexes," of molecules. Weakly bonded assemblies of proteins, sometimes with polynucleotides, of considerable biological significance abound in cells. The cell itself is such an assembly, composed of many constituent subassemblies and surrounded by a self-assembled membrane.

Because the self-assembly process utilizes weak bonds, it possesses some remarkable properties. The interaction of subunits is *specific*, when specificity is satisfied the interaction is relatively *stable*, and yet thermal mechanisms are sufficient to *break* the subunits apart. A weak bond with a strength of about $10k_BT_{300} = 4 \times 10^{-13}$ erg (~6 kcal/mole) has a thermal "time of expectation" for breaking of $t \sim 2 \times 10^{-9}$ sec (nanoseconds!), as was explained in Chapter 1. Nevertheless, three simultaneous weak bonds lead to $t \sim 1$ sec. This is rather a long time on the biochemical time scale and explains the relative stability of multiple weak bonds with as few as three bonds. To make three simultaneous weak bonds, the surfaces of two subunits must contain *complementarily* placed residues or groups that make the weak bonds [see (6.1)]. This obligatory complementarity confers specificity. Consequently, many interactions that occur between subunits involve only one weak bond, and these "nonspecific" interactions are neither stable nor specific. Whenever complementary surfaces become adjacent, the specific, relatively stable interaction of multiple weak bonds takes place. The stability is only relative. Thermal motions will continually break individual bonds in a multiple weak bond. However, the remaining weak bonds will keep the subunits together until the broken bond reforms. Simultaneous breaks will break the entire multiple bond [1-3].

Water molecules play a fundamental role in all biological assembly processes. The dispersed subunits are usually surrounded by organized H_2O structures because water hydrogen bonds readily with many protein residues as well as

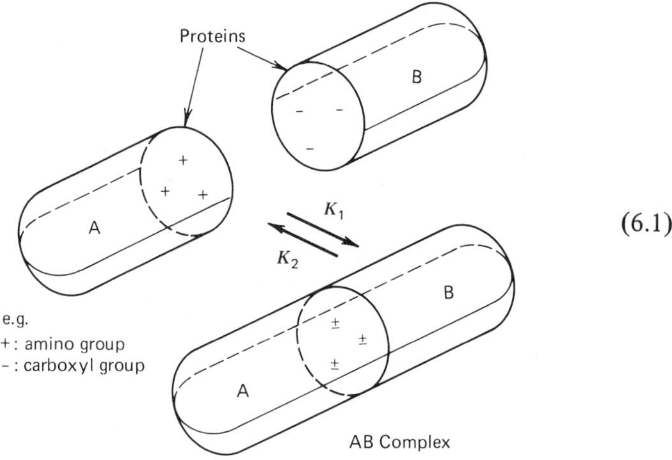

(6.1)

with other H_2O molecules. The various residues on a subunit that are involved in specific weak bonds with complementary residues on another subunit will be bonded to H_2O molecules that will have to be displaced when the specific, multiple weak bonding between the subunits takes place. Consequently, free energy considerations dealing with the ΔG for the formation of assemblies from dispersed subunits depend on an accounting of H_2O molecule effects on a detailed, molecular level. As in Chapter 4, $\Delta G \equiv G_{\text{final}} - G_{\text{initial}}$.

Experimentally, $\Delta G < 0$ has always been verified for the spontaneous self-assembly of an aggregate from its dispersed subunits. Nevertheless, observers often find the phenomenon of self-assembly to be paradoxical. The transition from the dispersed subunit state into the aggregated assembly state appears to be in the direction of more "organization," or increased *order.* In Chapter 3 this was associated with a decrease in entropy, or $\Delta S < 0$. It was already explained there that the second law of thermodynamics for an *isolated* system, $\Delta S > 0$, does not apply to biological systems in contact with temperature and pressure reservoirs, for which $\Delta G < 0$ is the second law. The identity

$$\Delta G = \Delta U + P\Delta V - T\Delta S \tag{6.2}$$

indicates that $\Delta G < 0$ and $\Delta S < 0$ are *not* incompatible, provided

$$|\Delta U + P\Delta V| > |T\Delta S| \tag{6.3}$$

and

$$\Delta U + P\Delta V < 0$$

It is instructive to show that the *entropy change for the subunits* does satisfy $\Delta S < 0$ during assembly, as suggested above. According to (2.17), the entropy is expressed, neglecting activity corrections,

$$S = N_A \overline{S}_A^0 + N_B \overline{S}_B^0 + N_{AB} \overline{S}_{AB}^0 - k_B N_A \ln \left(\frac{N_A}{e}\right) - k_B N_B \ln \left(\frac{N_B}{e}\right) - k_B N_{AB} \ln \left(\frac{N_{AB}}{e}\right) + \frac{3}{2} k_B (N_A + N_B + N_{AB}) \tag{6.4}$$

where $\overline{S}^0$, the shorthand expression introduced in Chapter 3, includes explicit contributions from the translational motion of the center of mass, rotational motion about the center of mass, and other internal state degrees of freedom:

$$\overline{S}^0 = k_B \ln \left[\left(\frac{2\pi M k_B T}{h^2} \right)^{3/2} V \right] + k_B \ln \left(\frac{8\pi^2 I k_B T}{h^2} \right) + S^0 \tag{6.5}$$

where M is the mass, I is the moment of inertia for a linear molecule, and S^0 accounts for the remaining internal degrees of freedom. [For an arbitrary molecule, the rotational inertial term is $k_B \ln [8\pi^2 (8\pi^3 ABC)^{1/2} / \sigma h^3 \times (k_B T)^{3/2}]$, where A, B, and C are the principal moments of inertia and σ is a symmetry number, which should also be present in the case of linear molecules to distinguish between homonuclear and heteronuclear diatomic molecules.] Overwhelmingly the biggest term in (6.5) is the translational entropy.

$$\left(\frac{2\pi M k_B T}{h^2} \right)^{3/2} V \sim 10^{26} \frac{V}{\text{cm}^3} \qquad (T \simeq 300^\circ\text{K}) \tag{6.6}$$

for M = mass of H_2O. For proteins the value is more than 50,000 times larger. By contrast, the rotational entropy has a smaller value,

$$\frac{8\pi^2 (8\pi^3 ABC)^{1/2}}{\sigma h^3} (k_B T)^{3/2} \simeq 10 \longrightarrow 10^5 \qquad (T \simeq 300^\circ\text{K}) \tag{6.7}$$

Typical small-molecule values for A, B, and C are $1 - 100 \times 10^{-40}$ g-cm^2. Proteins, however, may achieve moments of inertia in excess of $10^6 \times 10^{-40}$ g-cm^2. The ramaining internal degrees of freedom, such as vibrations and electronic levels, contribute partition functions with values in the range from 0 to 10. It is the translational entropy that is responsible for the impression of ordering, or growth of organization, when subunit assembly is observed.

Initially, imagine a volume of *1 liter of H_2O* in which subunits of species A

and B are dispersed. Let their initial concentrations both equal 10^{-3} molar. This means there are 6×10^{20} molecules of A and 6×10^{20} molecules of B, and zero molecules of AB before assembly takes place. The initial entropy is, again neglecting the γ terms for simplicity.

$$S_{\text{initial}} = \frac{3}{2} k_B(N_A + N_B) + N_A\bar{S}_A^0 + N_B\bar{S}_B^0 - k_BN_A \ln\left(\frac{N_A}{e}\right) - k_BN_B \ln\left(\frac{N_B}{e}\right) \tag{6.8}$$

where $N_A = N_B = 6 \times 10^{20}$. Permit assembly to occur. An equilibirum distribution of A, B, and AB will be produced, satisfying the constraint

$$N_A - N_A^{\text{eq}} = N_{AB}^{\text{eq}} = N_B - N_B^{\text{eq}} > 0 \tag{6.9}$$

in which the superscript "eq" denotes equilibrium value. These equilibrium populations must also satisfy the identity, in accord with (3.10),

$$\frac{N_{AB}^{\text{eq}}}{N_A^{\text{eq}}N_B^{\text{eq}}} = \frac{k_1}{k_2} = \exp\left[-\frac{1}{k_BT}(G_{AB}^0 - G_A^0 - G_B^0)\right] \tag{6.10}$$

where k_1 and k_2 are as defined in (6.1) and G_A^0, G_B^0, and G_{AB}^0 are given by

$$\begin{aligned} G_A^0 &= U_A^0 + PV_A^0 - T\bar{S}_A^0 \\ G_B^0 &= U_B^0 + PV_B^0 - T\bar{S}_B^0 \\ G_{AB}^0 &= U_{AB}^0 + PV_{AB}^0 - T\bar{S}_{AB}^0 \end{aligned} \tag{6.11}$$

The final entropy for the (subunit/complex) is

$$\begin{aligned} S_{\text{final}} = \frac{3}{2} k_B(N_A^{\text{eq}} + N_B^{\text{eq}} + N_{AB}^{\text{eq}}) + N_A^{\text{eq}}\bar{S}_A^0 + N_B^{\text{eq}}\bar{S}_B^0 + N_{AB}^{\text{eq}}\bar{S}_{AB}^0 - k_BN_A^{\text{eq}} \\ \ln\left(\frac{N_A^{\text{eq}}}{e}\right) - k_BN_B^{\text{eq}} \ln\left(\frac{N_B^{\text{eq}}}{e}\right) - k_BN_{AB}^{\text{eq}} \ln\left(\frac{N_{AB}^{\text{eq}}}{e}\right) \end{aligned} \tag{6.12}$$

The change in entropy, $\Delta S = S_{\text{final}} - S_{\text{initial}}$, can be expressed more compactly by using (6.9):

$$\begin{aligned} \Delta S = -\frac{3}{2} k_BN_{AB}^{\text{eq}} + N_{AB}^{\text{eq}}(\bar{S}_{AB}^0 - \bar{S}_A^0 - \bar{S}_B^0) - k_BN_A^{\text{eq}} \ln\left(\frac{N_A^{\text{eq}}}{N_A}\right) \\ - k_BN_B^{\text{eq}} \ln\left(\frac{N_B^{\text{eq}}}{N_B}\right) - k_BN_{AB}^{\text{eq}} \ln\left(\frac{eN_{AB}^{\text{eq}}}{N_AN_B}\right) \end{aligned} \tag{6.13}$$

Using a property of logarithms, the substitution

$$\ln\left(\frac{eN_{AB}^{\text{eq}}}{N_A N_B}\right) = 1 + \ln\left(\frac{N_{AB}^{\text{eq}}}{N_A^{\text{eq}} N_B^{\text{eq}}}\right) + \ln\left(\frac{N_A^{\text{eq}}}{N_A}\right) + \ln\left(\frac{N_B^{\text{eq}}}{N_B}\right) \qquad (6.14)$$

is allowed, and (6.13) becomes

$$\Delta S = -\frac{5}{2} k_B N_{AB}^{\text{eq}} + N_{AB}^{\text{eq}} \Delta \overline{S}^0 - k_B(N_A^{\text{eq}} + N_{AB}^{\text{eq}}) \ln\left(\frac{N_A^{\text{eq}}}{N_A}\right)$$
$$- k_B(N_B^{\text{eq}} + N_{AB}^{\text{eq}}) \ln\left(\frac{N_B^{\text{eq}}}{N_B}\right) - k_B N_{AB}^{\text{eq}} \ln\left(\frac{N_{AB}^{\text{eq}}}{N_A^{\text{eq}} N_B^{\text{eq}}}\right) \qquad (6.15)$$

The leading term is clearly negative. The second term can be written

$$N_{AB}^{\text{eq}} \Delta \overline{S}^0 \simeq -N_{AB}^{\text{eq}} k_B \ln\left[\left(\frac{2\pi k_B T}{h^2}\right)^{3/2} V \left(\frac{M_A M_B}{M_{AB}}\right)^{3/2}\right] \qquad (6.16)$$

because the translational entropy dominates other contributions in (6.5). For protein subunits such that $M_A = M_B = 20{,}000 \times 1.6 \times 10^{-24}$ g and $M_{AB} = M_A + M_B$, the value at $T = 300°\text{K}$ and $V = 10^3\ \text{cm}^3$ is

$$N_{AB}^{\text{eq}} \Delta \overline{S}^0 = -N_{AB}^{\text{eq}} k_B \ln(8 \times 10^{32}) = -76 k_B N_{AB}^{\text{eq}} \qquad (6.17)$$

The next two terms in (6.15), however, are positive because $N_A > N_A^{\text{eq}}$ and $N_B > N_B^{\text{eq}}$. Moreover, $N_A^{\text{eq}} + N_{AB}^{\text{eq}} = N_A$ and $N_B^{\text{eq}} + N_{AB}^{\text{eq}} = N_B$, which are the large initial values. To determine how big the entire term is requires finding the value of the ratios N_A^{eq}/N_A and N_B^{eq}/N_B. These can be determined from (3.10) and the constraint equation (6.9):

$$N_A^{\text{eq}} = \frac{N_A}{1 + \exp\left[-(1/k_B T)(G_{AB}^0 - G_A^0)\right]} \qquad (6.18)$$

and

$$N_B^{\text{eq}} = \frac{N_B}{1 + \exp\left[-(1/k_B T)(G_{AB}^0 - G_B^0)\right]}$$

Thus

$$-k_B(N_A^{eq} + N_{AB}^{eq}) \ln \left(\frac{N_A^{eq}}{N_A}\right) - k_B(N_B^{eq} + N_{AB}^{eq}) \ln \left(\frac{N_B^{eq}}{N_B}\right)$$

$$= +k_B N_A \ln \left\{1 + \exp\left[-\frac{1}{k_B T}(G_{AB}^0 - G_A^0)\right]\right\} \tag{6.19}$$

$$+k_B N_B \ln \left\{1 + \exp\left[-\frac{1}{k_B T}(G_{AB}^0 - G_B^0)\right]\right\} > 0$$

The inequality is valid for any values of $G_{AB}^0 - G_A^0$ and $G_{AB}^0 - G_B^0$. However, the largest positive value for these terms occurs when $G_{AB}^0 - G_A^0$ and $G_{AB}^0 - G_B^0$ are negative to a high degree. The last term in (6.13) can be rewritten by using (6.10)

$$-k_B N_{AB}^{eq} \ln \left(\frac{N_{AB}^{eq}}{N_A^{eq} N_B^{eq}}\right) = \frac{N_{AB}^{eq}}{T}(G_{AB}^0 - G_A^0 - G_B^0) \equiv N_{AB}^{eq} \frac{\Delta G^0}{T} \tag{6.20}$$

For the case at hand, depicted in (6.1) of this chapter, three weak bonds are involved in binding. This means that the size of ΔG^0 is less than roughly $-30k_BT$, or

$$N_{AB}^{eq} \frac{\Delta G^0}{T} \leqq -30 N_{AB}^{eq} k_B \tag{6.21}$$

Consequently this term is also negative. The total negative contribution to (6.15) is $-(\frac{5}{2} + 76 + 30)\ k_B N_{AB}^{eq}$; the total positive contribution is given in (6.19). Using the initial data $N_A = N_B = 6 \times 10^{20}$ and the equilibrium (6.10) with the constraints (6.9), it is found that

$$N_A^{eq} = N_B^{eq} = 7.7 \times 10^3 \qquad \text{and} \qquad N_{AB}^{eq} \simeq 6 \times 10^{20} \tag{6.22}$$

This means that there is essentially a complete conversion from dispersed subunits into dimeric AB complexes. These results also imply that

$$\frac{N_A^{eq}}{N_A} = \frac{N_B^{eq}}{N_B} = 1.3 \times 10^{-17} \tag{6.23}$$

which with (6.18) implies

$$\begin{aligned} G_{AB}^0 - G_A^0 = G_{AB}^0 - G_B^0 &= -k_B T \ln\left[(1.3 \times 10^{-17})^{-1}\right] \\ &= -39 k_B T \end{aligned} \tag{6.24}$$

In summation, the total negative contribution is less than or equal to

$$-\left(\frac{5}{2}+76+30\right)(6 \times 10^{20})\,k_B$$

whereas the total positive contribution is

$$+2(6 \times 10^{20})\,39k_B$$

The final result is that

$$\Delta S \leqq -(2 \times 10^{22})\,k_B \tag{6.25}$$

which surely justifies the impression that $\Delta S < 0$!

How can $\Delta G < 0$ also hold? From (6.2) and (6.3) it is necessary that

$$\Delta U + P\Delta V < T\Delta S < 0 \tag{6.26}$$

or

$$\Delta U + P\Delta V < -(2 \times 10^{22})\,k_B T$$

The $P\Delta V$ term is not large. In fact

$$P\Delta V = N_{AB}^{\text{eq}} P(V_{AB}^0 - V_A^0 - V_B^0) = N_{AB}^{\text{eq}} P\Delta V^0 \tag{6.27}$$

For proteins with a molecular weight of about 20,000, as is the case here, $V_A^0 = 25{,}000$ Å^3 $= 25 \times 10^{-24}$ liter. Surely ΔV^0 will be less than this because V_{AB}^0 is very nearly equal to $V_A^0 + V_B^0$. Using 25×10^{-24} liter as an upper bound for ΔV^0, it follows that

$$P\Delta V^0 \leqq 25 \times 10^{-15}\ \text{erg} = \frac{5}{8}\,k_B T_{300} \tag{6.28}$$

and

$$N_{AB}^{\text{eq}} P\Delta V^0 \leqq (3.75 \times 10^{20})\,k_B T_{300}$$

which is insignificant on the scale set by (6.25). Therefore, ΔU must account for the requirement (6.26). Because $\Delta U = N_{AB}^{\text{eq}}\Delta U^0 = (U_{AB}^0 - U_A^0 - U_B^0)$, it follows that

$$N_{AB}^{\text{eq}}\Delta U^0 < -(2 \times 10^{22})\,k_B T \tag{6.29}$$

or

$$\Delta U^0 < -33k_BT$$

This inequality guarantees (6.26), but it does not guarantee (6.21), because

$$\Delta G^0 = \Delta U^0 + P\Delta V^0 - T\Delta\overline{S}^0 = -30k_BT \tag{6.30}$$

requires, according to (6.17),

$$\Delta U^0 \leqslant -76k_BT - 30k_BT \tag{6.31}$$

for the subunit and complex. This requirement certainly includes (6.29).

How could such a large negative value for ΔU^0 be achieved? It results from the effect of solvent molecules, H_2O, on the energy states of the subunits A and B, and the complex AB. Before the complexes form, each subunit has H_2O associated with the residues which will eventually make up the weak bonds. This associated H_2O is released when the subunits aggregate into complexes. As H_2O effects were not included in the preceding analysis, they can be included in an "effective" internal energy contribution that is viewed as a solvent effect. Since each residue involved in a weak bond has at least one H_2O molecule (usually two to four) bound to it, the formation of AB releases at least $6 \times$ (two to four) H_2O molecules (three weak bonds). The entropy increase of the H_2O upon release is, according to (6.5) for H_2O and $V = 1$ liter,

$$\begin{aligned} \Delta S_{H_2O} &= 6k_B \ln (6 \times 10^{28}) N_{AB}^{\text{eq}} \times \text{(two to four)} \\ &= 66k_B N_{AB}^{\text{eq}} \times \text{(two to four)} \end{aligned} \tag{6.32}$$

Thus, the change in the Gibbs free energy that is associated with this H_2O is

$$\Delta G_{H_2O} = -66k_BTN_{AB}^{\text{eq}} \times \text{(two to four)} \tag{6.33}$$

or

$$-66k_BT/\text{complex} \times \text{(two to four)}$$

The weak bonds themselves will surely supply an additional $-10k_BT$ per complex, as would each additional bound H_2O, should there be more than one H_2O molecule per residue. The ΔG_{H_2O} can be incorporated into ΔU^0 for the subunit assembly, and this justifies (6.31). The factors 2 to 4 confirm (6.31) even more strongly.

In summary, it has been shown that, for the self-assembly of complexes from dispersed subunits, $\Delta G < 0$ and $\Delta S < 0$ for the (subunits/complex), and this is possible *because* of a solvent effect of bound H_2O on the subunits which is released upon assembly. The released H_2O contributes a large effect, as shown by (6.33), because the translational entropy of H_2O is large. This finally justifies all the claims at the beginning of this chapter, including the remark that detailed accounting of H_2O effects is essential.

Several examples of self-assembly are listed below. While each example involves more than two types of subunit and complexes of more than two constituents, the basic physics is essentially like that of the special case just analyzed detail.

Enzyme Complexes [4]

In Chapter 5 the transformation of pyruvate into acetyl-CoA was discussed. Equation (5.6) depicts the sequence of steps but does not indicate how these steps are organized by the enzymes involved. The enzyme complex isolated from *Escherichia coli* has the following properties: It is comprised of three different types of subunit: E_1 (pyruvate dehydrogenase), E_2 (dihydrolipoyltransacetylase), and E_3 (dihydrolipoyldehydrogenase). The TPP in (5.6) is bound to E_1 and the FAD in (5.6) is bound to E_3. The lipoate group is attached to a long lysine side chain of E_2, and this entire extending arm of E_2 can rotate freely. The complex contains 24 subunits of E_2, organized in eight groups of three. These eight "trimers" are arranged as a cube in (6.34). On each of the six

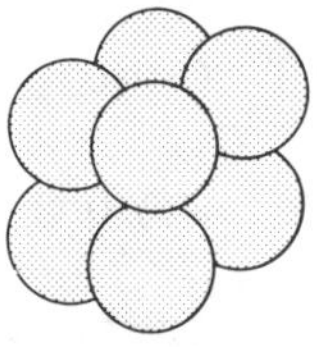

(6.34)

E_2 Trimer

faces of this cube there is a square of four E_3 subunits in (6.35). On each of the 12 edges of the cubic structure (6.34) there are two subunits of E_1, shown in (6.36). The entire complex contains 72 subunits [see (6.37)]. Adding to this complexity is the fact that E_1 subunits are actually $\alpha\beta$ dimers of two distinct (α and β) polypeptide chains. The molecular weights of E_1, E_2, and E_3 are 90,000, 40,000, and 55,000, respectively, and the entire complex has a molecular weight of 4,440,000!

The organizer of this complex is E_2. It spontaneously forms trimers, which

(6.35)

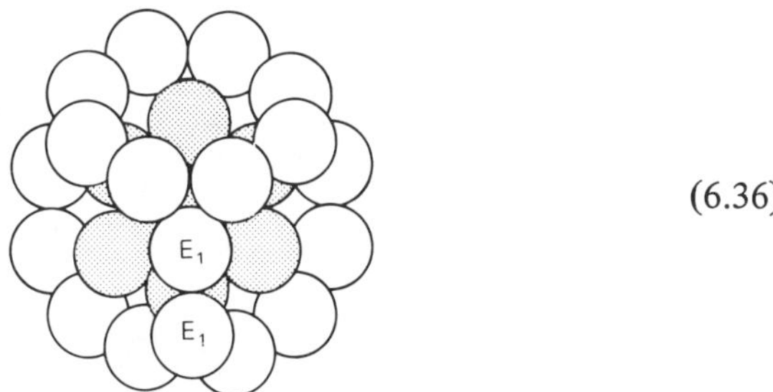

(6.36)

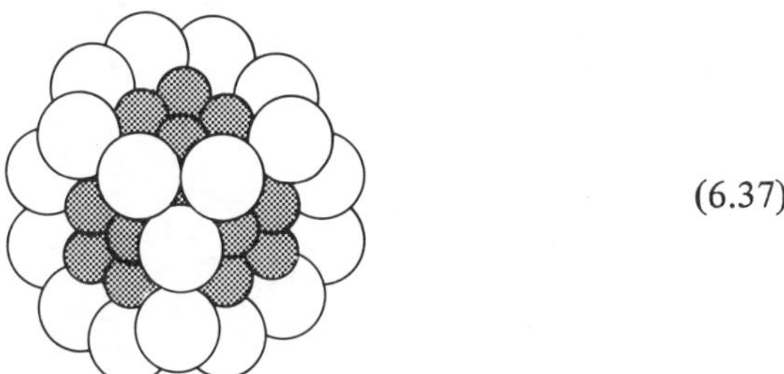

(6.37)

then spontaneously aggregate into the cubic structure (6.34). To this structure the subunits E_1 and E_3 spontaneously bind. This sequence of assembly, though short, is characteristic of complex formation. If one E_2 trimer is allowed to bind E_1 and E_3, a *functional* subcomplex is formed.

The functional properties of this complex can be depicted by considering just one corner of the cubic structure since there is an eightfold symmetry, or redundancy, to this function [see (6.38) and the discussion in Chapter 5]. The lipoate arm swings around as a result of thermal motion. This is truly a remarkable molecular machine.

There are other enzyme complexes, and some are partially characterized in Table 6.1.

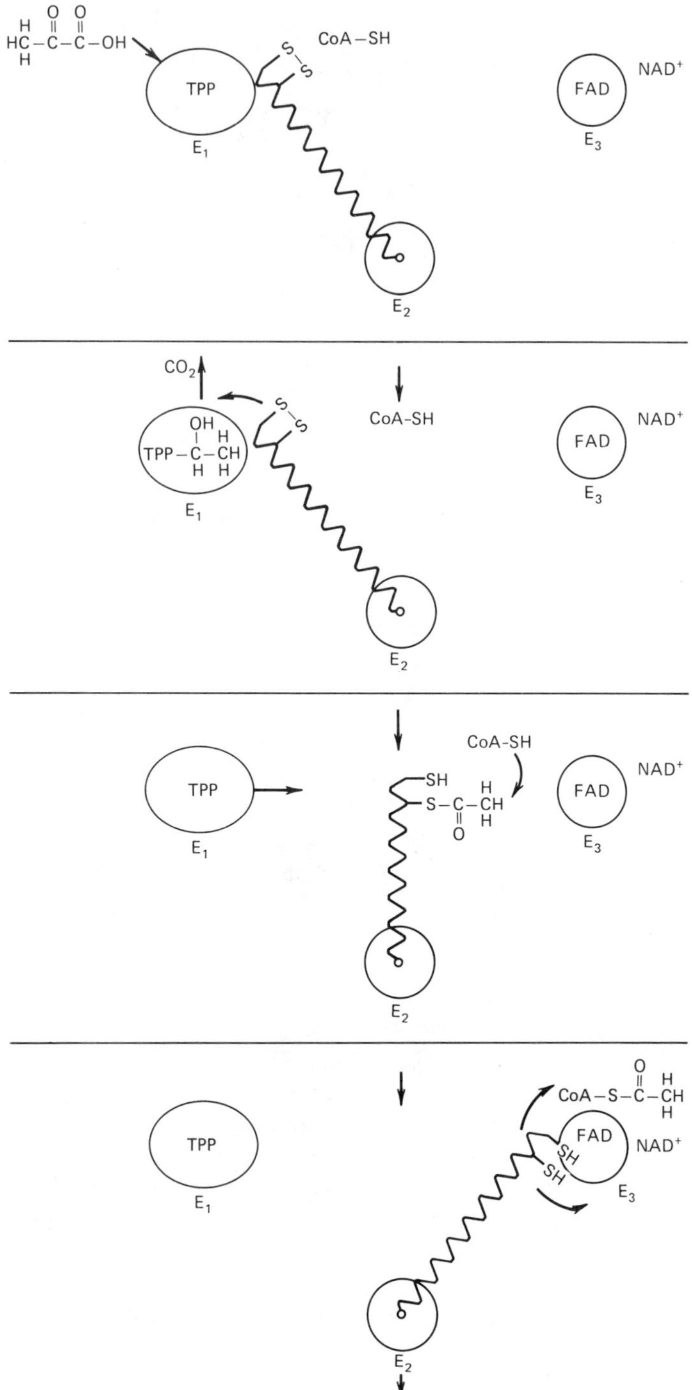

HC–C–C–OH
TPP
E_1
CoA–SH
FAD
NAD^+
E_3
E_2
CO_2
TPP–C–CH
CoA-SH
SH
S–C–CH
CoA–S–C–CH
SH

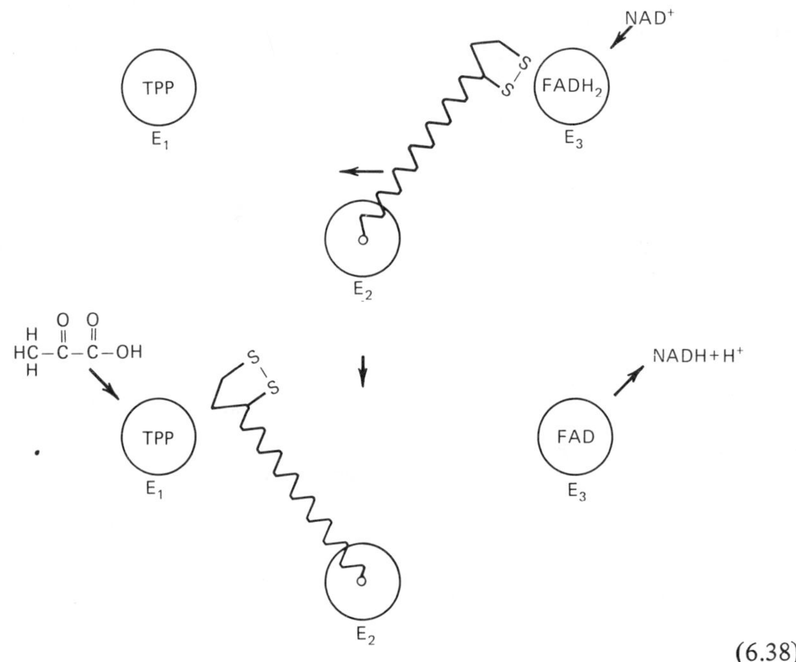

(6.38)

Table 6.1

Protein	Molecular Weight	Subunits Number	Subunits Molecular Weight
Insulin	11,466	2	5,733
β-Lactoglobulin	35,000	2	17,500
Neurospora malate dehydrogenase	54,000	4	13,500
Hemoglobin	64,500	4	16,000
Glycerol-1-phosphate dehydrogenase	78,000	2	40,000
Alkaline phosphatase	80,000	2	40,000
Creatine kinase	80,000	2	40,000
Enolase	82,000	2	41,000
Liver alcohol dehydrogenase	84,000	2	42,000
Procarboxypeptidase	87,000	1	34,500
		2	25,000
Hexokinase	96,000	4	24,000
Hemerythrin	107,000	8	13,500
Tryptophan synthetase A	29,000	1	29,000
Tryptophan synthetase B	117,000	2	60,000
Glyceraldehyde-3-phosphate dehydrogenase	140,000	4	37,000

Table 6.1 (*Continued*)

Protein	Molecular Weight	Subunits Number	Subunits Molecular Weight
Aldolase	142,000	3	50,000
Lactic dehydrogenase	150,000	4	35,000
Threonine deaminase	160,000	4	40,000
Cystathionine γ-synthetase	160,000	4	40,000
Fumarase	194,000	4	48,500
Tryptophanase	220,000	2	125,000
Pyruvate kinase	237,000	4	57,200
Catalase	250,000	4	60,000
Acetoacetate decarboxylase	260,000	8	30,000
Mitochondrial adenosine triphosphatase	284,000	10	26,000
Aspartyl transcarbamylase	310,000	2	96,000
		4	30,000
Phosphoenolpyruvate carboxytransphosphorylase	430,000	3–4	120,000
Apoferritin	480,000	20	24,000
Urease	483,000	6	83,000
Phosphorylase	495,000	4	125,000
β-Galactosidase	520,000	4	130,000
	130,000	3–4	40,000
Pyruvate carboxylase	660,000	4	165,000
	165,000	4	45,000
Propionyl carboxylase	700,000	4	175,000
Lipoic reductase-transacetylase	1,600,000	60	27,000
Glutamic dehydrogenase	2,000,000	8	250,000
	250,000	5	50,000
Turnip yellow mosaic virus	5,000,000	150	21,000
Poliomyelitis virus	5,500,000	130	27,000
Bushy stunt virus	9,000,000	120	60,000
Potato virux X	35,000,000	650	52,000
Tobacco mosaic virus	40,000,000	2130	17,500

From Dickerson and Geis [3].

Viruses [5]

Viruses are complexes of a few proteins and either DNA or RNA. There are double-stranded and single-stranded varieties of both RNA and DNA types. All of the viral components must be synthesized in a host cell that is replete with all of the necessary machinery for macromolecular synthesis. The assembly of the virus from its constituent polymers is spontaneous and can be achieved in a test tube (in vitro) as well as in vivo. The tobacco mosaic virus, (TMV), a single-stranded RNA virus, was the first virus to be reconstituted in vitro, in 1954 by H. Fraenkel-Conrat and R. C. Williams. The concept of self-assembly was given great impetus by this discovery. Parallel self-assembly studies in 1955 by F. O. Schmitt on the connective tissue protein collagen did likewise.

Of the many viruses that have been studied, only one is described below. It is the virus P22 of the host bacterium *Salmonella typhimurium*, and it is a double-stranded DNA virus. The self-assembly of this virus from its precursor DNA and its gene product proteins typifies the variety of functions and sequences of assembly of these macromolecules in viruses generally. Nevertheless, each type of virus involves assembly mechanisms unique to itself, so that the references must be consulted for details regarding other species.

Virus P22 contains a DNA chromosome that has a molecular weight of 27×10^6 and is confined inside an isometric "head" 6000 Å in diameter. Ten viral gene products, proteins, are required for the self-assembly of infectious particles. The notation "gp" is used [5] to designate a gene product protein, and the 10 proteins for P22 are labeled gp1, gp2, gp3, gp5, gp8, gp9, gp10, gp16, gp20, and gp26. Gene products gp2, gp3, and gp8 are not found in mature particles but are required during assembly.

Like some other viruses, for example, phages λ, T3, T4, and T7, P22 forms a "prohead" before its DNA is "encapsidated." Proteins gp5 and gp8 assemble into a double-shelled prohead. Protein gp8 has a molecular weight of 42,000, whereas gp5 has a weight of 55,000. About 250 molecules of gp8 catalyze the assembly of about 420 gp5 molecules into a double-shelled prohead that contains both proteins. Proteins gp16 and gp20 are not required for this process, but if DNA encapsidation is to subsequently proceed, they are also required. Proteins gp1, gp2, and gp3 are also required for the encapsidation of DNA by the prohead, after which gp2 and gp3 appear to be released.

The encapsidation of a "headful of DNA" by the prohead is concomitant with the release of all the gp8 proteins, which are apparently required as "scaffolding" for the assembly of the major coat protein gp5. Indeed, gp8 is called [5] the "scaffolding" protein. The released gp8 proteins can then engage in catalyzing the assembly of other gp5 proteins. Consequently, gp8 proteins are really quite scarce in a propagating culture of replicating virus particles. Cells infected with defective gp8 mutants accumulate unassembled coat gp5 proteins or aberrant aggregates of this protein. Cells infected with defective gp5 mutants accumulate unassembled gp8 proteins. It is therefore believed that in normal infections gp8 and gp5 proteins form a complex that then aggregates into the prohead. The encapsidation of a headful of DNA greatly alters the structure of the prohead, which forms a single shell as the gp8 proteins are released. At this stage the particles are unstable unless proteins gp10 and gp26 are added, after which the particle becomes stable. Without these proteins the heads lose the DNA both in vivo and in vitro. Proteins gp10 and gp26 provide the head-to-tail junction to which gp9 "tailplate" proteins attach to form the fully mature virus.

One step in this remarkable sequence stands out from the rest as truly remarkable. This step is the encapsidation of a DNA molecule with a molecular weight of 27×10^6 by a prohead with a volume of approximately 10^{11} Å^3 (10^{-16} liter). What causes such an immense molecule to confine itself to such a small volume? It would appear that this encapsidation is accompanied by a large decrease in translational entropy for the DNA. The concomitant release of 250 gp8 proteins will increase entropy, but this cannot be the explanation because this event is peculiar to P22, whereas in the λ virus, for example, proteins are *added* during encapsidation. The correct explanation at the molecular level for the process of DNA encapsidation remains an outstanding problem.

Ribosomes [6]

Ribosomes are very large complexes of ribosomal proteins and ribosomal RNAs (rRNAs). They are the essential "organelles" of bacterial cells on which protein biosynthesis occurs, a process described in detail in Chapter 10. The whole ribosome of *E. coli* consists of two major subunits that are labeled by their sedimentation coefficients in the centrifuge: the 30S subunit and the 50S subunit. The whole ribosome has sedimentation coefficient of 70S and a molecular weight of about 2.7×10^6. The 50S subunit consists of 1 23S RNA molecule, 1 5S RNA molecule, and 30 to 35 protein molecules; it has a molecular weight of about 1.8×10^6. The 30S subunit consists of 1 16S RNA molecule and about 20 protein molecules; it has a molecular weight of about 0.9×10^6. The rRNAs account for most of the mass with a combined molecular weight of about 1.8×10^6.

The dissociation and the subsequent "reconstitution" of 30S *E. coli* ribosomes with essentially full activity were accomplished in the late 1960s by Nomura and co-workers. Reconstitution experiments with the 50S *E. coli* ribosomal subunit have not been as successful, but success has been achieved with the comparable 50S subunit of the *Bacillus stearothermophilus* ribosome.

In reconstitution of the 30S subunit, the sequence in which the various proteins assemble with the 16S RNA has been worked out. Some proteins bind directly with the 16S RNA, while others bind to the already bound proteins afterward. Some proteins appear to play a purely assembly oriented role; others are clearly necessary for ribosomal functions in protein biosynthesis. Reconstitution experiments in which individual proteins are omitted have helped to identify their functions. For example, seven proteins become bound to the 16S RNA, whereas the rest bind only the first seven proteins. Proteins S16, S17, and possibly S9 have roles in assembly but no apparent ribosomal function roles. Without S16 and S17, 30S subunits assembled much more slowly, although fully functional 30S particles did eventually form.

Various types of test have shown strong evidence that the ribosomal subunits have a very precisely determined topological arrangement of their constituents that is faithfully reproduced in reconstitution experiments. It is now clearly demonstrated that the information for the correct assembly of ribosomal subunits is contained in the structure of their constituent proteins and rRNAs, and does not require nonribosomal factors in the cell.

References

1 A. Lehninger, *Biochemistry*, 1st ed. (Worth, New York, 1970), Chapter 33.

2 J. Watson, *Molecular Biology of the Gene*, 2nd ed. (Benjamin, New York, 1970), Chapter 4.

3 R. E. Dickerson and I. Geis, *The Structure and Action of Proteins* (Harper and Row, New York, 1969), Chapter 5.

4 M. Calvin, *Chemical Evolution* (Oxford, New York, 1969).

Chapter 9 has some beautiful electron micrographs of several enzyme complexes.

5 S. Casjens and J. King, "Virus Assembly," *Annual Review of Biochemistry*, **44**, 555–611 (1975).

An outstanding, comprehensive review of self-assembly in many types of virus.

6 M. Nomura, "Assembly of Bacterial Ribosomes," *Science*, **179**, 864 (1973).

CHAPTER

7

Cell Membranes and the Membrane Potential

A self-assembly process of considerable interest occurs when lipids are dissolved in H_2O. When the phosphoglyceride phosphatidylcholine is used, spherical vesicles spontaneously assemble. These vesicles have a bilayer structure that strikingly resembles the bilayer structure of the bacterial membrane as well as the membranes of several organelles such as the mitochondrion and the chloroplast. For this reason, it is believed that the basic structure of biological membranes is the lipid bilayer.

It is not difficult to explain the tendency of phosphoglycerides to segregate themselves from H_2O in the manner in which they do. The explanation depends on their structure (see Chapter 1). These molecules can be represented schematically as shown in (7.1), in which a charged polar head exists at one end and two

(7.1)

acyl hydrocarbon chains extend in the opposite direction. In (7.1) phosphatidylcholine is schematized, whereas other phosphoglycerides will possess different charge groups at their polar end. The charged ends are hydrophilic because of their strong tendency to form hydrogen bonds with H_2O molecules. The acyl chains are hydrophobic because of their strong tendency to associate with other acyl chains instead of with H_2O molecules. The geometrical way to maximize

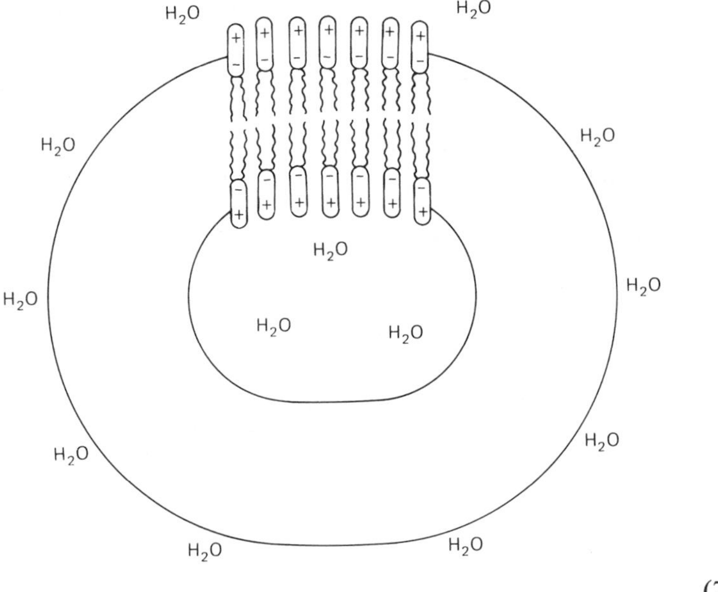

(7.2)

these two tendencies is for lipid molecules to assemble into a bilayer spherical vesicle. The radius of curvature in (7.2) has been exaggerated and is much too small in comparison with the length of a lipid molecule.

Lipids do not totally comprise biological membranes. Numerous membrane proteins are associated with the lipids, either extending entirely through the bilayer or on either the inner or outer surface. Such proteins usually contain on their exteriors regions of polar charged residues and regions of hydrophobic residues. The polar residues tend to interface with H_2O, while the hydrophobic residues tend to associate with the acyl chains of the lipids. In (7.3) the radius

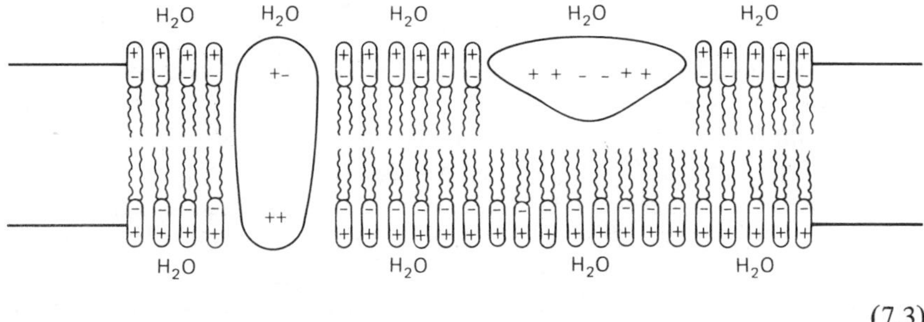

(7.3)

of curvature is given a more realistic perspective and only a portion of the membrane is depicted.

None of the interactions between lipid molecules and between lipids and proteins are covalent. Consequently, the thermal perturbations cause these molecules to engage in considerable motion. This has led to the modern view of the lipid bilayer membrane as a *fluid* system in which the associated membrane-bound proteins and individual lipid molecules are freely mobile. In addition, the acyl chains consist of entirely or mostly saturated hydrocarbons arranged in a linear sequence of single carbon–carbon bonds that can freely rotate within the confines of steric hindrance. Consequently, the acyl interior of the lipid bilayer is very much like the interior of a bulk sample of hydrocarbon fluid and possesses the hydrodynamic viscous properties of bulk fluid lipid.

While the lipid bilayer is permeable to H_2O molecules, it is virtually impermeable to ions and small, as well as large, organic molecules. The biological membranes are able to selectively transport these nonpermeants either by means of membrane-bound transport proteins or by means of freely mobile membrane-soluble organic carriers, some of which are polypeptides and many of which are not. A number of these carriers and transporter systems are discussed in detail in Chapter 11 [1-3].

A phenomenon of fundamental importance for the discussion of energy transduction in Part 2 of this book is the generation of a transmembrane electrochemical potential. For many years the importance of the transmembrane potential was overlooked because of the analytic tendency of researchers to break apart the membrane into its constituents. Once the intact membrane has been disrupted, the membrane potential is obliterated and thereby rendered unobservable. Only after it was realized that the intact membrane has properties not to be found in its constituents was the full significance of the membrane potential appreciated.

In order to discuss the generation of the transmembrane potential, a special system is considered. It involves the transport of the potassium ion, K^+, across the membrane by the lipid-soluble "ionophore" valinomycin. Valinomycin is a small molecule that binds K^+ by surrounding it and coordinating the K^+ in its interior with six carbonyl oxygens. The exterior of this complex is covered with the hydrophobic groups of D-valine and L-valine as well as some simple methyl groups that render it lipid soluble. The complex traverses the membrane by diffusion in the lipid interior of the membrane; on reaching the lipid–H_2O interfaces, it either releases K^+ into, or takes up a K^+ ion from, the aqueous medium.

Consider a preparation of lipid-bilayer vesicles containing an aqueous solution of a potassium salt K^+A^-, where A^- is the counteranion for K^+ that guarantees overall electroneutrality. Since the vesicles are suspended in an aqueous medium of low K^+ concentration, there exist unequal concentrations of K^+ on

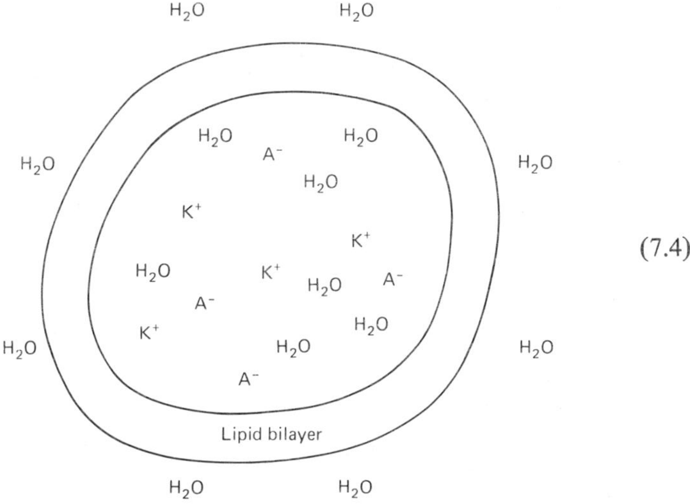

(7.4)

the inside and the outside of the vesicles [see (7.4)]. To this system valinomycin is added. The valinomycin molecules mostly enter the lipid phase because they are hydrophobic (lipophilic) and begin to shuttle K^+ cations out of the vesicles. The directionality of the K^+ flux is solely a result of the imbalance of K^+ concentration inside and outside the vesicles. The diffusion process that accounts for the motion of valinomycin in the membrane takes these molecules in both directions equally likely, but there is a greater likelihood of K^+ cations being carried outward rather than inward because there are more inside per unit volume than outside. As K^+ cations are transported, their counteranions remain behind and a net negative electric charge builds up in the interior of the vesicle. Simultaneously, the exterior medium accumulates a net positive charge as the K^+ cations come out. The problem here is to determine the equilibrium state that will eventually result.

In addition to being a thermodynamics problem, this problem is also an electrostatics problem. In equilibrium the interior of the vesicle must be an equipotential region, as must be the exterior medium because both regions are *conductors*. Because the interior and exterior potentials are generally not equal, there must be an electric field across the membrane, which in this case points from outside to inside. According to the elementary theory of conductors and insulators (dielectrics) in electrostatics, these conditions can be met only if all the K^+ ions that come out of the vesicles end up bounded electrostatically to the *exterior surface* of the vesicle, while the counteranions bind to the *inner surface*. This situation is depicted in (7.5). The vesicle behaves just like a capacitor because the lipid membrane is a good dielectric. In fact, its capacitance (in Gaus-

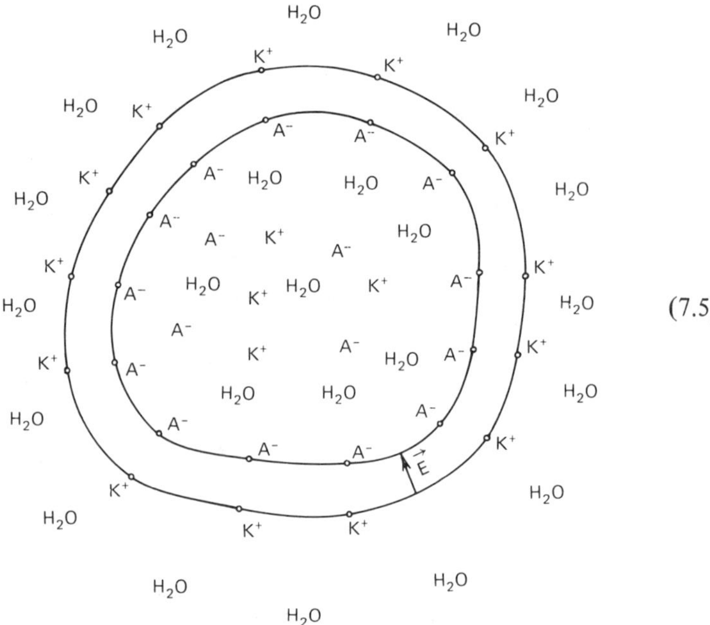

(7.5)

sian units) is given by

$$C = \epsilon\left(\frac{1}{r_{\text{in}}} - \frac{1}{r_{\text{out}}}\right)^{-1} \tag{7.6}$$

where ϵ is the dielectric constant for lipids, r_{in} is the inner radius of the bilayer, and r_{out} is the outer radius. Bulk lipid values for ϵ work well even for the thin membrane and justify using $\epsilon = 3$ or 4 (in Gaussian units). The potential difference across a capacitor of capacitance C is

$$\Delta\psi = \frac{N^*_{\text{K}^+}|e|}{C} \tag{7.7}$$

where $N^*_{\text{K}^+}$ is the number of K$^+$ ions on the outer surface (and equals the number of counterions on the inner surface), C is given in (7.6), and $|e|$, the value of one charge, is 4.8×10^{-10} esu (1.6×10^{-19} C). This completes the electrostatic analysis and leaves only the thermodynamic consideration of how big N_{K^+} is.

In Chapter 4 it was seen that equilibrium is determined by the equality of the chemical potential μ (the Gibbs free energy per molecule) on the inside and the outside of the vesicle. When activity coefficients are neglected, the total

Gibbs free energy for K^+ inside the vesicle is given by

$$G_{in} = N_{K^+}^{in}(U_{K^+}^0 + PV_{K^+}^0 - T\bar{S}_{K^+}^{0in}) + N_{K^+}^{in}|e|\psi_{in} + N_{K^+}^{in}k_BT \ln\left(\frac{N_{K^+}^{in}}{e}\right) \quad (7.8)$$

in which the electrical potential term has been explicitly separated from the rest of the internal energy, as in (4.82), and

$$\bar{S}_{K^+}^{0in} = S_{K^+}^0 + k_B \ln\left[\left(\frac{2\pi M_{K^+}k_BT}{h^2}\right)^{3/2} v\right] \quad (7.9)$$

where M_{K^+} is the mass of K^+ and v is the volume of the vesicle. The total Gibbs free energy for K^+ outside the vesicle is given by

$$G_{out} = N_{K^+}^{out}(U_{K^+}^0 + PV_{K^+}^0 - T\bar{S}_{K^+}^{0out}) + N_{K^+}^{out}|e|\psi_{out} + N_{K^+}^{out}k_BT \ln\left(\frac{N_{K^+}^{out}}{e}\right) \quad (7.10)$$

Here $\bar{S}_{K^+}^{0out}$ is

$$\bar{S}_{K^+}^{0out} = S_{K^+}^0 + k_B \ln\left[\left(\frac{2\pi M_{K^+}k_BT}{h^2}\right)^{3/2} V\right] \quad (7.11)$$

V being the volume of the external medium. In both (7.9) and (7.11), $S_{K^+}^0$ refers to the entropy of K^+ derived from rotation and formation, but the translational entropy has been explicitly separated and differs in its volume dependence. Dividing (7.8) by $N_{K^+}^{in}$ and (7.10) by $N_{K^+}^{out}$ yields μ_{in} and μ_{out}, respectively. Their equality, required for equilibrium, implies that

$$|e|(\psi_{out} - \psi_{in}) = -k_BT \ln\left(\frac{N_{K^+}^{out}/V}{N_{K^+}^{in}/v}\right) \quad (7.12)$$

Writing $\Delta\psi \equiv \psi_{out} - \psi_{in}$, this can be expressed

$$\Delta\psi = \frac{RT}{F} \ln\left(\frac{[K^+]_{in}}{[K^+]_{out}}\right) \quad (7.13)$$

This is just (4.84) in another context. *Initially* the interior concentration of K^+ was some value, say $C_{K^+}^{in}$, and the outside concentration was $C_{K^+}^{out}$. Because $N_{K^+}^*$

ions have ended up on the outer surface of the vesicle after equilibrium is achieved,

$$[\mathrm{K}^+]^{\mathrm{in}} = C^{\mathrm{in}}_{\mathrm{K}^+} - \frac{N^*_{\mathrm{K}^+}}{v} \tag{7.14}$$

On the other hand, the exterior concentration has not really changed, even though $N^*_{\mathrm{K}^+}$ ions have come out of the vesicle because they have ended up bound to the external surface by electrostatic forces. Because of an exchange between these bound K^+ ions and those in the external medium both populations have equal chemical potentials, but $C^{\mathrm{out}}_{\mathrm{K}^+}$ remains $C^{\mathrm{out}}_{\mathrm{K}^+}$. Therefore

$$[\mathrm{K}^+]^{\mathrm{out}} = C^{\mathrm{out}}_{\mathrm{K}^+} \tag{7.15}$$

Finally, unlike in (4.84), where $\Delta\psi$ is *externally applied*, in (7.13) $\Delta\psi$ is determined by (7.7), so that together they imply

$$\frac{N^*_{\mathrm{K}^+}|e|}{C} = \frac{RT}{F} \ln\left[\frac{C^{\mathrm{in}}_{\mathrm{K}^+} - (N^*_{\mathrm{K}^+}/v)}{C^{\mathrm{out}}_{\mathrm{K}^+}}\right] \qquad C^{\mathrm{in}}_{\mathrm{K}^+} > C^{\mathrm{out}}_{\mathrm{K}^+} \tag{7.16}$$

This is a transcendental equation that determines a unique value for $N^*_{\mathrm{K}^+}$. This entire consideration assumed at the outset that $C^{\mathrm{in}}_{\mathrm{K}^+} > C^{\mathrm{out}}_{\mathrm{K}^+}$. A parallel argument can be made for $C^{\mathrm{in}}_{\mathrm{K}^+} < C^{\mathrm{out}}_{\mathrm{K}^+}$ if desired. Equation (7.16) can be rewritten

$$N^*_{\mathrm{K}^+} = C^{\mathrm{in}}_{\mathrm{K}^+}v - C^{\mathrm{out}}_{\mathrm{K}^+}v \exp\left[\frac{|e|^2}{k_BT}\,\frac{1}{\epsilon}\left(\frac{1}{r_{\mathrm{in}}} - \frac{1}{r_{\mathrm{out}}}\right)N^*_{\mathrm{K}^+}\right] \tag{7.17}$$

The quantity $|e|^2/k_BT$ has the units (in the Gaussian system of units) of capacitance (cm):

$$\frac{e^2}{k_BT} = \frac{23.04 \times 10^{-20}}{4 \times 10^{-14}}\ \mathrm{esu}^2/\mathrm{erg} = 5.76 \times 10^{-6}\ \mathrm{cm} \tag{7.18}$$

The quantity $|e|/C$ is the change in electrical potential caused by one ion. For a vesicle with

$$r_{\mathrm{in}} = 10{,}000\ \text{Å} \qquad \text{and} \qquad r_{\mathrm{out}} = 10{,}070\ \text{Å} \tag{7.19}$$

which reflect the usually quoted value for the thickness of a lipid bilayer, 70 Å, and with $\epsilon \simeq 4$,

$$\frac{|e|}{C} = \frac{|e|}{\epsilon}\left(\frac{1}{r_{\text{in}}} - \frac{1}{r_{\text{out}}}\right) = (4.8 \times 10^{-10})\,\frac{1}{4}\,(7 \times 10^{-7}\ \text{esu/Å})$$

$$= 8.4 \times 10^{-17}\ \text{esu/Å} = 8.4 \times 10^{-9}\ \text{esu/cm}$$

$$= 8.4 \times 10^{-9}\ \text{statvolt} = 25.2 \times 10^{-7}\ \text{V} \tag{7.20}$$

(1 statvolt ≡ 300 V). This is only *microvolts*. Typically, measured values of transmembrane potentials are on the order of 100 mV. From (7.7) and (7.20) this would require $N^*_{\text{K}^+}$ to be about 4×10^4. In order for $C^{\text{in}}_{\text{K}^+} > C^{\text{out}}_{\text{K}^+}$ to be satisfied and for $N^*_{\text{K}^+} \simeq 4 \times 10^4$ to hold, it must follow that

$$C^{\text{in}}_{\text{K}^+} v > 4 \times 10^4 \tag{7.21}$$

if it is to be possible to achieve a 100-mV potential. For $r_{\text{in}} = 10{,}000$ Å, $v = 4.2 \times 10^{-12}$ cm^3, so that (7.21) implies

$$C^{\text{in}}_{\text{K}^+} > 10^{16}\ \text{ions/cm}^3 = 10^{19}\ \text{ions/liter} = 1.6 \times 10^{-5}\ \text{molar} \tag{7.22}$$

This is an easily achieved requirement! In addition, from (7.19) it follows that

$$C = 4\,\frac{10^8}{70}\ \text{Å} = 5.7 \times 10^{-2}\ \text{cm} \tag{7.23}$$

so that

$$\exp\left[\frac{e^2}{k_B T}\,\frac{1}{\epsilon}\left(\frac{1}{r_{\text{in}}} - \frac{1}{r_{\text{out}}}\right) N^*_{\text{K}^+}\right] = \exp\left(\frac{5.76 \times 10^{-6}}{5.7 \times 10^{-2}}\,N^*_{\text{K}^+}\right)$$

$$= \exp(10^{-4} N^*_{\text{K}^+}) \tag{7.24}$$

This term becomes significant only for $N^*_{\text{K}^+}$ as large as 10^4, which is compatible with the requirement for a 100-mV potential. Thus the entire analysis is self-consistent.

It may be asked whether there is room on the exterior surface of the vesicle for 10^4 to 10^5 K^+ ions. The surface area for vesicles with $r_{\text{out}} \simeq 10^4$ Å is about 12.5×10^8 Å^2, which allows $(100\ \text{Å})^2$ per ion if 10^5 ions are involved. Such an area is certainly large enough to accommodate all these ions.

The capacitance for the vesicles just described was found to be 0.057 cm for a vesicle with a radius of 10,070 Å. The conversion to the mks unit for capacitance, the farad, is achieved with the factor 9×10^{11}, the longest conversion factor for any electromagnetic quantity:

$$0.57 \text{ cm} \longrightarrow \frac{1}{9 \times 10^{11}} \times 0.057 \text{ farads} = 6.3 \times 10^{-14} \text{ farads} \qquad (7.25)$$

The surface area is $4\pi \times 10^{-8}$ cm^2. Therefore the capacitance density is

$$\frac{6.3 \times 10^{-14}}{4\pi \times 10^{-8}} \text{ farad/cm}^2 = 0.5 \text{ microfarad}(\mu\text{F})/\text{cm}^2) \qquad (7.26)$$

This is only half the value usually quoted as the capacitance density for general biological membranes. From (7.6) it is seen that the value of the capacity density can be increased if $r_{out} - r_{in}$ is decreased. There is some reason to believe that $r_{out} - r_{in}$ may be as small as 40 Å. Hydrodynamic analysis and electron density profiles for egg phosphatidylcholine vesicles support this smaller value. These studies were done with vesicles only 100 Å in radius, however, and therefore it is not definite that this smaller value is also accurate for the bigger vesicles, 10,000 Å in radius, discussed above. It is amusing to note that, as early as 1923, Fricke [4] had measured the capacitance density of red blood cell membranes as 0.80 μF/cm^2. He assumed that the membrane might be an oil with a dielectric constant equal to 4 and thereby estimated that the thickness of the membrane was about 40 Å, a value *not* taken too seriously since then, because it seemed too small by a factor of 2. From the analysis above, it is seen that this value may not be so bad after all!

The Cell Theory [5]

Part 1 of this book has ended with a discussion of the largest macromolecular structure to be discussed in this book, the cell membrane. At the end of Chapter 5 it was suggested that the cell itself is the minimal macromolecular structure that can replicate itself entirely self-sufficiently. The idea that the *cell* is the basic unit of life is an old idea that originally was reached from the other direction. Usually the German botanist Schleider and the zoologist Schwann are credited with enunciating in 1838 the view that all organic tissues are composed of cells and that the cell is the basic unit of all living things. They had no detailed molecular view within which to consider the cell from the self-replicative viewpoint of molecular biology. Nevertheless, looking at the constitution of tissues in higher organisms led them to this view. In 1824 the French biologist Dutrochet wrote [5]–

> All organic tissues are actually globular cells of exceeding smallness, which appear to be united only by simple adhesive forces; thus all tissues, all animal organs are actually only a cellular tissue variously modified.

He recognized that growth is the result of increases in cellular volume *and* the addition of new small cells produced by cellular replication. Somewhat later, in 1858, Virchow [5] also recognized that cells arise from cells and that this is the universal basis of all animal and plant tissues.

Recently Jacob [6] has written an elegant account of the history of heredity and the cell theory. He has developed the idea of the "integron." In describing the hierarchical structure of organisms from the molecular level on up, he says (p. 302):

> At each level, units of relatively well defined size and almost identical structure associate to form a unit of the level above. Each of these units formed by the integration of sub-units may be given the general name "integron." An integron is formed by assembling integrons of the level below it; it takes part in the construction of the integron of the level above.

He later concludes (p. 306):

> And yet biology has demonstrated that there is no metaphysical entity hidden behind the word "life." The power of assembling, of producing increasingly complex structures, even of reproducing, belongs to the elements that constitute matter. From particles to man, there is a whole series of integration, of levels, of discontinuities. But there is no breach either in the composition of the objects or in the reactions that take place in them; no change in "essence." So much so that investigation of molecules and cellular organelles has now become the concern of physicists. Details of structure are now defined by crystallography, ultracentrifugation, nuclear magnetic resonance, fluorescence and other physical techniques. This does not at all mean that biology has become an annex of physics, that it represents, as it were, a junior branch concerned with complex systems. At each level of organization, novelties appear in both properties and logic. To reproduce is not within the power of any single molecule by itself. This faculty appears only with the simplest integron deserving to be called a living organism, that is, the cell.

References

1 M. Eigen and L. DeMaeyer, *Carriers and Specificity in Membranes* (Neurosciences Research Program Bulletin, Vol. 9, No. 3, 1971).

2 Ken Cole, *Membranes, Ions and Impulses* (University of California Press, Berkeley, 1968).

This book contains an account of Hugo Fricke's work.

3 J. T. Mason and C. Huang, "Hydrodynamic Analysis of Egg Phosphatidylcholine Vesicles," in *Liposomes and Their Uses in Biology and Medicine*, D. Papahadjopoulos, ed. (New York Academy of Sciences, New York, 1978), Vol. 308, p. 29.

4 Hugo Fricke, "The Electric Capacity of Cell Suspensions," *Phys. Rev.*, 21, 708–709 (1923).

5 G. G. Simpson, C. S. Pittendrigh, and L. H. Tiffany, *Life* (Harcourt, Brace, and World, New York, 1957).

Here one can find a brief introduction to the cell theory.

6 F. Jacob, *The Logic of Life* (Pantheon, New York, 1973).

This book contains a scholarly account of the history of heredity and the cell theory. It is in this book that Jacob develops his idea of the integron.

PART

2

Energy Transduction in Organisms

"Great knowledge sees all in one.
Small knowledge breaks down into the many."

Chuang Tzu

Part 2 of this book is devoted to a review of the principal energy-generating and energy-utilizing processes in organisms. Chapter 8 outlines the central pathways of anabolic and catabolic energy metabolism, exhibiting the pathways for sunlight-driven carbohydrate synthesis and the pathways by which carbohydrate is oxidized, such that its energy content is transduced into ATP energy and into the redox energy of electron flow. Chapter 9 examines a variety of control, or regulation, processes, with emphasis on the importance of proteins for these functions. It also shows how the byproducts of energy metabolism are involved at several different levels of integration in control processes.

Chapter 10 discusses the biosynthesis of proteins, one of the major energy (ATP) utilization processes in cells; it explains where and how all the protein catalysts required for energy metabolism are manufactured in the cell. Chapter 11 is devoted to the presentation of macromolecular "rotors," "wheels," or "motors." Its purpose is to exhibit the remarkable protein complexes that are found in cells, so that the membrane-associated complexes diagrammed at the end of the chapter are more easily accepted as additional examples of the remarkable inventiveness of self-assembly. Chapter 12, the last chapter of this part, contains a detailed quantitative account of how ATP is manufactured by the membrane-associated protein complexes. It is in this chapter that the chemiosmotic hypothesis is described and explained.

Two major revolutions in biological understanding have taken place during the last 30 years. The first concerns the structure of DNA and the detailed elucidation of gene-directed protein biosynthesis. The second concerns the chemiosmotic theory of ATP synthesis, which underlies the energy requirements for protein biosynthesis. The former revolution led to the detailed description of the most complex process ever described by biochemistry or molecular biology. It gave rise to knowledge about the genetic code and the associated machinery for protein biosynthesis, including ribosomes, tRNAs, synthetases, and polymerases. The second required recognition of the fundamental importance of the intact, whole membrane and its transmembrane electrochemical proton potential in the overall scheme of energy transduction.

While each of these revolutions in understanding has produced spectacular insights and voluminous accounts of the detailed mechanisms, major unsolved problems remain for both. The "mystery of the second code," introduced in Chapter 10, is an unsolved, often underemphasized, research problem whose solution is absolutely essential for a *complete* picture of protein biosynthesis. This mystery is elaborated in greater detail in Chapter 14. The elucidation of the mechanism of proton translocation across membranes by the electron transport chain, as described in Chapter 12, is still incomplete in a very important way, and the structure of the active site of the membrane-associated ATPase enzyme has yet to be determined.

CHAPTER

8

Intermediary Energy Metabolism [1-3]

The pathways for energy metabolism can be divided into two categories: the anabolic portion, during which the energy of sunlight is used to convert CO_2 and H_2O into carbohydrate $(CHOH)_n$ *and* O_2; and the catabolic portion, during which the energy stored in carbohydrate *and* O_2 is transduced into ATP energy *and* other energy forms while releasing CO_2 and H_2O. The anabolic portion occurs in photosynthetic organisms and involves two major pathways: the light reactions, which produce ATP and NADPH; and the carbon cycle, in which the energy in ATP and NADPH is used to manufacture carbohydrate from CO_2 and H_2O. The catabolic portion may occur in photosynthetic organisms in the dark and in aerobic organisms; part of it also occurs in anaerobes. The major pathways are the fermentation of glucose to pyruvate, the conversion of pyruvate to acetyl-CoA, the citric acid cycle, and electron transport. Anaerobes use only the fermentative pathway and, for example, convert pyruvate into ethanol as an end product. Aerobes use all of these pathways because the oxidant O_2 appears in the last step of the last segment, electron transport.

There are other pathways as well, such as the phosphogluconate pathway and the generation of energy starting with fats or proteins instead of with carbohydrates. In this book only the central pathways, which provide a minimal basis for a fully integrated view of energy metabolism, are reviewed in detail. The reader may refer to any of the previously referenced biochemistry books for a detailed account of these other pathways.

The light reactions of photosynthesis are shown in the energy figure (8.1). The energy scale is given in terms of the redox electrical potential $E^{0\prime}$ of the substance indicated for pH = 7. Since electrons are being transported by this pathway, the negative values of $E^{0\prime}$ times the negative charge of an electron yield positive energies. Therefore, in the diagram, positive energies are in the upper portion, whereas lower energies are below. The components of this pathway are as follows: pigment systems I and II are arrays of chlorophyll molecules that absorb sunlight. In pigment system II the excitation energy of absorbed

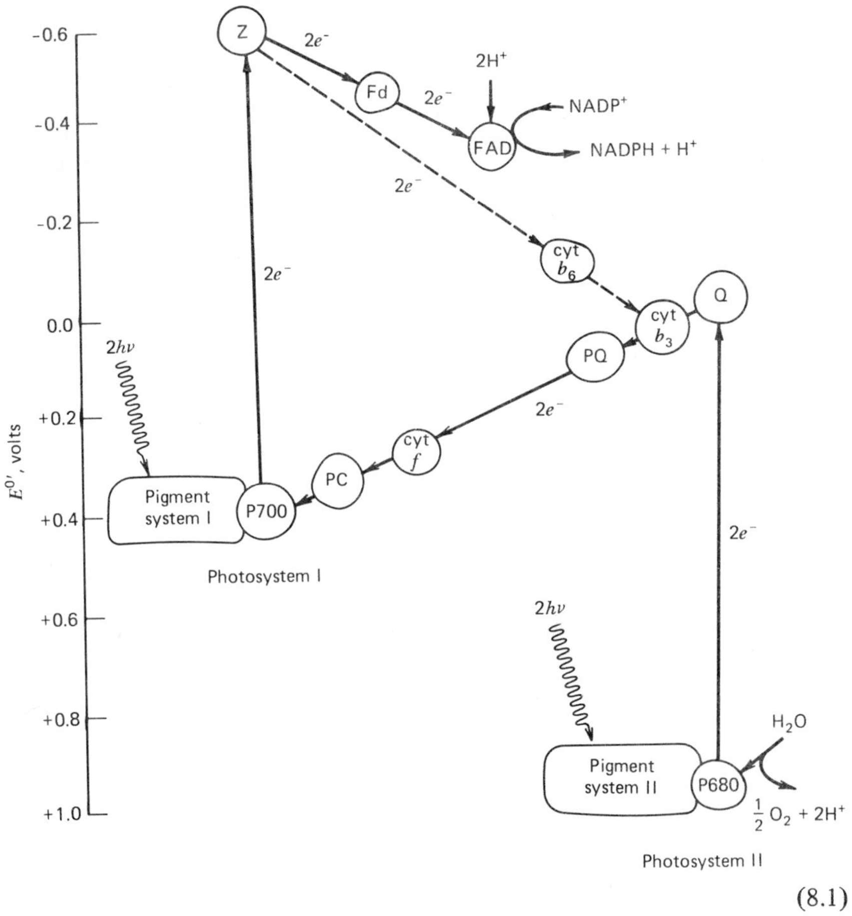

(8.1)

light is shuttled to a chlorophyll, designated P680, from which excited electrons arise. Two photons give rise to two electrons that reach the acceptor Q. From Q these electrons are transported to photocenter P700 via cytochrome b_3, (Cyt.b_3), plastoquinone (PQ), cytochrome f, (Cyt.f), and plastocyanine (PC). Two more photons absorbed by the chlorophyll array of pigment system I are transduced as excitations that excite the two electrons at P700 up to acceptor Z, an iron-sulfur protein. From this energy pinnacle the electron pair flows downhill, energywise, through the iron-sulfur protein ferrodoxin (Fd), to a flavin-adenine dinucleotide bound protein (FAD), which along with two protons from the medium is reduced to $FADH_2$. $FADH_2$ finally reduces $NADP^+$ to NADPH. The original electrons excited at photosystem II are derived from H_2O, releasing $\frac{1}{2}O_2$ and two protons ($2H^+$).

The pathway just described uses four photons to carry two electrons from H_2O all the way to $NADP^+$. This is called "noncyclic" electron transport by the light reactions. A "cyclic" electron transport involving only photosystem I also occurs. In this case two photons excite two electrons to acceptor Z, from which they return to P700 via cytochrome b_6, cytochrome b_3, plastoquinone, cytochrome *f*, and plastocyanine. The energy derived from this cyclic process shows up as ATP, which is generated during both types of electron transport from Q to P700. The mechanism of this transduction is discussed in Chapter 12. There is no direct molecular, or "chemical," coupling between the molecular components of the electron transport system between Q and P700, and the components involved in making ATP from its precursors, even though *the energy* for ATP synthesis derives from this portion of the pathway. This perplexing situation is explained in Chapter 12.

Two metal atoms are important in this pathway, iron and copper. Plastocyanine contains copper and is a protein of relatively low molecular weight with two copper atoms per molecule. Iron is far more prevalent and varied in its appearance in the pathway. The cytochromes are proteins that contain heme groups. The heme groups are based on the porphyrin structure shown in Chapter 1; they contain one iron (Fe) atom each, as shown in (8.2). The various cyto-

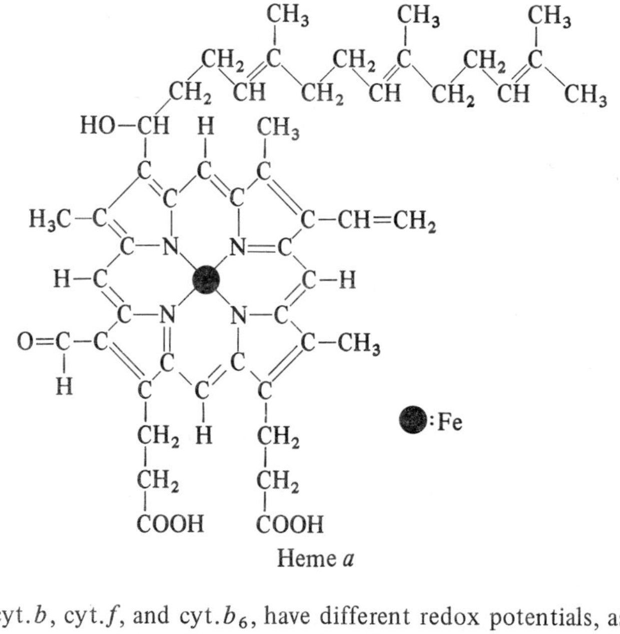

(8.2)

Heme *a*

chromes, cyt.*b*, cyt.*f*, and cyt.b_6, have different redox potentials, as indicated in (8.1), because of their different proteins. Their hemes are virtually identical. Ferrodoxin and Z are iron-sulfur proteins that do not contain porphyrin; they are based on an iron-sulfur structure that appears to be rather primitive. This structure is shown in (8.3), where R designates the rest of the protein, specifically the cysteine residues.

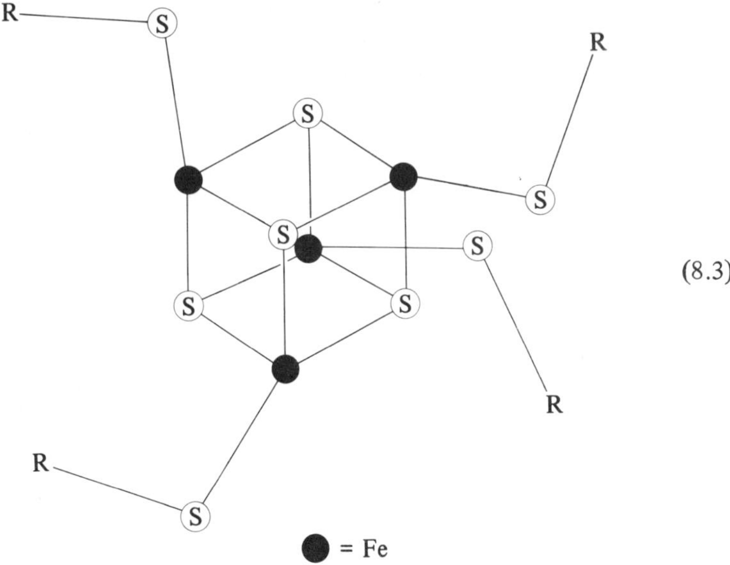

The quinone, Plastoquinone (PQ), is closely related to ubiquinone (CoQ), which was depicted in Chapter 1. It differs in that its benzenoid structure is slightly different [see (8.4)].

H O H
HCO CH
H H
H
HCO R
H O

Ubiquinone

H O
HC H
H
H
HC R
H O

Plastoquinone

(8.4)

The chlorophyll molecules are structurally derived from a modified porphyrin like the cytochromes, except that it contains magnesium rather than iron and there is no protein in chlorophyll as there is in cytochromes. Instead, chlorophyll has a long phytol side chain [see (8.5)].

The energy content of a photon of frequency ν is $h\nu$, where h is Planck's constant. The frequency ν is related to the wavelength λ by $\lambda\nu = c$, the speed of light ($c = 3 \times 10^{10}$ cm/sec). Chlorophyll absorption is largest around $\lambda = 5500$ Å,

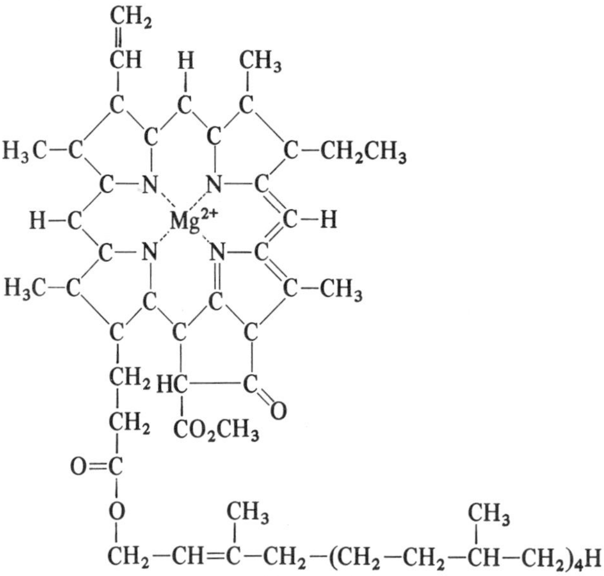

```
          CH2
          ||
          CH     H     CH3
          |      |      |
          C      C      C
        /   \  /   \  /   \
H3C—C         C      C        C—CH2CH3
        \     |      |      /
          C—N        N—C
        /     \    /     \\
  H—C          Mg2+        C—H
        \     /    \     //
          C—N        N—C
        /     |      ||     \\
H3C—C         C      C        C—CH3
        \   /   \  /   \  /
          C       C       C
          |       |       |
          CH2    HC———————C
          |       |        \\
          CH2    CO2CH3      O
          |
       O=C
          |
          O            CH3                      CH3
          |            |                        |
          CH2—CH=C—CH2—(CH2—CH2—CH—CH2)4H
```

Chlorophyll (8.5)

which corresponds to an energy of $h(c/\lambda) = 6 \times 10^{-27}\ (3 \times 10^{10})/(5.5 \times 10^{-5})$ erg $= 3.3 \times 10^{-12}$ erg. This is about $80k_BT_{300}$, a sizable energy per photon. In (8.1) it is seen that each electron traverses about 1 V in photosystem II and about 1 V in photosystem I. The energy per electron would be 1 V $(1.6 \times 10^{-19}$ C$) = 1.6 \times 10^{-19}$ J $= 1.6 \times 10^{-12}$ erg (i.e., one electron volt, 1 eV). Thus, there is ample energy available in green sunlight ($\lambda = 5500$ Å) to excite electrons the required amount.

On the other hand, the energy released during electron transport from Q to P700 is about 0.4 Volt $\times$ 1.6×10^{-19} C $= 6.4 \times 10^{-13}$ erg per electron ($\sim 16k_BT_{300}$). Later it will be seen whether this will satisfy the energy requirements for making ATP.

In Chapter 11 the structure of the light reaction pathway within the chloroplast membrane is elucidated.

The carbon cycle of photosynthesis has already been shown, for another purpose, in Chapter 5 [see (5.13).]. There it is shown how the energy of ATP and NADPH is used to transform CO_2 and H_2O into carbohydrate $(CHOH)_n$. As carbohydrate accumulates, it is removed from the carbon cycle by the conversion of fructose-6-phosphate into glucose-6-phosphate, which is "stored" or further processed ultimately into polysaccharides.

The overall balance sheet for the anabolic portion of energy metabolism can be written

$$nh\nu + 6CO_2 + 6H_2O \longrightarrow (CHOH)_6 + 6O_2 \tag{8.6}$$

In the carbon cycle (5.13) it is seen that this conversion requires six molecules of NADPH and six molecules of ATP. Figure 8.1 shows that to get six molecules of NADPH, six *pairs* of electrons must traverse the entire pathway. This requires 12 *pairs* of photons. Simultaneously, at least six molecules of ATP are generated from the energy released between Q and P700. Therefore, n in (8.6) is at least 24. The Gibbs free energy of formation for each molecular species at

Gibbs free energy (kcal/mole)

0	• O_2(gas) 0.00	• H_2O(liquid) -56.69	
-100			• CO_2(gas) -94.26
-200	• Glucose -219.22		
-300	• $6H_2O$(liquid) -340.14		
-400			
-500	• $6CO_2$(gas) -565.56		
-600			

(8.7)

300°K is shown in (8.7), from which it is seen that glucose lies (-219.22) - (-340.14 - 565.56) = 905.70 - 219.22 = 686.48 kcal/mole above six molecules of H_2O and six molecules of CO_2 together. This is equivalent to about 2.74×10^{-12} erg *per molecule* of glucose, or about $69k_BT_{300}$.

The catabolic portion of energy metabolism can be reckoned from glucose as the starting point. Many organisms possess precursor pathways for the generation of glucose from polysaccharide storage molecules. These pathways are not elucidated in this book. The outline of how fats and proteins feed into the pathways described here is given later.

Glucose is processed according to the sequence of reactions which comprises the pathway of "glycolysis" which was diagrammed in Chapter 5 [see (5.12)]. There the byproduct, pyruvate, is shown being reduced by NADH to lactate. For the perspective of this chapter, the pathway ends with pyruvate instead of lactate, and the reduced NADH generated during an earlier step in the sequence is not recycled as in (5.12) but instead accumulates, to be processed later. The earlier, NADH-generating step is an extraordinary step because it is an oxidative-phosphorylation step that is mechanistically achieved by "chemical coupling." This means that the inorganic phosphate is combined with ADP to make ATP through an intermediary, activated, phosphorylated chemical, 1,3-diphosphoglyceric acid, which has been generated from glyceraldehyde-3-phosphate and inorganic phosphate through an *oxidation* reaction driven by NAD^+, the oxidant. Historically, oxidative phosphorylation by chemical coupling was the earliest known mechanism for making ATP, and this fact inhibited the discovery of other mechanisms for a long time. More will be said on that score in Chapter 12.

The fact that glyceraldehyde-3-phosphate already contains a phosphate group may cause confusion. This phosphate is incidental in the mechanism of oxidative phosphorylation. It is the inorganic, free phosphate that is oxidatively attached to glyceraldehyde-3-phosphate that is important.

For each glucose entering the glycolysis pathway, two pyruvate molecules are generated, as are two NADH, two H_2O, and four ATP. The ATP balance is a little bit subtle. Two ATP molecules are used to "activate" the pathway; so there is only a net gain of two ATP molecules. This requires an input of two phosphate and two ADP molecules. The NADH generated during the pathway require an input of two NAD^+.

The subsequent fate of the pyruvate and the NADH involves the remaining pathways in the catabolic portion of energy metabolism. Pyruvate is converted into "active acetate," acetyl-CoA, by a pathway already diagrammed for another purpose [see (5.6)]. The accumulating pyruvate drives the pathway. For each pyruvate molecule (two pyruvates per glucose molecule), one molecule of CO_2 is released, one molecule of NAD^+ is reduced to NADH, and one molecule of coenzyme A (CoA) is transduced into active acetate, acetyl-CoA.

The fate of acetyl-CoA is now known as the citric acid cycle (Kreb's cycle *or* tricarboxylic acid cycle (TCA)). This pathway is given in (8.8). Notice that two steps involve $\Delta G^{0\prime} > 0$. This should be compared with (5.4). It is estimated that reaction 2 has $\Delta G^{0\prime} \simeq +2.0$ kcal/mole and reaction 3 has $\Delta G^{0\prime} \simeq -0.5$ kcal/mole. Nevertheless, in the steady-state functioning condition of the cycle, isocitrate is rapidly removed by reactions 4 and 5, thereby favoring reactions 2 and 3 in the direction of isocitrate production because $\Delta G < 0$ under these conditions, in spite of $\Delta G^{0\prime} = +1.5$ kcal/mole. Similarly, reaction 10 has a relatively large positive $\Delta G^{0\prime}$ (+7.1 kcal/mole). Again, however, oxaloacetate is rapidly removed through conversion to citrate by reaction 1 whenever acetyl-CoA is plentiful. Therefore, $\Delta G_{10} < 0$ is achieved for this reaction in spite of $\Delta G_{10}^{0\prime} = +7.1$ kcal/mole. Reaction 1, with $\Delta G_1^{0\prime} = -7.7$ kcal/mole, provides a very strong impetus for the *removal* of oxaloacetate as the product of reaction 10. What really counts overall is that $\Delta G^{0\prime} \simeq -12.8$ kcal/mole for all of the steps of the cycle together. This is a strong thermodynamic stimulus for the cycle to proceed as indicated in (8.8), in the clockwise direction.

Although reactions 2 and 3 and reactions 4 and 5 are indicated in the figure as though each pair involved a pair of enzymes, in fact only *one* enzyme is involved in each case, and *two* distinct reaction steps occur in each case. The intermediates, *cis*-aconitate and oxalosuccinate, respectively, are enzyme bound; that is, they are not freely occurring, as are the other components of the cycle.

After one complete cycle, a number of changes have occurred. The CoA-SH utilized during the transduction of pyruvate to acetyl-CoA has been regenerated for another cycle as acetate carrier. Coenzyme A is therefore a *catalytic* component; it is also used and regenerated in the sequence from α-ketoglutarate to succinate. This sequence is noteworthy for another property: it exhibits an

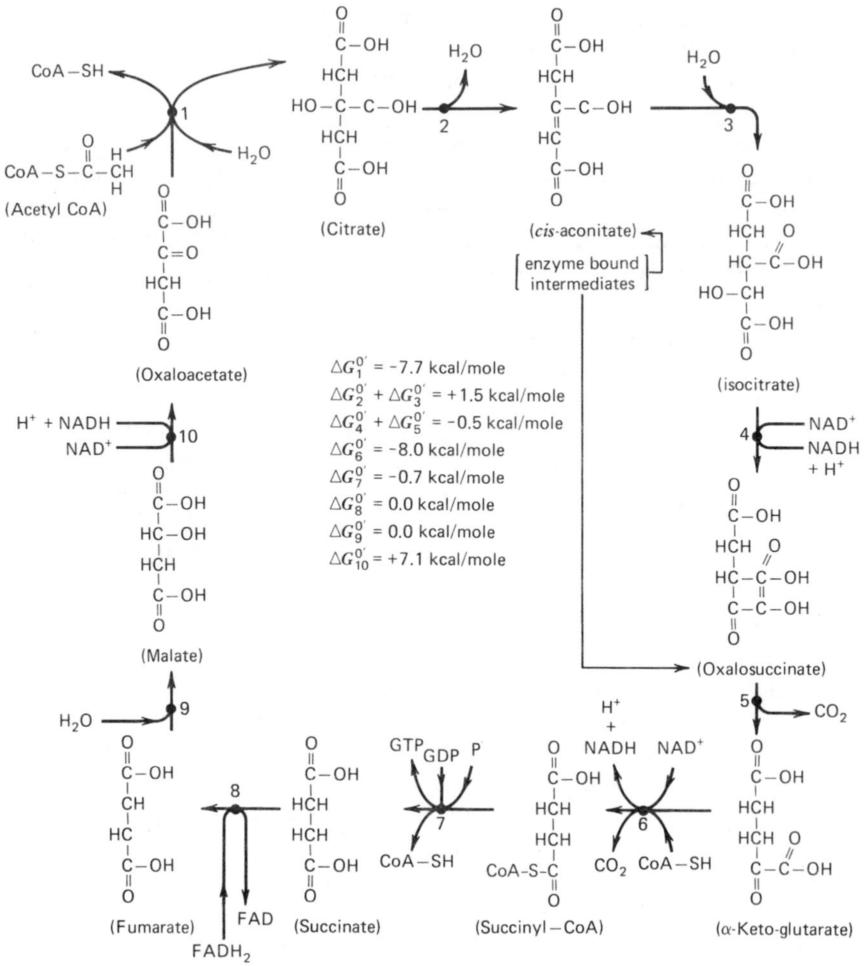

CoA−SH
(Acetyl CoA)
H2O
(Oxaloacetate)
(Citrate)
(cis-aconitate)
enzyme bound intermediates
(isocitrate)
NAD+
NADH + H+
(Oxalosuccinate)
CO2
(α-Keto-glutarate)
NADH NAD+
CO2 CoA−SH
(Succinyl−CoA)
GTP GDP P
CoA−SH
(Succinate)
FAD
FADH2
(Fumarate)
(Malate)
H+ + NADH
NAD+
$\Delta G_1^{0'} = -7.7$ kcal/mole
$\Delta G_2^{0'} + \Delta G_3^{0'} = +1.5$ kcal/mole
$\Delta G_4^{0'} + \Delta G_5^{0'} = -0.5$ kcal/mole
$\Delta G_6^{0'} = -8.0$ kcal/mole
$\Delta G_7^{0'} = -0.7$ kcal/mole
$\Delta G_8^{0'} = 0.0$ kcal/mole
$\Delta G_9^{0'} = 0.0$ kcal/mole
$\Delta G_{10}^{0'} = +7.1$ kcal/mole

(8.8)

oxidative decarboxylation in reaction 6, followed by an oxidative phosphorylation in reaction 7 in which succinate-CoA is the oxidant, being reduced to CoA-SH and succinate while *GDP* is phosphorylated to *G*TP with inorganic phosphate (P_i). Water is released once, in reaction 2, but is required twice, in reactions 3 and 9. Carbon dioxide is released twice, in reactions 5 and the aforementioned 6. In reactions 4, 6, and 10, NAD^+ is reduced to NADH and H^+, while FAD is reduced to $FADH_2$ in reaction 8.

Each acetate carbon is released in a CO_2 molecule. The fate of each oxygen and hydrogen atom will be deferred for a while and then followed in detail.

The fate of the reduced dinucleotides NADH and $FADH_2$ is known as the electron transport chain pathway. It involves O_2 as the ultimate oxidant. In between the dinucleotides and O_2 there is a sequence of iron–sulfur proteins, quinone, and several cytochromes, much like in the Q to P700 segment of the light reactions in photosynthesis. The specific types of these components are somewhat different in bacteria and in mitochondria, the electron transport pathway containing organelles of higher, aerobic organisms. For bacteria it is drawn in (8.9).

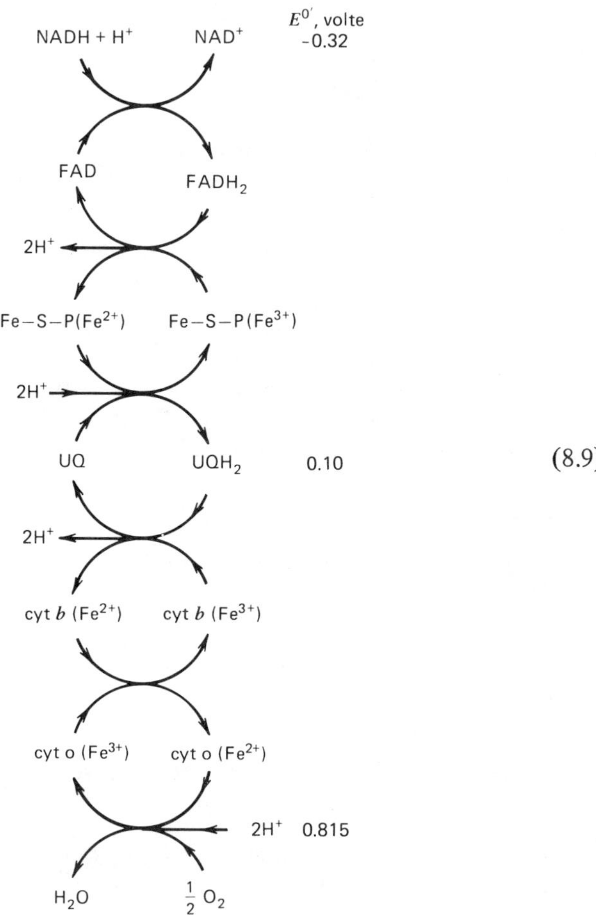

(8.9)

The components of this chain involve NADH generated during the citric acid cycle, FAD bound to a protein in the chain, iron–sulfur protein (Fe-S-P), ubiquinone (UQ), and cytochromes *b* and *o*. The final oxidant is $\frac{1}{2}O_2$. The redox potentials of three members of this chain are given for pH = 7. When multiplied by the *negative* charge of the electron, they yield the energy, which is relatively positive at the top of the figure.

There are several peculiar features associated with this sequence of steps. In (8.9) there are four steps at which two H^+ are either eliminated or taken up. These steps mark the boundary between redox reactions that involve a whole hydrogen atom and those involving only the electrons. This is the same as those not involving iron and those involving iron, respectively. What is the significance of the H^+ steps? The answer to this question is given in Chapter 12. How can the energy gleaned from a pair of electrons going from NADH to $\frac{1}{2}O_2$ to form H_2O be used to make ATP, the known energy product of this transduction? Are there chemical coupling intermediates? The answer to this question is also given in Chapter 12.

The overall balance sheet for the catabolic portion of energy metabolism can be written

$$(CHOH)_6 + 6O_2 \longrightarrow 6CO_2 + 6H_2O + \text{energy} \qquad (8.10)$$

$$\Delta G^{0\prime} = -686 \text{ kcal/mole}$$

which is essentially the reverse of (8.6), except that the energy harvested by these pathways is not in the form of photons–it is in the form of ATP generated from ADP and phosphate by this energy. Each pathway discussed above is shown below along with the overall balance of molecular species.

Glycolysis

$$1 \text{ glucose} + 2NAD^+ + 2ADP + 2P \longrightarrow 2 \text{ pyruvate} + 2H_2O + 2(NADH + H^+) + 2ATP \qquad (8.11)$$

$$\Delta G^{0\prime} = -19.2 \text{ kcal/mole}$$

Acetyl-CoA connection

$$2 \text{ pyruvate} + 2\text{CoA-SH} + 2NAD^+ \longrightarrow 2 \text{ acetyl-CoA} + 2CO_2 + 2(NADH + H^+) \qquad (8.12)$$

$$\Delta G^{0\prime} = -16.0 \text{ kcal/mole}$$

Citric acid cycle

$$2 \text{ acetyl-CoA} + 6NAD^+ + 2FAD + 2GDP + 2P \longrightarrow 2\text{CoA-SH} + 6(NADH + H^+) + 2FADH_2 + 2GTP + 4CO_2 - 4H_2O \qquad (8.13)$$

$$\Delta G^{0\prime} = -25.6 \text{ kcal/mole}$$

Electron transport chain

$$10(NADH + H^+) + 2FADH_2 + 6O_2 \longrightarrow 12H_2O + \text{energy} + 10NAD^+ + 2FAD \quad (8.14)$$

$$\Delta G^{0\prime} = -611 \text{ kcal/mole}$$

The balance sheet for glycolysis follows directly from (5.12) and the remarks made earlier in this chapter regarding stopping with pyruvate and NADH rather than proceeding to lactate as in (5.12). Similarly, (5.6) makes (8.12) obvious, although the overall factor of 2 in (8.12) reflects the need to consider two molecules of pyruvate since one molecule of glucose gave rise to two pyruvates. Glucose has six carbons, pyruvate has three, and acetate has two. In the citric acid cycle, (8.8), H_2O is eliminated in reaction 2 but is taken up in reactions 1, 3, and 9, for a net uptake of $2H_2O$ per acetate or $4H_2O$ for two acetate molecules. This explains the $-4H_2O$ in (8.13). All of the other terms in (8.13) are obvious from (8.8). All of the NADH and $FADH_2$ produced in any of these pathways is processed by the electron transport chain, and this explains (8.14). Note that all 10 NAD^+ utilized in (8.11), (8.12), and (8.13) are regenerated in (8.14). Similarly, the FAD used in (8.13) is regenerated in (8.14) as well. It is also seen that six molecules of CO_2 are released overall: two during the transformation of pyruvate to acetyl-CoA and four during the citric acid cycle. The CoA-SH used in (8.12) is regenerated in (8.13). All that remains to be accounted for is the H_2O. Counting up all H_2O molecules in these pathways yields a net of $10H_2O$. This is four molecules too many! However, if the results in (8.11) through (8.14) are tabulated overall, they yield

$$1 \text{ glucose} + 2ADP + 2P + 2GDP + 2P + 6O_2 \longrightarrow 6CO_2 + 10H_2O + 2ATP + 2GTP + \text{energy} \quad (8.15)$$

$$\Delta G^{0\prime} = -672 \text{ kcal/mole}$$

and the "excess" of $4H_2O$ can be accounted for through the *dehydration condensations*:

$$\begin{aligned} 2ADP + 2P &\longrightarrow 2ATP + 2H_2O \qquad \Delta G^{0\prime} = -14.6 \text{ kcal/mole} \\ 2GDP + 2P &\longrightarrow 2GTP + 2H_2O \qquad \Delta G^{0\prime} = -14.6 \text{ kcal/mole} \end{aligned} \quad (8.16)$$

Put the other way around, the $4H_2O$ can be used to hydrolyze 2ATP and 2GTP, yielding a final total of $6H_2O$, as desired, and also yielding 2ADP, 2GDP, and four molecules of phosphate to cancel out the same on the left-hand side of (8.15). The final, net result is indeed (8.10), and the overall $\Delta G^{0\prime}$ is -699 kcal/mole, not too far off from -686 kcal/mole ($<2\%$ error).

NAD^+, FAD, and CoA-SH are "catalytic" components. They are continuously recycled. The ATP and GTP produced account for only a small portion of the available energy. It was seen in (8.7) that about 686 kcal/mole is released in (8.10). Most of this energy is harvested during electron transport. However, so far the mechanism of this transduction has not been described. It is known that ATP is the energy product of this transduction, but the mechanism remained a fundamental mystery for decades. The direct chemical coupling mechanism that accounts for the 2ATP formed during glycolysis and for the 2GTP formed during the citric acid cycle was taken as the basis for other phosphorylations during the electron transport chain. Nevertheless, no chemical intermediates were ever found. The enzymes and the components of the electron transport chain were found to be associated with membranes, which were dissociated, fractionated, and otherwise scrutinized for chemical coupling intermediates analogous to 1.3-diphosphoglycerate and succinyl-CoA, but without success. This mystery was resolved when it was finally realized that the answer lay in the *intact* membrane. This answer is explored in detail in Chapter 12.

The integration of fatty acid and protein metabolism with the pathways for *catabolic* energy metabolism is of basic importance in the overall regulation of cellular processes. Proteins are hydrolyzed into their constituent amino acids,

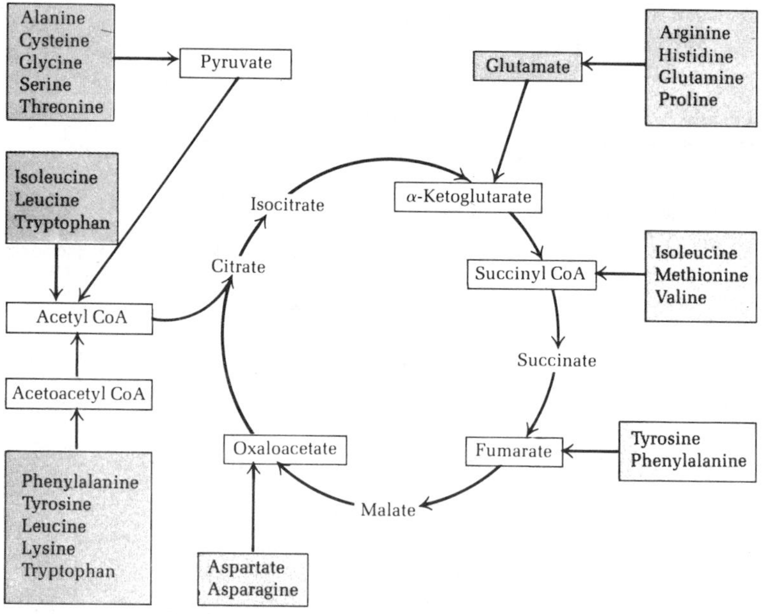

Reproduced, with permission, from Ref. 1, 1st ed.

(8.17)

which are then transformed enzymatically, usually in only a very few steps, into intermediates of the citric acid cycle. This is illustrated in (8.17). The catabolic metabolism of fatty acids leads ultimately to the production of acetyl-CoA in which the two-carbon acetate derives from two contiguous carbons in the fatty acid hydrocarbon chain. This acetyl-CoA is then processed by the citric acid cycle just like the acetyl-CoA derived from pyruvate.

References

1 A. Lehninger, *Biochemistry*, 2nd ed. (Worth, New York, 1975).

2 E. Baldwin, *Dynamic Aspects of Biochemistry*, 4th ed. (Cambridge University Press, Cambridge, England, 1965).

3 E. E. Conn and P. K. Stumpf, *Outlines of Biochemistry* (Wiley, New York, 1963).

CHAPTER

9

Control Processes: Molecular Cybernetics

In each step of each of the pathways discussed in the preceding chapter, an enzyme is involved as a catalyst. Each enzyme is specific for the particular step it catalyzes. Many of these enzymes are actually complexes of several polypeptide chains that self-assemble to form the complex. Not all of the subunits in a complex have a catalytic function: some are specialized for regulatory functions. In some instances both regulatory and catalytic functions reside in a single polypeptide chain. This chapter examines the operation of these regulatory, or control, functions.

Another type of regulation can occur with respect to the *production* of the enzyme within the cell. This kind of control has to do with the biosynthesis of proteins according to the gene-directed mechanism of protein synthesis. The details of how a protein is constructed according to the message contained in the gene are described in the next chapter. This chapter describes some of the control processes associated with protein synthesis.

A prevalent kind of control process occurs in the anabolic pathway that leads to the synthesis of amino acids. Whenever the end product of this pathway, an amino acid, accumulates in excess, it interacts with an enzyme that catalyzes an early step in the pathway and renders the enzyme nonfunctional. This is the case for arginine, histidine, proline, leucine, isoleucine, lysine, serine, threonine, tryptophan, and valine [see (9.1)]. In each case the pertinent enzyme contains a specific site at which the amino acid is bound. The binding of the amino acid involves "weak" bonds but is "strong" enough to induce, in the whole enzyme, a conformational change that causes the catalytic function to be inhibited partly or completely. If the regulatory binding site is on one subunit and the catalytic site is on another, a conformational change in the regulatory subunit is communicated to the catalytic subunit through their mutually specific interface of multiple weak bonds, and the conformation of the catalytic subunit is also altered. If both functions reside in a single polypeptide, they nevertheless

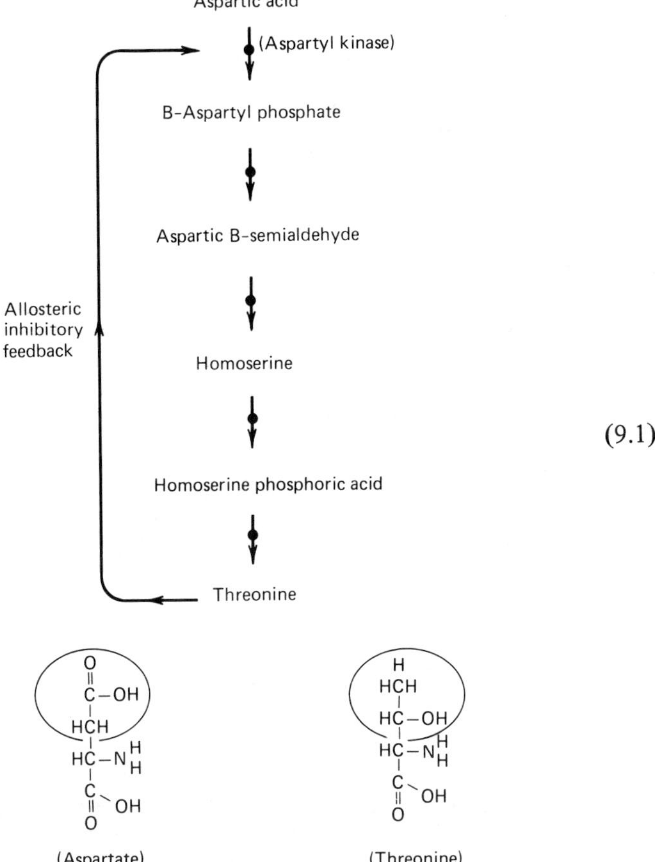

have always been found to be located at distinct sites. Correlative with this situation is the simple fact that the end product molecule that produces regulation by binding to an enzyme located several steps earlier in its synthesis pathway is usually only remotely similar structurally to the substrate molecule for that enzyme. Thus regulation is achieved by a structure *different* from that of the substrate. For this reason the general phenomenon is called "allostery" from the Greek for "other structure" [1-6].

The idea of allostery was stimulated by the detailed work on aspartate transcarbamylase by Gerhart and Pardee [5], and was given a broader theoretical and experimental basis by Monod and co-workers in 1963 [6]. Aspartate transcarbamylase is a complex composed of six catalytic and six regulatory subunits. It is regulated by cytidine triphosphate (CTP), which bears little resemblance to its substrates aspartic acid and carbamyl phosphate (see Chapter 1).

Escherichium coli contains an enzyme complex, glutamine synthetase, that carries the mechanism of allostery to an extreme! This complex contains 12 identical subunits that can be inhibited by each of 6 different end products arising in the many diverse pathways stemming from glutamine, as well as by alanine and glycine, which arise independently.

The molecular species that binds an enzyme as a regulator, the "allosteric effector," remains bound until its cellular concentration subsides through use or degradation, whichever occurs earlier. As the cellular concentration subsides, the binding equilibrium shifts until it is mostly unbound, and this reactivates the catalytic function of the enzyme. Thus allosteric inhibition is entirely reversible. This state of affairs is the basis for relatively rapid regulatory processes.

An example of this kind of rapid regulation is the control of energy metabolism in glycolysis and the citric acid cycle. In addition to inhibition by pathway end products, there is also positive control, or excitation of the pathway, by precursors of pathway end products. In this example of energy metabolism, the ultimate end product is ATP, which exerts numerous inhibitory allosteric actions on enzymes of the integrated pathways for its synthesis, while its precursors, AMP and ADP, are positive effectors in some of the same enzymatic steps. Some enzymes may therefore be inhibited or excited by different specific effectors. All this is diagrammed in (9.2) [7], where the solid lines indicate reaction pathways, the dotted lines are the inhibitory allosteric interactions of the molecule at the tail of an arrow with the enzyme at the head of an arrow, and the broken lines represent excitatory allosteric interactions of the molecule with the enzyme. The electron transport chain between NADH and H_2O has been schematized very simply, and in particular, the transduction of the energy harvested by this pathway into ATP energy has been expressed by the vague symbolic form "energy." As was pointed out earlier, not until Chapter 12 is a detailed account of this transduction given.

On a relatively slower time scale, the other regulatory mechanism alluded to in the second paragraph of this chapter is functional. This mechanism is described by the "operon" model of the gene, which emerged in the early 1960s from the work of Jacob and Monod. It should be borne in mind that this insight, to be described below, arose at a time when the genetic code was only just being deciphered and the function of transfer and messenger RNAs had been barely established. These equally important matters are discussed in detail in the next chapter.

In the operon model [8] there is more to DNA than just genes coding for the synthesis of specific proteins. There are also regions of the DNA that are the sites at which the obligatory polymerase enzyme binds and commences the transcription of DNA into a particular RNA, messenger RNA, from which the protein is made. The details of these processes, especially protein synthesis, are deferred until the next chapter. For the present, it is sufficient to realize that

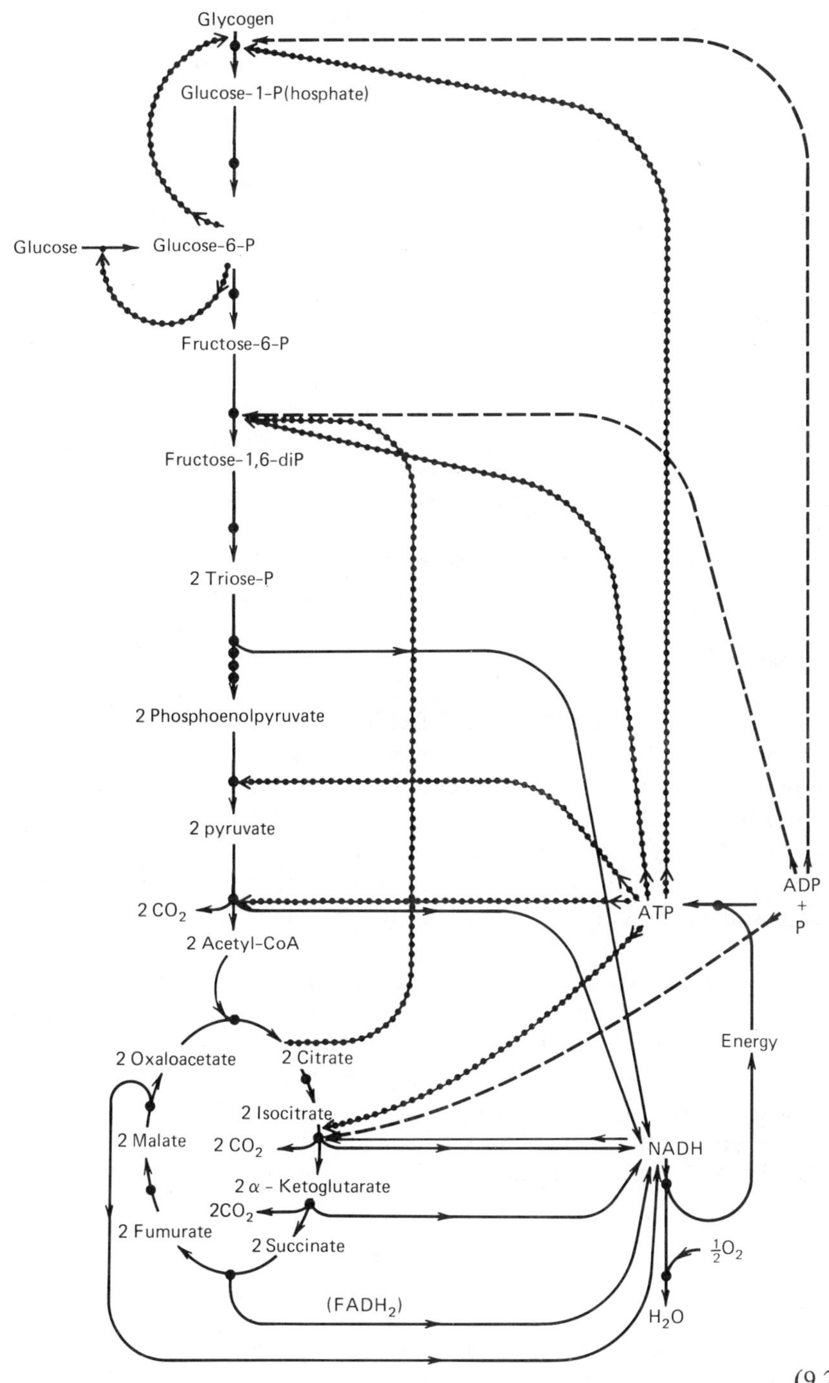
Glycogen
Glucose-1-P(hosphate)
Glucose
Glucose-6-P
Fructose-6-P
Fructose-1,6-diP
2 Triose-P
2 Phosphoenolpyruvate
2 pyruvate
2 CO_2
2 Acetyl-CoA
ATP
ADP
+
P
2 Oxaloacetate
2 Citrate
Energy
2 Isocitrate
2 Malate
2 CO_2
NADH
2 α - Ketoglutarate
$2CO_2$
2 Fumurate
2 Succinate
$\frac{1}{2}O_2$
($FADH_2$)
H_2O

(9.2)

the polymerase enzyme "recognizes" a "promoter" site on the DNA and commences to catalyze the synthesis of an RNA molecule, complementary to one of the two DNA strands, from ribose triphosphate precursors. This process is called the "transcription" of DNA into RNA. The polymerase moves along the DNA, perhaps through *several* genes, before it becomes detached at a specific site and in this way transcribes several genes in a coordinated fashion. This is illustrated

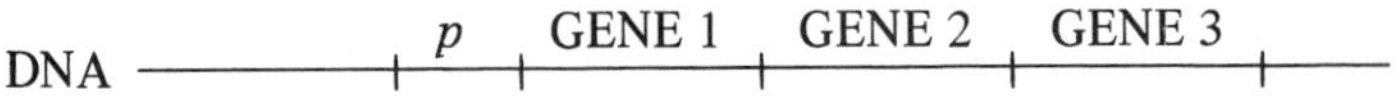

```
                     p     GENE 1     GENE 2     GENE 3
DNA  ---------------+---+----------+----------+----------+-------
```

(9.3)

schematically in (9.3), where p denotes the promoter site and the polymerase is imagined to move from left to right, coming off beyond gene 3. An example of such a system of coordinated genes occurs in *E. coli*. The lactose operon contains three genes: gene 1 (β-galactosidase), gene 2 (galactoside permease), and gene 3 (galactoside acetylase).

Lactose is a disaccharide comprised of D-glucose and D-galactose, two aldose isometes, linked by a glycosidic bond. β-Galactosidase catalyzes the hydrolysis of the glycosidic bond, freeing D-glucose and D-galactose, the latter of which is enzymatically isomerized into D-glucose, for energy metabolism. Thus lactose is an energy food for *E. coli*. Getting lactose inside the cell is one problem faced by *E. coli* cells during "energy gathering." β-Galactosidase permease is a carrier protein "catalyzing" the transmembrane transport of lactose into the cell from the external medium. This protein-induced permeability of the membrane to lactose is an example of how the cell can selectively enhance its natural permeability, which may be very low, to a variety of molecules.

As shown in detail in Chapter 10, the making of these two gene products, as well as the gene 3 product, whose function is not yet fully appreciated, consumes energy. In the absence of any environmental lactose, it would be convenient for the cellular energy economy if the cell had a mechanism for totally inhibiting the transcription of the lactose operon: no energy would be wasted making messenger RNA and proteins that do no good for the cell. The *E. coli* cells do possess such a mechanism, and it involves only one protein, the lactose *repressor*.

Between the promoter region and gene 1 of the lactose operon is a region called the "operator" region, as shown in (9.4). The repressor protein can bind

```
                    p     o     GENE 1     GENE 2     GENE 3
DNA  --------------+-----+-----+----------+----------+----------+-------
```

(9.4)

very "strongly" to the operator region marked by o. When it does, the polymerase protein can bind to p, the promoter, but is unable to proceed with the transcription of the three operon genes. That o is to the right of p in the *E. coli* lactose operon suggests that the polymerase simply cannot get past the operator region because the repressor protein blocks the way. This may well be so, but there are also some operons that are to the *left* of the promoter site even though the polymerase transcribes to the *right*. In these cases the repressor's presence perhaps induces a conformational change in the adjacent polymerase, which is then unable to function properly.

The lactose repressor is itself a gene product made from the *E. coli* chromosome (DNA). Its gene and gene promoter are located some distance to the left of the lactose operon in (9.5). The repressor gene is called the "*i*-gene," perhaps for

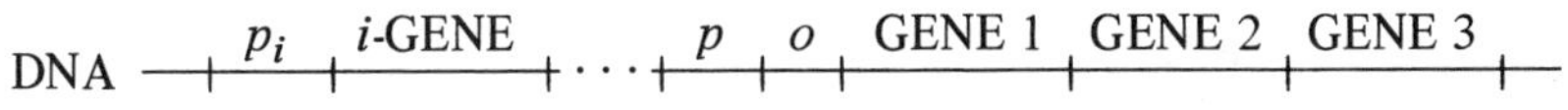

$$\text{DNA} \;\; -\!\!\!+\!\!\!\overset{p_i}{-\!-}\!\!\!+\!\!\!\overset{i\text{-GENE}}{-\!-\!-\!-\!-}\!\!\!+\cdots+\!\!\!\overset{p}{-\!-}\!\!\!+\!\!\!\overset{o}{-}\!\!\!+\!\!\!\overset{\text{GENE 1}}{-\!-\!-\!-}\!\!\!+\!\!\!\overset{\text{GENE 2}}{-\!-\!-\!-}\!\!\!+\!\!\!\overset{\text{GENE 3}}{-\!-\!-\!-}\!\!\!+\!- \tag{9.5}$$

"inhibitor gene," and its promoter is designated p_i. There is *no* operator region for the *i*-gene! Therefore the repressor is always made and will consequently bind the operator region of the lactose operon. How, then, does *E. coli* ever get to use its lactose operon?

The lactose repressor is an allosteric protein. When lactose is around, it binds tightly to the lactose repressor, even when the repressor is tightly bound to the lactose operator. When lactose is bound to repressor, the repressor changes conformation so that it is no longer able to bind the lactose operator region. The changed repressor comes off the DNA, and the polymerase is free to transcribe the operon. The key here is that *lactose* allosterically inhibits the repressor's ability to inhibit the lactose operon. This will happen whenever there is enough lactose in the environment of the cell for some of it to leak into the cell and bind the repressor. Once "derepressed," the lactose operon is transcribed and the gene products catalyze the transport of more lactose into the cell as well as begin its energy-yielding metabolism. As long as lactose concentrations remain relatively high, the repressors are inhibited and lactose is metabolized. As soon as lactose is exhausted in the medium, the repressors are reversibly freed of lactose and can again bind the lactose operator to shut off the operon. This is truly a splendidly conceived molecular mechanism. Monod [3] called this kind of control "molecular cybernetics" and characterized the function of lactose, the lactose repressor, and the lactose operon in terms of "logical operators." To wit, the lactose-induced derepression is a double negative ≡ positive ($- \; - = +$), with lactose ($-$) *inhibiting* the repressor ($-$), which *inhibits* the operon.

The specificity of polymerase proteins for promoter regions of DNA, and of repressor proteins for operator regions of DNA, is an outstanding example of a molecular phenomenon not very well understood in molecular detail. "Protein-polynucleotide recognition" in general is an actively discussed topic at present.

Much research is being done in this area to increase the understanding of the molecular basis for the specificity.

The regulation of the lactose operon is important for the cell because the cell needs to maintain a good ATP economy in order to perform innumerable functions requiring energy (ATP), including especially the synthesis of macromolecular constituents. Thus, if ATP levels fall perilously low in the cell, it is to the cell's advantage to be able to rapidly transcribe the lactose operon when lactose is found in the environment. There is a protein in the cell called adenyl cyclase. This enzyme catalyzes the conversion of ATP into cyclic AMP [see (9.6)]. Since

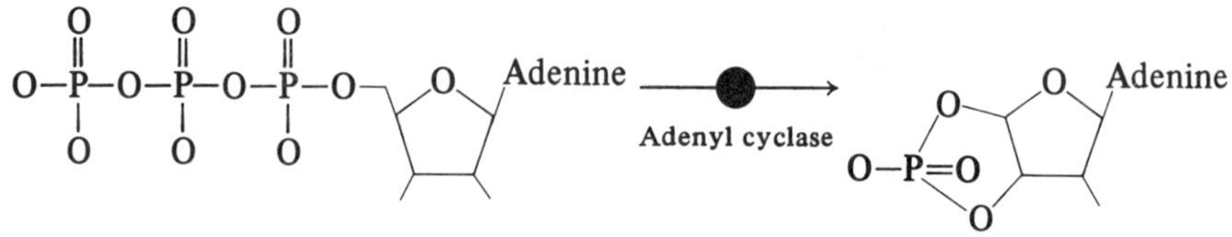

$$\text{O-}\overset{\overset{\text{O}}{\|}}{\underset{\underset{\text{O}}{|}}{\text{P}}}\text{-O-}\overset{\overset{\text{O}}{\|}}{\underset{\underset{\text{O}}{|}}{\text{P}}}\text{-O-}\overset{\overset{\text{O}}{\|}}{\underset{\underset{\text{O}}{|}}{\text{P}}}\text{-O-(ribose)-Adenine} \xrightarrow[\text{Adenyl cyclase}]{} \text{O-P(=O)(-O-)}_2\text{(ribose)-Adenine}$$

ATP

Cyclic AMP

+ Pyrophosphate PP

(9.6)

cyclic AMP functions *catalytically*, it can be produced even when ATP levels are low because only a few molecules of ATP are required to make a few molecules of cyclic AMP. The cyclic AMP allosterically binds another protein, the "catabolite gene activator protein," called CAP for short. With bound cyclic AMP, CAP binds a region of the *E. coli* DNA between the *i*-gene and the promoter region, as shown in (9.7). With CAP bound to the CAP region of the DNA and

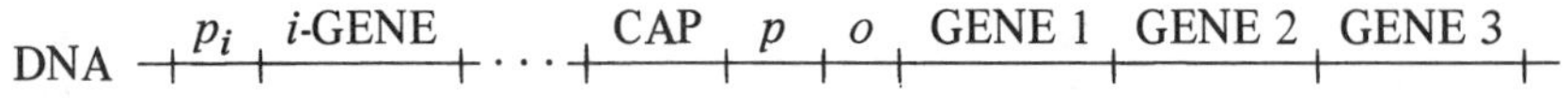

DNA —| p_i | *i*-GENE | . . . | CAP | p | o | GENE 1 | GENE 2 | GENE 3 |—

(9.7)

the repressor inhibited from binding the operator region by lactose, the lactose operon is transcribed by polymerase protein at a rate up to 20 times greater than it is when no CAP is bound.

When ATP levels are normal, there is no reason to increase the rate of lactose operon transcription above its normal, unexcited, rate, and *E. coli* has a mechanism for this kind of delicate control as well: ATP, the substrate for adenyl cyclase, is also an *allosteric* effector, which binds to adenyl cyclase at a site other than the catalytic site, to which ATP is also bound. When bound to the allosteric site, ATP inhibits adenyl cyclase, and this reduces the production of cyclic AMP. Still another enzyme, cyclic AMP phosphodiesterase, degrades cyclic AMP to ordinary AMP, which does not bind CAP. Thus, if ATP is inhibiting adenyl cyclase *and* cyclic AMP is being degraded to an inactive AMP, CAP is no longer activated.

All of these mechanisms are diagrammed together in (9.8). The seven proteins are 1, polymerase; 2, repressor; 3, CAP; 4, adenyl cyclase; 5, galactoside permease; 6, β-galactosidase; and 7, cyclic AMP phosphodiesterase.

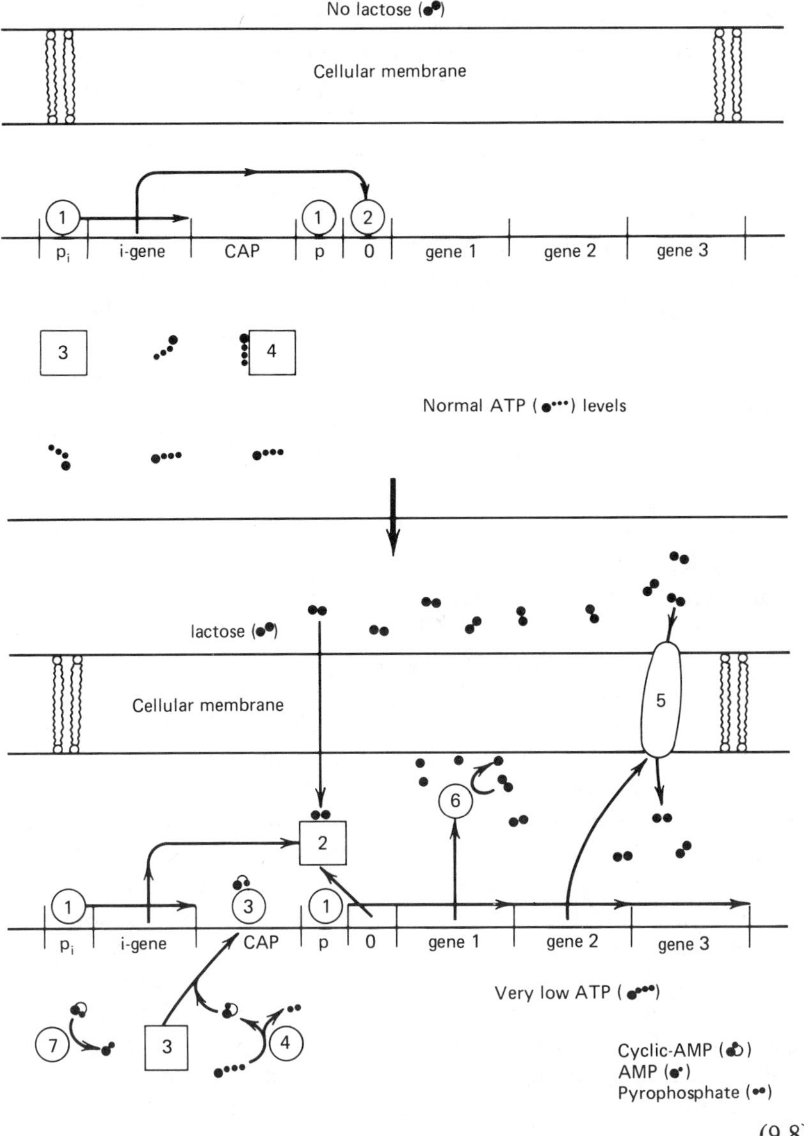

(9.8)

As a final example of control processes, the *phosphorylase cascade* mechanism of carbohydrate metabolism in higher organisms is described [9]. It is of interest because it controls a pathway that ultimately affects ATP levels. The control effectors are ATP and cyclic AMP, as is the case with the lactose operon, and some of the enzymes involved are activated by phosphorylation steps that use ATP.

The phosphorylase cascade is the mechanism by which excess glucose is polymerized into glycogen for storage or by which glycogen is degraded to glucose for glycolysis when energy is needed. Polymerization requires the activation of glucose monomers with UTP to form UDP-glucose, which can be added to the growing glucose polymer, glycogen. This activation is similar to the activation of amino acids required in order to polymerize them into proteins, as described in Chapter 10. The polymerization step is catalyzed by the enzyme synthase I. An enzyme, synthase I kinase, phosphorylates synthase I, using ATP as phosphate donor, and produces the enzyme synthase D, which is much less active than synthase I. Another enzyme, synthase phosphatase, removes the phosphate on synthase D, thereby regenerating synthase I. Glycogen, the product of synthase I, inhibits synthase phosphatase when glycogen levels are high. This means that most of the synthase will end up as synthase D, which catalyzes glycogen synthesis very little. The synthase I kinase is allosterically controlled by cyclic AMP, having an active form when cyclic AMP is bound and an inactive form when cyclic AMP is absent. Cyclic AMP, in turn, is made from ATP by adenyl cyclase, an enzyme that is activated by hormones like epinephrine and glucagon. These relations are shown in (9.9), where SI is synthase I; SD is synthase D; SIK and SIK_a are the inactive and active forms of synthase I kinase, respectively; AC and AC_a are the inactive and active forms of adenyl cyclase, respectively; SP is synthase phosphatase; and GP is UDP-glucose pyrophosphorylase. The dotted line denotes the inhibiting effect of glycogen.

While the preceding scheme for glycogen synthesis is reasonably complex, it involves only one kinase and one phosphatase. The degradation of glycogen to glucose-1-phosphate, on the other hand, is even more elaborate, involving two distinct kinases and two distinct phosphatases. The degradation of glycogen proceeds as a "phosphorolysis" in which glucose-1-phosphate is cleaved from glycogen through the incorporation of phosphate (hence phosphorolysis) by the enzyme complex phosphorylase *a*. Phosphorylase *a* is a tetramer with a molecular weight of 380,000. Each of its four subunits contains a phosphorylated serine residue. The enzyme phosphorylase phosphatase will remove these phosphates, thereby converting the enzyme complex into two dimeric molecules of phosphorylase *b*, which is inactive as a glycogen degradation enzyme. The active form, phosphorylase *a*, can be restored by rephosphorylating the serine residues, as a result of which the tetrameric complex reaggregates, and this phosphorylation is catalyzed by phosphorylase *b* kinase. So far, this group of

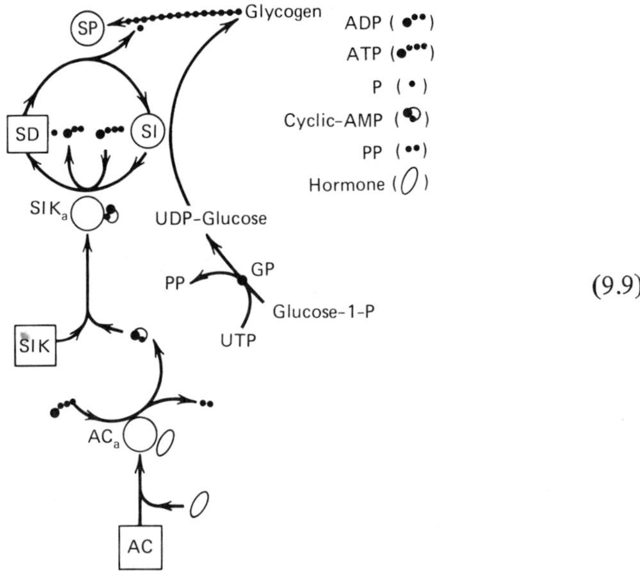

(9.9)

enzymes parallels the synthase enzyme group with its kinase and phosphatase. In addition, phosphorylase *b* kinase has both active and inactive forms, just as does the synthase kinase. The similarity ends here. The active form of phosphorylase *b* kinase is made from its inactive form by phosphorylation, using ATP as phosphate donor, and this is catalyzed by phosphorylase kinase kinase. Phosphorylase kinase kinase in turn has active and inactive forms. The active form is produced when cyclic AMP binds the inactive form. Cyclic AMP is made, as usual, from ATP by adenyl cyclase, which is converted from its inactive form to its active form on the binding of hormones like epinephrine and glucagon. All of these relationships are depicted in (9.10), in which the following abbreviations are used: PP_a, phosphorylase *a*; PP_b, phosphorylase *b*; PPP, phosphorylase phosphatase; PP_bK_a, phosphorylase *b* kinase (active form); PP_bK, phosphorylase *b* kinase (inactive form); PPKP, phosphorylase kinase phosphatase; $PPKK_a$, phosphorylase kinase kinase (active form); PPKK, phosphorylase kinase kinase (inactive form); AC_a, adenyl cyclase (active form); and AC, adenyl cyclase (inactive form). The effect of this "cascade" of activations is that a small amount of hormone leads to a large amount of glycogen degradation, as if the system amplified the initial signal.

Typically, hormones bind on the outside of cells, and the cyclic AMP that is produced by the activation of adenyl cyclase is on the inside of the cells. Thus, the adenyl cyclase depicted in (9.8), (9.9), and (9.10) is usually a complex with a subunit that binds specific hormones while *another* subunit is thereby allo-

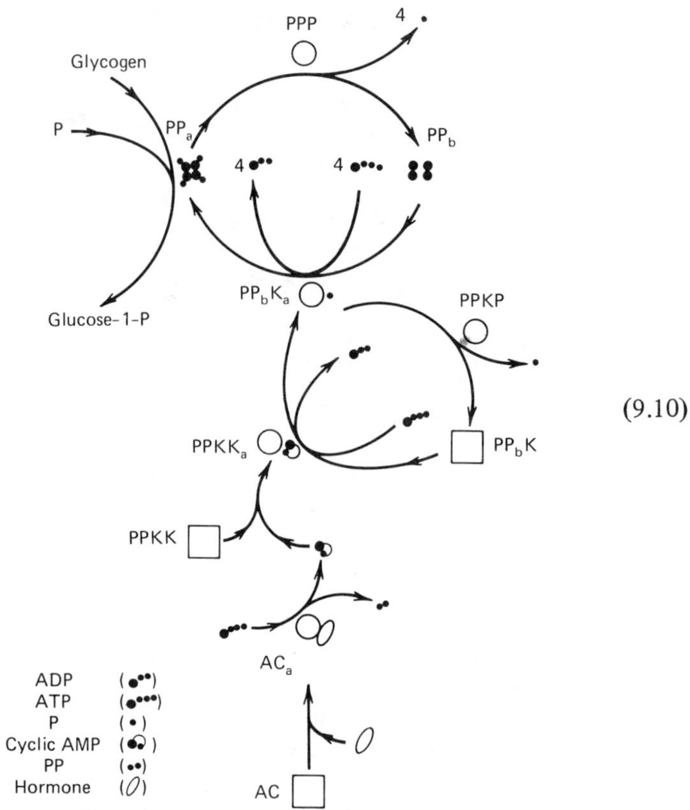

(9.10)

sterically altered into an active adenyl cyclase. This constitutes the "second messenger" mechanism of hormone action.

References

1 J. Watson, *Molecular Biology of the Gene*, 2nd ed. (Benjamin, New York, 1970).

2 J. Miller and W. Reznikoff, editors, *The Operon* (Cold Spring Harbor Laboratory, 1978).

Many operons, including the lactose operon, are described in great detail in this volume. Detailed molecular accounts of nearly current understanding of protein-polynucleotide recognition problems are provided.

3 J. Monod, *Chance and Necessity* (A. A. Knopf, New York, 1971).

4 R. E. Dickerson and I. Geis, *Proteins* (Harper and Row, New York, 1969).

5 J. C. Gerhart and A. Pardee, *J. Biol. Chem.*, **237**, 391 (1962).

6 J. Monod, J. P. Changeux, and F. Jacob, *J. Mol. Biol.*, **6**, 306 (1963).

7 A. Lehninger, *Biochemistry*, 2nd ed. (Worth, New York, 1975).

8 B. Lewin, *Gene Expression*, Vol. I (Wiley, New York, 1974).

This book contains detailed accounts of many operon control circuits.

9 J. Lerner, *Intermediary Metabolism and Its Regulation* (Prentice-Hall, Englewood Cliffs, N.J., 1971).

CHAPTER 10

The Monomer-to-Polymer Transition

The pathways for energy metabolism require a large number of enzymes to catalyze the reactions in each step. Enzymes are also needed for each of the synthetic pathways leading to amino acids, nucleotides, and other molecular components of the cell. In addition, there are numerous degradative pathways that require enzymes. Each of these many enzymes is synthesized by the cell in the complicated process of protein biosynthesis. This process involves several species of polynucleotides that must also be synthesized. One of the major energy-*utilizing* functions of the cell is the synthesis of proteins and polynucleotides. Thus, as is typical of biological organization, energy metabolism depends on macromolecules, whose synthesis depends on energy metabolism.

This chapter presents an up-to-date account of macromolecular synthesis. It shows that the energy requirements are always fulfilled by ATP, whose synthesis is described in detail in Chapter 12. It also shows that, even though an incredible amount of detailed knowledge exists regarding protein biosynthesis, there still remains an outstanding unsolved problem, the "mystery of the second code," which has not been sufficiently emphasized in the literature.

Before delving into the specific details of polynucleotide and protein synthesis in cells, a review of the thermodynamics of polymer synthesis is useful. As was observed in Chapter 1, the biological polymers–polynucleotides, proteins, and polysaccharides–are dehydration condensates of their monomeric constituents. This means that the linkage between adjacent monomers in the polymer–whether it is a peptide bond, a phosphodiester bond, or a glycosidic bond–is hydrolyzed by H_2O with an OH (or O^-) becoming bonded to one monomer while the remaining H (or H + H^+) bonds to the other [see (10.1)].

$$\underset{\text{Glycylglycine}}{\mathrm{H\overset{H}{\underset{H}{N^+}}-\overset{H}{\underset{H}{C}}-\overset{O}{\overset{\|}{C}}\overset{H_2O}{\swarrow}\underset{H}{N}-\overset{H}{\underset{H}{C}}-\overset{O}{\overset{\|}{C}}-O^-}} \rightarrow \underset{\text{Glycine}}{\mathrm{H\overset{H}{\underset{H}{N^+}}-\overset{H}{\underset{H}{C}}-\overset{O}{\overset{\|}{C}}-O^-}} + \underset{\text{Glycine}}{\mathrm{H\overset{H}{\underset{H}{N^+}}-\overset{H}{\underset{H}{C}}-\overset{O}{\overset{\|}{C}}-O^-}} \qquad (10.1)$$

In (10.1) the dipeptide glycylglycine as well as the amino acid glycine are depicted as zwitterions, which are their charged forms at pH 7. In this case H_2O is split into O^- and $H + H^+$ when it hydrolyzes the peptide bond. A similar situation holds for polynucleotides and polysaccharides. Consequently, the synthesis of these polymers ultimately corresponds to the reaction equation

$$M_1 + M_2 + M_3 \cdots + M_n \rightleftharpoons P_n + (n - 1)\, H_2O \tag{10.2}$$

This reaction need not be imagined to occur as an n-molecule event, but rather as a sequence of successive additions of monomers M_i, until the final polymer, P_n, is produced concomitantly with $(n - 1)$ H_2O molecules. As shown below, this equation represents only the final balance, much as (8.10) did for energy metabolism, and in actuality a great many intermediate steps are required for the production of the polymer P_n.

Several considerations lead to the same thermodynamic conclusion: polymer synthesis does not occur spontaneously because the equilibrium for (10.1) lies far on the side of the monomers, or put another way, the hydrolysis of polymers is thermodynamically favored. Three important reasons are (1) the measured enthalpy change of the bonds in going from monomers to polymers plus H_2O is positive; (2) the biological concentration of H_2O, 55.55 molar, greatly favors hydrolysis; and (3) the entropy change in going from monomers to polymer and H_2O is negative. To overcome these considerable barriers to polymer production, the cell has evolved synthetic schemes that use the energy harvested by energy metabolism and transduced into the energy of ATP, to "activate" the monomers. The activated monomers are then condensed into polymers with an overall decrease in Gibbs free energy, a thermodynamically favorable disposition. Moreover, the intervention of ATP for activation of the monomers removes the elimination of H_2O to a more remote step in the overall pathway sequence, thereby lessening the inhibitory thermodynamic effect of high H_2O molarity. For example, the formation of a dinucleotide from guanosine monophosphate (GMP) and cytosine monophosphate (CMP) as a "dehydration" is depicted in (10.3), which will be thermodynamically inhibited in the right direction and

[GMP: O^-–P(=O)(OH)–O–sugar(O)–G, 2' OH, 3' OH] + [CMP: O^-–P(=O)(OH)–O–sugar(O)–C, OH, OH] $\rightleftharpoons$ [O^-–P(=O)(OH)–O–sugar(O)–G, O–P(=O)(O^-)–O–sugar(O)–C, OH; OH, OH] + H_2O (10.3)

thermodynamically favored in the left, hydrolytic, direction. However, ATP can be used to activate CMP by converting it through an enzyme-dependent sequence of steps into CTP:

$$\begin{aligned} \text{ATP} + \text{CMP} &\rightleftharpoons \text{ADP} + \text{CDP} \\ \text{ATP} + \text{CDP} &\rightleftharpoons \text{ADP} + \text{CTP} \end{aligned} \tag{10.4}$$

$$\text{O}^-\text{–P(=O)(OH)–O–[ribose(OH)(OH)]–G} + \text{O}^-\text{–P(=O)(O}^-\text{)–O–P(=O)(OH)–O–P(=O)(OH)–O–[ribose(OH)(OH)]–C} \rightleftharpoons \text{O}^-\text{–P(=O)(O}^-\text{)–O–[ribose(OH)–G]–O–P(=O)(OH)–O–[ribose(OH)(OH)]–C} + \text{O–P(=O}^-\text{)(O)–O–P(=O)(OH)–OH} \tag{10.5}$$

GMP and CTP will now react according to (10.5). Note that pyrophosphate is the byproduct of this reaction, and not H_2O! In fact, *using* one H_2O molecule to split pyrophosphate as in (10.6) provides the two phosphates needed by the cell to regenerate the ATP used in (10.4):

$$\text{O}^-\text{–P(=O)(O}^-\text{)–O–P(=O)(OH)–OH} + H_2O \rightleftharpoons \text{O}^-\text{–P(=O)(OH)–OH} + \text{O}^-\text{–P(=O)(OH)–OH} \tag{10.6}$$

$$2\text{ADP} + 2\text{P} \rightleftharpoons 2\text{ATP} + 2H_2O \tag{10.7}$$

It is this regeneration of ATP that *releases* H_2O and *requires* the energy generated by energy metabolism. The *two* molecules of H_2O in (10.7) minus the *one* molecule of H_2O in (10.6) yield the *one* molecule of H_2O associated with (10.3). This overall process is depicted in (10.8).

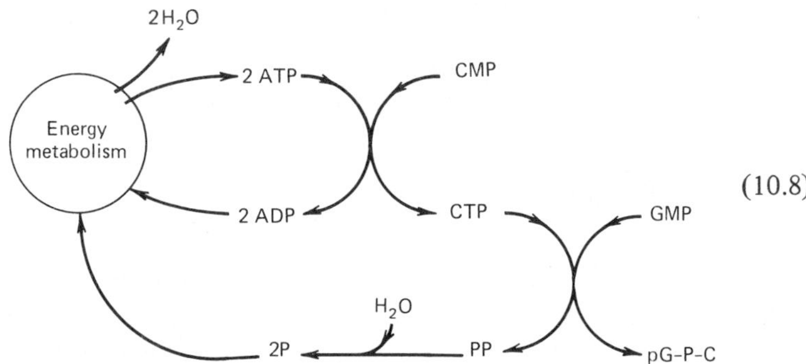

(10.8)

Now that the main aspects of the problem of the monomer-to-polymer conversion have been indicated, each of these aspects will be developed in greater detail.

A quantitative assessment of the thermodynamic barrier to polymer synthesis is possible. Although the *enthalpy* change measured in forming a dipeptide varies with the specific amino acids, and their order (glycylalanine is different from alanylglycine), a value of about 1.5 kcal/mole ($\sim 2.5 k_B T_{300}$ per bond) is representative; in contrast, for the phosphodiesterase bonds of polynucleotides values on the order 3 kcal/mole are typical. The entropy contributions, which are *negative*, tend to increase the Gibbs free energy change:

$$\Delta G = \Delta H - T\Delta S \tag{10.9}$$

That the entropy change on polymerization would be negative is seen as follows. Before polymerization the entropy for the reaction in (10.2) is, according to (2.17) and neglecting activities,

$$S_{\text{before}} = \frac{3}{2} k_B \sum_{i=1}^{n} N_{\mathrm{M}_i} + \sum_{i=1}^{n} N_{\mathrm{M}_i} S^0_{\mathrm{M}_i} + k_B \sum_{i=1}^{n} N_{\mathrm{M}_i} \ln\left[\left(\frac{2\pi m_{\mathrm{M}_i} k_B T}{h^2}\right)^{3/2} V\right] - k_B \sum_{i=1}^{n} N_{\mathrm{M}_i} \ln\left(\frac{N_{\mathrm{M}_i}}{e}\right) \tag{10.10}$$

where $S^0_{\mathrm{M}_i}$ includes rotational, vibrational, and electronic contributions to the entropy while the translational entropy has been explicitly separated. After polymerization, there is only the polymer P and $(n - 1)$ H_2O molecules with an entropy

$$S_{\text{after}} = \frac{3}{2} k_B (N_{P_n} + N_{H_2O}) + N_{P_n} S^0_{P_n} + N_{H_2O} S^0_{H_2O}$$
$$+ k_B N_{P_n} \ln\left[\left(\frac{2\pi m_{P_n} k_B T}{h^2}\right)^{3/2} V\right] + k_B N_{H_2O} \ln\left[\left(\frac{2\pi m_{H_2O} k_B T}{h^2}\right)^{3/2} V\right]$$
$$- k_B N_{P_n} \ln\left(\frac{N_{P_n}}{e}\right) - k_B N_{H_2O} \ln\left(\frac{N_{H_2O}}{e}\right) \tag{10.11}$$

In addition, stipulate, for example, $N_{H_2O} = n - 1$ when $N_p = 1$ and $N_{M_i} = 1$ for each i. Moreover, $m_{P_n} + (n - 1) m_{H_2O} = \Sigma_i m_{M_i}$ exhibits mass conservation for this case. Therefore the entropy change is (see appendix for more details)

$$\Delta S \equiv S_{\text{after}} - S_{\text{before}} = S^0_{P_n} + (n - 1) S^0_{H_2O} - \sum_{i=1}^{n} S^0_{M_i}$$
$$+ k_B \ln\left[\left(\frac{m_{P_n} m^{n-1}_{H_2O}}{\prod_{i=1}^{n} m_{M_i}}\right)^{3/2}\right] - k_B \ln\left[(n - 1)^{n-1}\right] \tag{10.12}$$

To estimate the magnitude of this entropy change, it is assumed that (1) the intrinsic entropy change resulting from rotation, vibration, and electronic levels is relatively small in comparison with the mass term and the entropy of mixing term; (2) the mass term can be replaced by a simpler expression if each monomer has a mass about equal to a common mass, m_M; and (3) n is large in comparison with 1. Then it follows that

$$\Delta S \simeq \frac{3}{2} k_B \ln\left[n \left(\frac{m_{H_2O}}{m_M}\right)^{n-1}\right] - n k_B \ln(n)$$
$$\simeq -\frac{3}{2} k_B (n - 1) \ln\left(\frac{m_M}{m_{H_2O}}\right) - n k_B \ln(n) < 0 \tag{10.13}$$

where the ratio m_M/m_{H_2O} is about 10 for amino acids and about 20 for nucleotide monophosphates. Therefore, for proteins, the mass term in (10.13) contributes about $+3k_B T_{300}$ to the Gibbs free energy *per bond*; for polynucleotides, it contributes about $+5k_B T_{300}$. These values are comparable to the enthalpy values already quoted, and the minus sign in (10.9) has been included.

As an example of how *badly* the equilibrium ratio for (10.2) "favors" the formation of polymer, consider the case of a protein composed of $n = 100$ amino acids, with an average Gibbs free energy input of about 3 kcal/mole of bonds. Polynucleotides are usually much longer and each bond requires more energy, so that this example is a conservative one. The equilibrium constant is

$$K_{eq} = \frac{[P_n][H_2O]^{n-1}}{\prod_{i=1}^{n} [M_i]} = \exp\left[-(n-1)\frac{1}{RT} 3 \text{ kcal/mole}\right] \quad (n = 100) \quad (10.14)$$

where $R = 6 \times 10^{23} k_B$ and $RT_{300} = 600$ cal/mole. Therefore, the right-hand side of (10.14) is $\exp[-(n-1)5] = 10^{-(n-1)2.2} = 10^{-215}$ for $n = 100$. Even if the reaction starts with 1 molar concentrations of the monomers M_i, concentrations that are well above any cellular levels, the concentration of polymer works out to be

$$\begin{aligned}[P_n] &= 10^{-215} \times [H_2O]^{-(n-1)} = 10^{-215} \times (55.55)^{-99} \\ &= 10^{-215} \times 10^{-173} \text{ molar} \\ &= 10^{-388} \text{ molar} \end{aligned} \quad (10.15)$$

What does a concentration of 10^{-388} molar mean? It means that the volume required to guarantee the presence of *one* polymer *molecule*–not a mole (6×10^{23}) of them, but just *one*–is 1.6×10^{364} liters = 1.6×10^{361} m^3. This is not really properly called an "astronomical" volume because it is way *too big*! The nearest galaxy to our own is the Andromeda Nebula, which is about 2×10^6 light years away. A light year is about 10^{16} m. The volume of a sphere with a radius as long as the distance between us and the Andromeda Nebula would be only 4×10^{48} m^3! The volume above is 4×10^{312} times bigger! This is much, much bigger than the volume of the entire universe in the "big bang" theory! (The big bang theory gives a volume $\ll 10^{100}$ m^3.) It is therefore safe to say that thermodynamics does not favor the production of polymers from monomers *without* energy coupling!

On the other hand, by simply paying the price per bond in energy, the cell overcomes this immense barrier. The molecular details of how the cell utilizes ATP to pay for bond formation in polymers occupies the rest of this chapter. The description of this process represents one of the major achievements of twentieth century science, rivaling the theory of relativity and the quantum theory in its significance. Curiously, these triumphs of physics represent the work of very few men, whereas the accomplishments of molecular biology represent the achievements of literally hundreds of major contributors.

Payment for polymer bonds with ATP works for the following reasons:

The reaction in (10.2), when specialized to nucleotide polymerization according to the activation mechanisms for nucleotides depicted in (10.4), should instead read

$$M_1 + M_2 + M_3 + \cdots + M_n + 2(n-1)\text{ATP} \rightleftharpoons P_n + 2(n-1)\text{ADP} + (n-1)\text{PP} \tag{10.16}$$

where PP is pyrophosphate. This involves a net release of energy from $2(n-1) - (n-1) = (n-1)$ phosphodiester bonds, at an energy content of roughly 7.3 kcal/mole ($12k_BT_{300}$ per linkage). This value will be explained in Chapter 12. The coupling of ATP to the reaction, through activation of the monomeric nucleotide monophosphates, couples some of this energy into the polymerization process. The analog of (10.14) becomes

$$K_{\text{eq}} = \frac{[P_n][\text{ADP}]^{2(n-1)}[\text{PP}]^{n-1}}{\prod_{i=1}^{n} [M_i][\text{ATP}]^{2(n-1)}}$$

$$= \exp\left[-(n-1)\frac{1}{RT}(3 \text{ kcal/mole} - 7.3 \text{ kcal/mole})\right] \simeq e^{7n} \gg 1 \tag{10.17}$$

In this exponential the 3 kcal/mole is the $\Delta G^{0\prime}$ for the transition from monomers to polymers while the 7.3 kcal/mole is the $\Delta G^{0\prime}$ for the phosphodiester bond change. From (10.17) it follows that

$$[P_n] = \frac{\prod_{i=1}^{n} [M_i][\text{ATP}]^{2(n-1)}}{[\text{ADP}]^{2(n-1)}[\text{PP}]^{n-1}} e^{7n} \tag{10.18}$$

which can be made at least of order unity. By reducing [PP] through the hydrolysis of PP to $2P_i$, more energy is released, driving the overall tendency toward more P_n. Moreover, the removal of PP decreases the likelihood of backreactions that free previously linked up monomers, and this shows up as a decrease in [PP] in the denominator of (10.18), which increases $[P_n]$. Ultimately, as described by (10.8), this expenditure of ATP is paid for by energy metabolism, which regenerates ATP from its degradation products, ADP and P. Thus (10.8) can be generalized for polynucleotide synthesis as shown in (10.19). The nucleotide monophosphates (NMP) may be either oxy or deoxy varieties, depending on whether RNA or DNA is the polynucleotide product P_n, respectively. Enzyme activity is indicated by stippling. The most studied enzymes are the polymerase

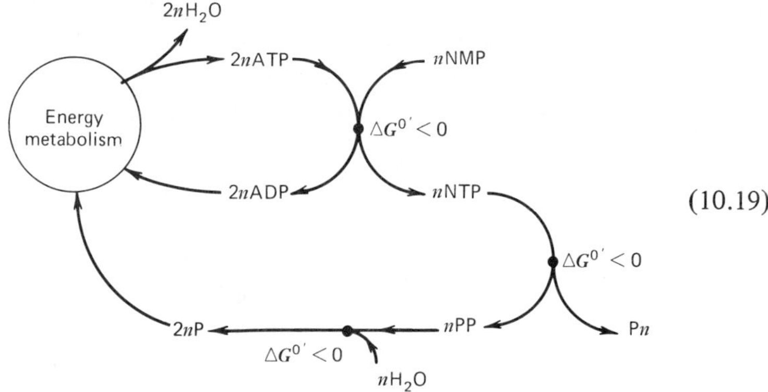

(10.19)

enzymes, which catalyze the condensation of the nucleotide triphosphates (NTP) into P_n.

A description of each of the two types of polynucleotide synthesis, DNA replication and RNA transcription of DNA, follows [1-4].

DNA Replication in *E. coli*

The two most profound scientific papers written in the twentieth century are each only one journal page long: (1) Einstein's paper in the 1905 volume of *Annalen der Physik*, in which he deduces the earth-shaking consequence of relativity, $E = mc^2$; and (2) the 1953 paper in *Nature* by J. Watson and F. Crick, in which the structure of DNA is elucidated. Watson and Crick wrote:

> It has not escaped our notice that the specific pairing we have postulated immediately suggests a possible copying mechanism for the genetic material.

Indeed, the replicative ability of DNA made it the quintessential molecule of all of biology, the *macromolecule embodiment of life itself*. Nevertheless, only after 20 years of painstaking research have scientists finally understood the molecular mechanism of DNA replication, and during much of that period, the accepted mechanism was wrong! It is now realized that, while the structure of DNA clearly explains *why* it can be replicated, it in no way explains *how* this takes place. It turns out that several large proteins are required for replication, and while much is now known about the process, much is still being learned about it.

As was indicated in Chapter 1, double-stranded DNA consists of two antiparallel polynucleotide chains bound together by a multitude of hydrogen bonds between purine and pyrimidine bases on opposite strands, according to the base-pairing rules A=T and G≡C. Schematically, this structure is represented in (10.20), in which the symbolism of (10.21) is schematic for (10.22), which explains the $5' \rightarrow 3'$ polarity.

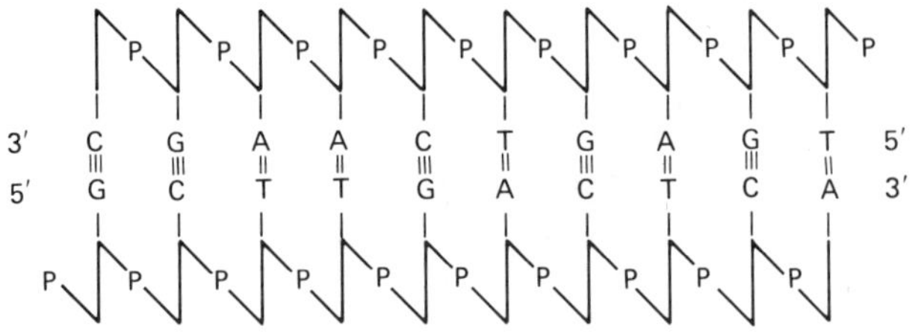

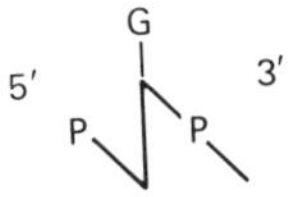

3′ C G A A C T G A G T 5′
5′ G C T T G A C T C A 3′ (10.20)

G
5′ P P 3′ (10.21)

G
O 1′ 2′ 3′ 4′ 5′
O^-—P(=O)—O— ; —H ; O—P(=O)—O ; O^- ; O^- (10.22)

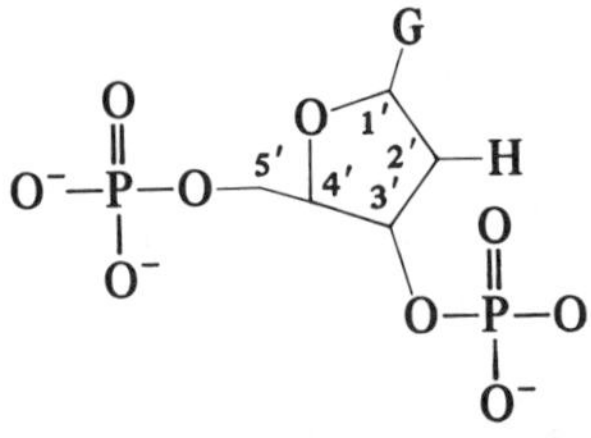

plains the $5' \rightarrow 3'$ polarity. The replication of this structure simply requires that the two strands become separated, activated nucleotide triphosphates arrange themselves according to the base-pairing rules along each of the two separated strands, and phosphodiester linkages form between adjacent activated nucleotides when they are base paired to the DNA strand "template." If all of this happened spontaneously when DNA is incubated with nucleotide triphosphates, DNA would be truly *self-replicating*. However, it does not happen without the considerable aid of several protein catalysts [3, 4].

The two DNA strands will not separate under cellular conditions if the DNA is more than a few dozen base pairs long (a typical gene is on the order of 1000 base pairs long) because of the multiple hydrogen bonds holding the two strands together. Thermal motions are insufficient to break enough of these bonds simultaneously for strand separation. Thermal motions will produce occasional "bubbles" in the DNA [see (10.23)] in which the DNA is now schematized in an even more simplified manner than in (10.21), and the helical twist of the structure is omitted from the drawing for the sake of clarity. These bubbles are

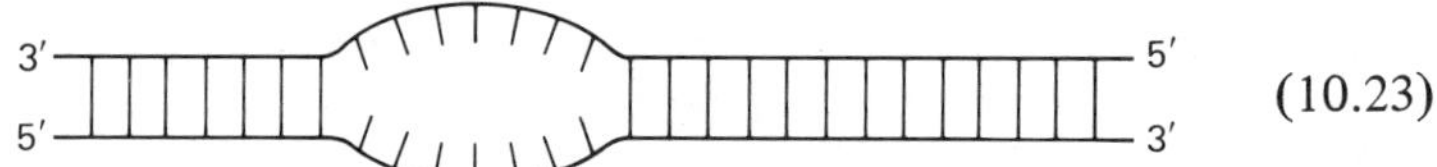

(10.23)

used by the replication machinery in the following way, according to the current, still tentative, model mechanism.

1 At a specific region of the DNA which is usually found to be about 10 bases long and rich in pyrimidines, "DNA-directed" RNA polymerase binds. Yes, RNA polymerase, a pentameric complex of molecular weight 490,000, initiates DNA replication. This protein complex has been successfully dissociated and then allowed to self-assemble into fully active complexes. One of the five subunits of the $\alpha\alpha\beta\beta'\sigma$ complex, the σ factor, is required for initiation but is eventually released, while the remaining four subunits, $\alpha\alpha\beta\beta'$, comprise the functional polymerase during the elongation process that follows initiation.

2 As bubbles or loops spontaneously form adjacent to the RNA polymerase on its 5′ side, "unwinding proteins" bind a strand of the DNA and maintain these open conformations of the DNA so that activated nucleotides may enter the base pair with the bases on the separated strands [see (10.24)].

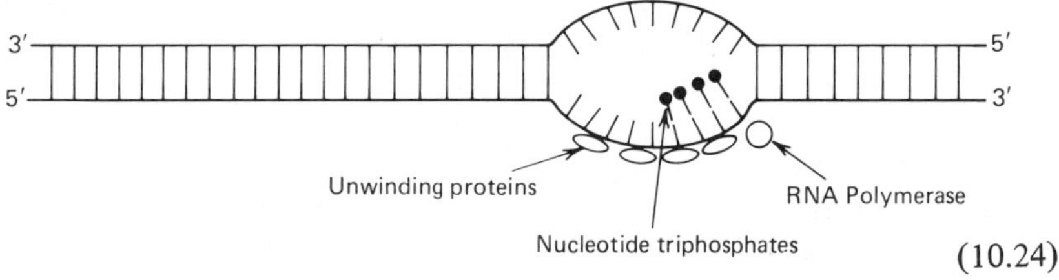

(10.24)

3 The nucleotides used by RNA polymerase are ribonucleotides, and the polymerase catalyzes the formation of phosphodiester bonds between the ribonucleotide triphosphates that are base paired to the DNA strand in the bubble region. Another RNA polymerase commences RNA polymerization on the opposite strand, in the opposite direction. The direction of synthesis is from 3′ to 5′ along the DNA strand templates, but from 5′ to 3′ for the newly synthesized RNA strands because the pairing of bases results in antiparallel strands [see (10.25)].

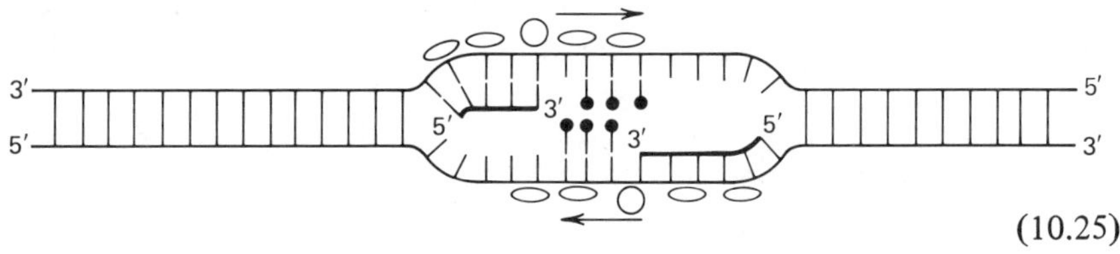

(10.25)

4 These growing RNA strands reach 50 to 100 bases in length. As the bubble region grows longer, there may arise several disjoint stretches of "primer RNA" of this kind. At this point DNA polymerase enters the picture. This enzyme complex, called pol III*, is the dimer of a protein with a molecular weight of 180,000. To commence DNA synthesis on the 3′ end of the RNA primer, an auxiliary dimer, copol III*, and ATP are required. Approximately 1000 or so bases of deoxyribonucleotides are added to the RNA primer by DNA polymerase III*, and the copol III* proteins are released shortly after this elongation commences [see (10.26) for details].

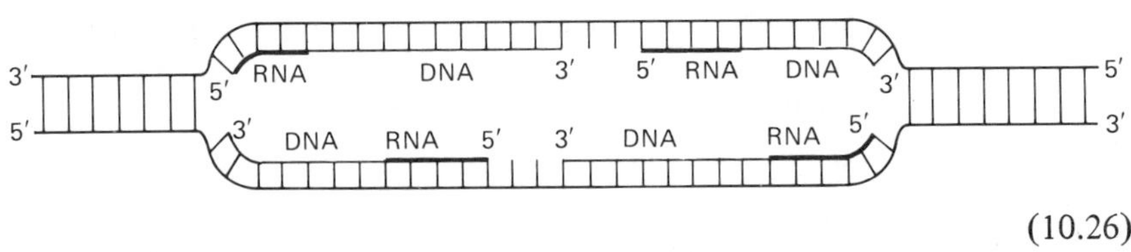

(10.26)

5 The RNA primers are excised, residue by residue, from their 5′ ends by an RNA endonuclease, another specific protein. The resulting DNA stretches, complementary to the original DNA strands, are called Okazaki fragments after their discoverer. These fragments are finally made continuous by DNA polymerase I (pol I), which fills in the gaps between the Okazaki fragments, and DNA ligase, which connects the adjacent 5′ and 3′ termini of adjacent Okazaki fragments, once they have been fully extended by pol I. Both of these additional enzymes are specific proteins. DNA polymerase I has a molecular weight of 109,000 and is a single polypeptide chain of about 1000 amino acids, an unusually big protein. From 1956 until 1972, it was believed that pol I was *the* DNA polymerase, and only since 1973 has it been realized that pol III* is really the main polymerase, as described above, whereas pol I is used only to fill out the gaps between Okazaki fragments. DNA ligase is also a single polypeptide chain with a molecular weight of 77,000.

Altogether, six types of protein are required for DNA replication: DNA-dependent RNA polymerase, unwinding protein, RNA endonuclease, DNA polymerase III*, DNA polymerase I, and DNA ligase. Each of these proteins is synthesized by the DNA-directed protein biosynthesis machinery, to be described later in this chapter. Thus, as another illustration of biological organization: DNA replication depends on proteins, whose synthesis depends on DNA. Consequently, *DNA* is *not* self-replicating. The *cell* is self-replicating, and it contains all of the concatenated components for DNA replication.

DNA Transcription in *E. coli* [3]

The transcription of DNA into RNA is catalyzed by RNA polymerase, an enzyme complex that was partially described in the preceding section. In vitro studies of viral, bacterial, and animal DNA polymerases have never found one that can start a polynucleotide chain. RNA polymerase can start new chains, apparently without the aid of any other proteins, in the case of *E. coli*. The difference between DNA replication initiation and DNA transcription, both catalyzed by RNA polymerase, may be due to differing structures of the RNA complex in the two cases, and this difference may be due to additional small protein subunits. Much work is still going on which will help to clarify this conundrum.

The transcribed RNA is synthesized in the 5′ to 3′ direction from one strand of double-stranded DNA. This is from 3′ to 5′ on the DNA template. The σ subunit of RNA polymerase appears to be a regulatory subunit responsible for stabilizing an RNA polymerase conformation that engages in the initiation of transcription. Two or more protein factors may play a role in chain termination, which is triggered by an as yet not fully elucidated termination sequence on the DNA.

The transcription products are of three types. Ribosomal RNA, rRNA, is made this way, although the final 16S and 23S pieces of the ribosomal subunits are made as one single transcript that is cleaved and modified by "post-transcriptional" enzymes (see Chapter 6 about ribosomes). Similarly, the 5S ribosomal RNA is processed from a larger precursor transcript. Transfer RNA, tRNA, is also processed post-transcriptionally by a variety of enzymes that trim its length and modify many of its bases. The tRNAs are essential in protein synthesis. The third type of RNA is messenger RNA, mRNA, which is the transcript of the genes that code for proteins. It is clearly essential for protein synthesis as well. How these three RNA transcripts are involved in protein synthesis is discussed below.

Protein Biosynthesis in *E. coli* [5]

The problem addressed at the beginning of this chapter regarding the thermodynamic requirements for protein synthesis is solved by the cell by using ATP energy for the activation of the amino acid monomers. While this process obeys the general principles already applied to DNA and RNA synthesis, the details of the mechanism of ATP coupling are necessarily distinct. Indeed, protein synthesis proves to be decidedly more complicated and involves a polymer-synthesizing machinery considerably more massive than the polymerases, discussed above.

The activation of amino acids is not a phosphorylation of the simple type in nucleotide activation (NMP → NDP → NTP); it is instead an adenylation, depicted in (10.27). The adenylates possess activated carboxyl phosphate esters (10.28):

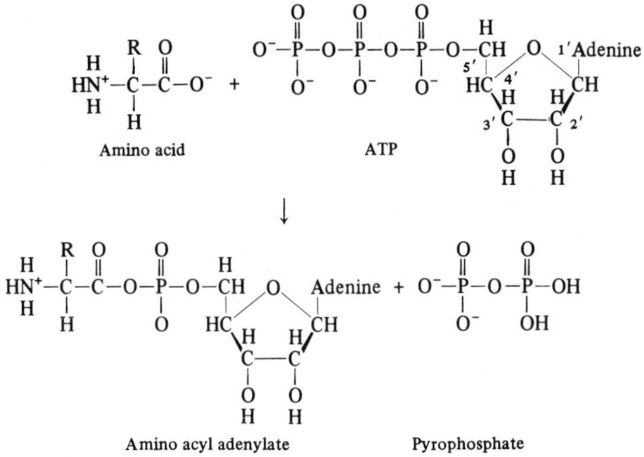

$$\mathrm{HN^+H_2{-}\overset{R}{\underset{H}{C}}{-}\overset{O}{\overset{\|}{C}}{-}O^-} \;+\; \mathrm{O^-{-}\underset{O^-}{\overset{O}{\overset{\|}{P}}}{-}O{-}\underset{O^-}{\overset{O}{\overset{\|}{P}}}{-}O{-}\underset{O^-}{\overset{O}{\overset{\|}{P}}}{-}O{-}\overset{H}{C}H{-}(ribose:\ 5',4',3',2',1')\text{-}Adenine}$$

Amino acid ATP

$$\downarrow \qquad (10.27)$$

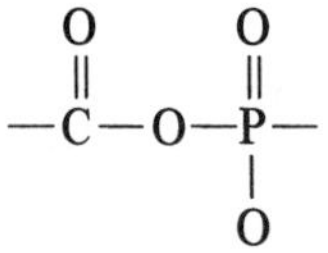

$$\mathrm{HN^+H_2{-}\overset{R}{\underset{H}{C}}{-}\overset{O}{\overset{\|}{C}}{-}O{-}\underset{O}{\overset{O}{\overset{\|}{P}}}{-}O{-}\overset{H}{C}H{-}(ribose){-}Adenine} \;+\; \mathrm{O^-{-}\underset{O^-}{\overset{O}{\overset{\|}{P}}}{-}O{-}\underset{OH}{\overset{O}{\overset{\|}{P}}}{-}OH}$$

Amino acyl adenylate Pyrophosphate

$$-\overset{O}{\overset{\|}{C}}-O-\underset{O}{\underset{|}{\overset{O}{\overset{\|}{P}}}}- \qquad (10.28)$$

An amino acyl adenylate and another amino acid will "condense" to form a dipeptide [see (10.29)]. Steps (10.27) and (10.29) are parallel to (10.4) and (10.5). The balance sheet for this reaction sequence is

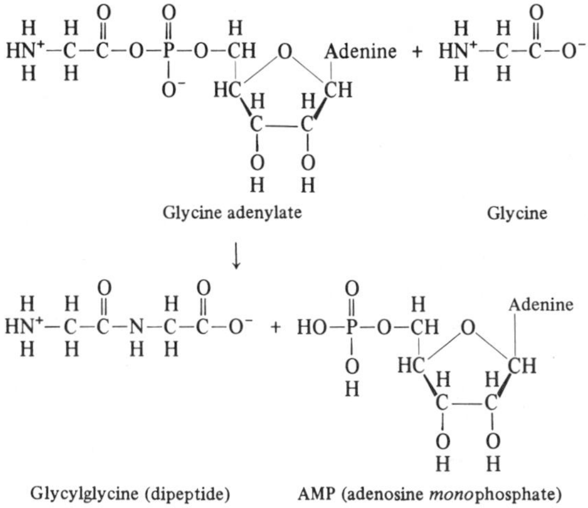

$$\mathrm{HN^+H_2{-}CH_2{-}\overset{O}{\overset{\|}{C}}{-}O{-}\underset{O^-}{\overset{O}{\overset{\|}{P}}}{-}O{-}\overset{H}{C}H{-}(ribose){-}Adenine} \;+\; \mathrm{HN^+H_2{-}CH_2{-}\overset{O}{\overset{\|}{C}}{-}O^-}$$

Glycine adenylate Glycine

$$\downarrow \qquad (10.29)$$

$$\mathrm{HN^+H_2{-}CH_2{-}\overset{O}{\overset{\|}{C}}{-}\underset{H}{N}{-}CH_2{-}\overset{O}{\overset{\|}{C}}{-}O^-} \;+\; \mathrm{HO{-}\underset{OH}{\overset{O}{\overset{\|}{P}}}{-}O{-}\overset{H}{C}H{-}(ribose){-}Adenine}$$

Glycylglycine (dipeptide) AMP (adenosine *mono*phosphate)

$$\begin{aligned} \text{glycine} + \text{ATP} &\rightleftharpoons \text{glycine adenylate} + \text{pyrophosphate} \\ \text{glycine} + \text{glycine adenylate} &\rightleftharpoons \text{glycylglycine} + \text{AMP} \end{aligned} \tag{10.30}$$

Therefore

$$\text{ATP} + \text{glycine} + \text{glycine} \rightleftharpoons \text{glycylglycine} + \text{AMP} + \text{pyrophosphate}$$

As in the nucleoside example of (10.5), H_2O is not directly involved. However, energy metabolism coupling to ATP regeneration produces

$$\text{AMP} + \text{pyrophosphate} + \text{energy} \rightarrow \text{ATP} + H_2O \tag{10.31}$$

which finally manifests the dehydration condensation byproduct, H_2O. Thus, the analog of (10.19) is (10.32), which illustrates the energy coupling between

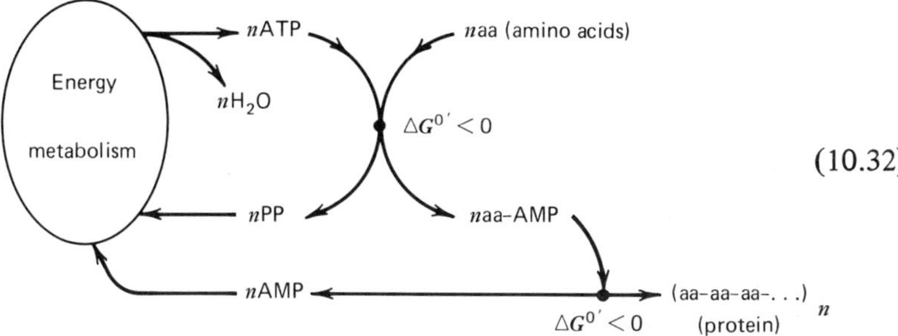

(10.32)

metabolically generated ATP energy and the thermodynamic requirement for energy by the amino acid polymerization process. It in no way begins to exhibit the complex macromolecular apparatus used by the cell to achieve this coupling, indicated in the figure by stippled spots. Prominent among these macromolecular components are the three transcription products, rRNA, tRNA, and mRNA; ribosomes; activation enzymes called "amino acyl tRNA synthetases"; and several other "protein factors."

Both DNA and proteins are linear sequences of a fixed number of basic units. There are four bases for DNA sequences: A (adenine), G (guanine), C (cytosine), and T (thymine) (see Chapter 1 for structures); and there are 20 amino acids for protein sequences (see Chapter 1 again). The evolutionary significance of DNA in cells is as a repository of genetic heritage accumulated over the eons in a codified linear sequence of bases. This heritage is made functional by a translation mechanism that reads the messages codified in DNA and translates them into linear sequences of amino acids by making proteins. The base-pairing properties of bases explain how DNA can be replicated so that progeny cells stemming from an antecedent cell may inherit the genetic information of the DNA. Yet this precious molecular heritage, the DNA, only possesses *latent* functional abilities that only become manifested by the cell when they have been expressed as proteins. The functional proteins interface with the possibly hostile outside environment of the cellular cytoplasm and the external medium; they are often destroyed, damaged, and degraded. The DNA is sequestered and protected from intimacy with the chemical milieu. Only the replication machinery and the DNA-dependent RNA polymerase interact catalytically with the DNA, exposing it momentarily to destructive tendencies. Indeed, damage to DNA does occur; and it also seems reasonable that some inherent "mutability," perhaps merely thermal "noise," is useful to the cell, evolutionarily.

The amino acid activation scheme in (10.32) is the process that enables the cell to couple the information content of the DNA base sequence into the functional-protein sequence of amino acids. The chemical problem that the evolution of the cell had to solve was how to couple the linear sequence of DNA bases to the polymerization mechanism for protein synthesis. How an *E. coli* cell does this is described below. This process is one of the most amazing processes ever revealed by science! Its evolution is the subject of Chapter 14.

To protect DNA from too much chemistry, the sequence of bases in DNA is transcribed into RNA, which has a complementary base sequence [with uracil (U) in place of thymine (T)] as a consequence of base pairing [see (10.33)].

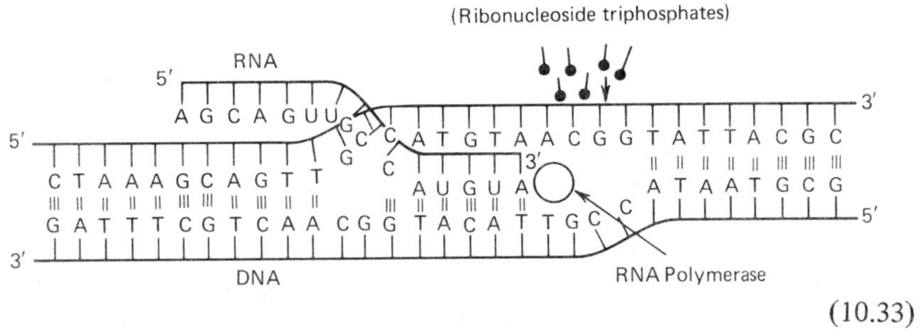

(10.33)

When the genetic message for a protein is transcribed in this manner, the resulting RNA is mRNA, and it is the base sequence of mRNA that engages directly in the protein synthesis mechanism, not the DNA. As will be seen below, the mRNA is translated into protein in the $5' \longrightarrow 3'$ direction. This corresponds to reading the DNA gene from $3'$ to $5'$. The protein, "colinear" with this, is synthesized in the direction amino terminus to carboxyl terminus.

While making mRNA from DNA protects the integrity of the DNA because mRNA is to be translated into protein instead of the DNA gene itself, it does not change the problem of how to associate an amino acid sequence with a nucleotide base sequence. This problem has merely been shifted to an RNA.

The evolutionary solution to this problem is to use an "adaptor" molecule that can read mRNA *and* couple to amino acids [6].

The adaptor molecules are the "transfer" RNAs, tRNAs, that are transcribed from tRNA genes on the DNA. Their structure is discussed a little later. At one end of all tRNAs there is a sequence of three bases, called the "anticodon," which is capable of base pairing with a sequence of three bases on the mRNA, called the "codon." Thus, the reading of mRNA is done by using base pairing. At the other end of the tRNA, about 75 to 80 Å away, the appropriate amino acid is attached by a covalent bond. This arrangement is schematized in (10.34).

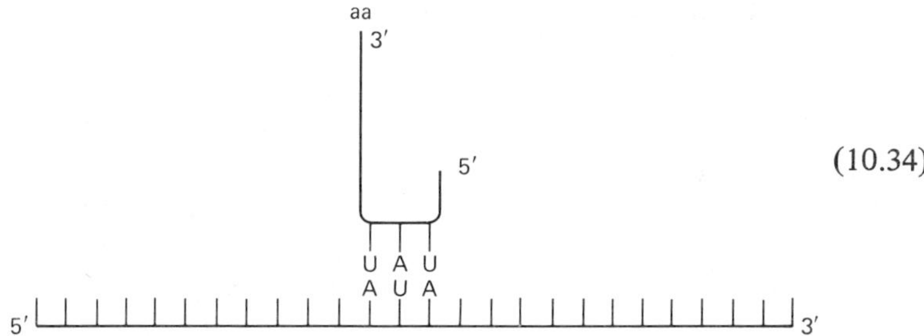

(10.34)

The base pairing of the tRNA to the mRNA is "antiparallel"—as is the base pairing between the two strands of double-stranded DNA and as is the base pairing of RNA transcripts to DNA during transcription.

This adaptor scheme shows how to "associate" an amino acid with a triplet sequence of bases. However, it does not *explain* how the "cognate" amino acid gets attached to the tRNA, whose anticodon is the "correct" anticodon. In other words, for this adaptor scheme to work, it is necessary that for each amino acid there be one, or perhaps a few, specific codons that correspond to it. This means that the cell must guarantee that tRNAs with the appropriate anticodons must become attached to only those amino acids to which they correspond. Because there are only four bases and consequently a possible 64 triplets, and because there are only 20 amino acids for making proteins, there are from one to six codons specific for a specific amino acid, depending on the amino acid. For example, tryptophan has a unique codon, whereas serine has six. Thus tryptophan requires a tRNA with a specific anticodon, whereas serine needs enough tRNAs to read six codons. For reasons to be explained in detail later, the six serine codons can be read by five tRNAs.

The hooking up of the right amino acid with the right tRNAs is achieved by a class of enzymes called amino acyl synthetases [7]. There is one synthetase that is specific for each amino acid and all of the associated tRNAs. In some cells there may be two of these enzymes with the same specificities. This multiplicity is not fully understood. The synthetases are doubly important because they not only guarantee that the cognate tRNAs and amino acids are joined covalently but also couple in the energy of ATP, which is required for protein synthesis.

The amino acyl tRNA synthetases are pure proteins, containing no polynucleotide components or prosthetic groups like heme. They are made by the protein biosynthesis machinery that is being described and are translated from DNA genes specifically for them. They recognize and bind three distinct substrates: ATP, a specific amino acid, and a specific tRNA from a small class of cognate tRNAs. The synthetase catalyzes the activation of the amino acid by forming a synthetase-bound amino acyl adenylate. Studies show that only after the enzyme-bound amino acyl adenylate has formed does the tRNA become attached. This selective, sequential binding of substrate reflects allosteric conformation changes in the synthetase enzyme. Once the tRNA has bound, the synthetase then catalyzes the "transacylation" from amino acyl AMP to the 3′ terminal ribose residue of the tRNA. The AMP is then released. This is followed by the binding of ATP and another amino acid, adenylate formation, and finally the release of amino acyl tRNA. What remains is a synthetase enzyme already primed with amino acid AMP (adenylate), ready to bind another tRNA. This is depicted for isoleucine (Ile) activation in (10.35).

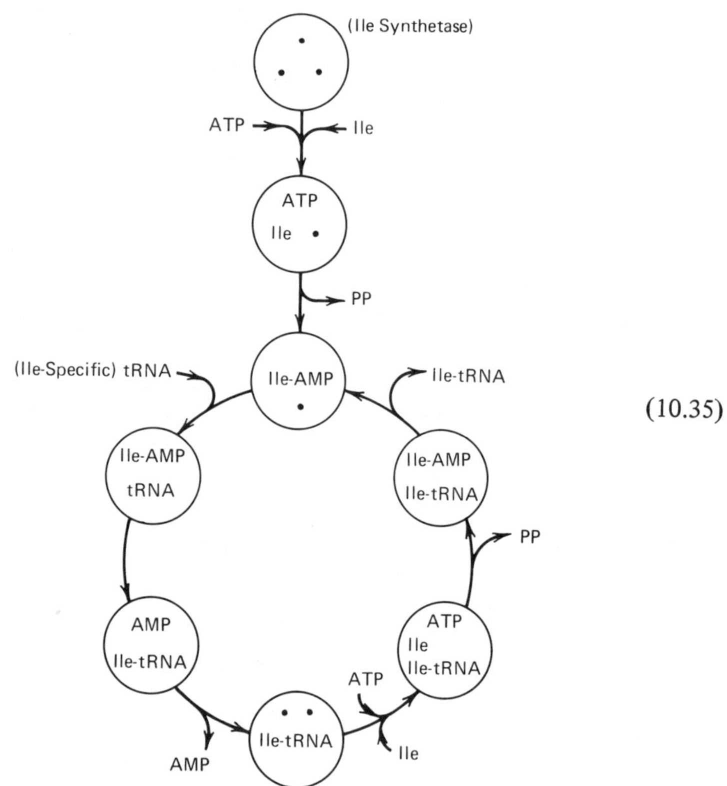

(10.35)

All tRNAs have identical 3′ acceptor stems for amino acid attachment with a terminal adenine ribonucleoside residue. The resulting amino acyl ester has the structure given in (10.36). A 2′OH carboxyl-ribose ester is shown in (10.36).

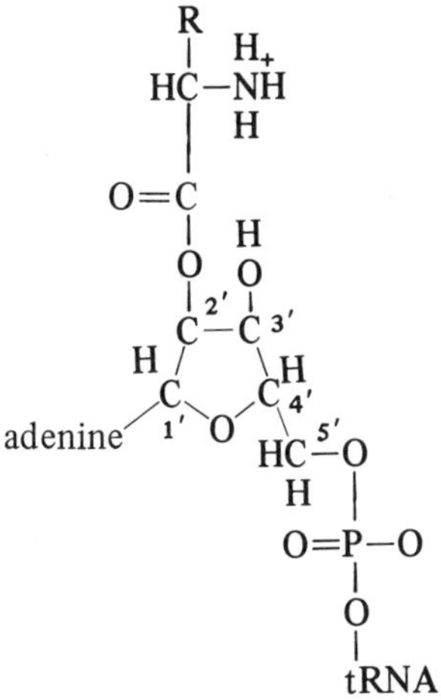

```
              R
              |  H+
             HC—NH
              |  H
              |
           O=C
              |  H
              O  O
              |2′|
              C—C 3′
           H /   \H
            C      C 4′
  adenine/ 1′ \O/   \ 5′
                    HC—O
                     H  |
                    O=P—O
                        |
                        O
                        |
                       tRNA
```

(10.36)

This is what forms initially on the synthetase, but it is believed that subsequently an equilibrium develops between 2′-OH and 3′-OH esters on the 3′ terminus of the tRNA. The remarkable feature about this bond is that it contains almost precisely as much energy as did the activated amino acid adenylate ester linkage between the carboxyl group and the phosphate of AMP [see (10.27)]. As a consequence of its energy content, the tRNA-*activated* amino acid is poised for peptide bond formation, which remains thermodynamically favored as it was for the adenylates depicted in (10.29).

The mRNA and tRNAs are cumbersome structures. It is necessary to organize their interactions felicitously if efficient synthesis of protein is to occur. Such organizational requirements are fulfilled by the *ribosome.* The ribosome is the locus at which the mRNA is read in an orderly way by properly coordinated and arranged amino acyl tRNAs. As already stated in Chapter 6, ribosomes consist of two major pieces, the 30S and the 50S subunits, each of which is itself a complex of rRNA and ribosomal proteins. The ribosomal proteins are synthesized by the protein biosynthesis machinery described here. Ribosomal protein genes in the DNA must be translated into protein, just as is the case for the amino acyl tRNA synthetases. The rRNAs are also transcribed from DNA genes for rRNAs, by a process mentioned earlier which involves some post-transcriptional processing before the rRNA components for ribosomes are completed. The mRNA is bound to the 30S ribosomal subunit, whereas the amino acyl tRNAs that "read" the mRNA are coordinated by the 50S subunit of the ribosome. The 50S subunit also catalyzes the formation of the peptide bond between adjacent amino acyl tRNAs, freeing the tRNA adaptor to be recycled as an amino acid carrier again.

The initiation of mRNA translation begins with the mRNA codon AUG (or GUG sometimes). This triplet codes for the amino acid methionine and determines the initiation locus on the mRNA for protein synthesis. The initiation process is complex, involving three protein factors, guanosine triphosphate (GTP), and a tertiary conformational change in ribosomal structure [5].

Nonfunctioning 70S ribosomes are in equilibrium with their dissociated major subunits, the 30S subunit and the 50S subunit. Protein factor F_3 binds to the 30S subunit, which then binds the mRNA [see (10.37) and (10.38)].

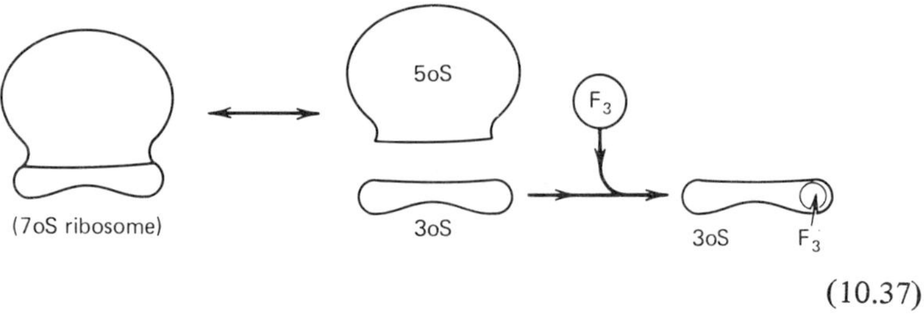

(10.37)

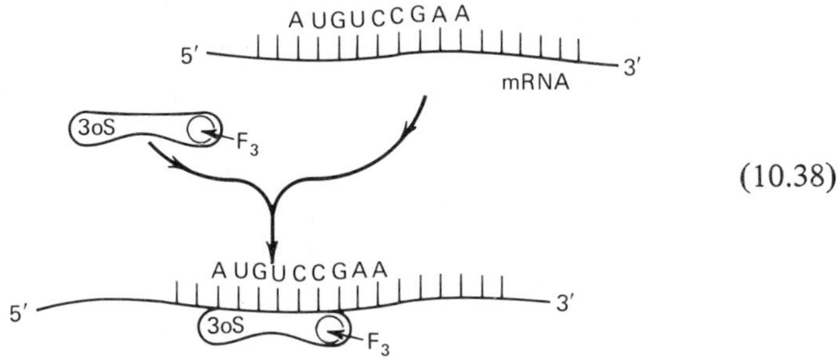

(10.38)

Methionine tRNA$_f$, activated to methionyl-tRNA$_f$, is *N*-formylated by an enzyme. Two types of methionine tRNA exist: met-tRNA and met-tRNA$_f$. The latter one is specifically recognized by the "transformylase" enzyme, which adds a formyl group to the free amino group of met-tRNA$_f$. The other methionine tRNA, met-tRNA, is not formylated and remains available for reading the AUG codons that occur somewhere in the gene sequence, coding for methionine residues somewhere internally in the protein's amino acid sequence. The *N*-formyl-met-tRNA$_f$ then combines with protein factor F_2 and with a molecule of GTP while a third protein factor, F_1, enables this complex to combine with the mRNA-F_3-30S subunit complex. Some disagreement still exists regarding the actual sequence of binding to form the final initiation complex depicted in (10.39). Finally, the 30S initiation complex combines with the 50S

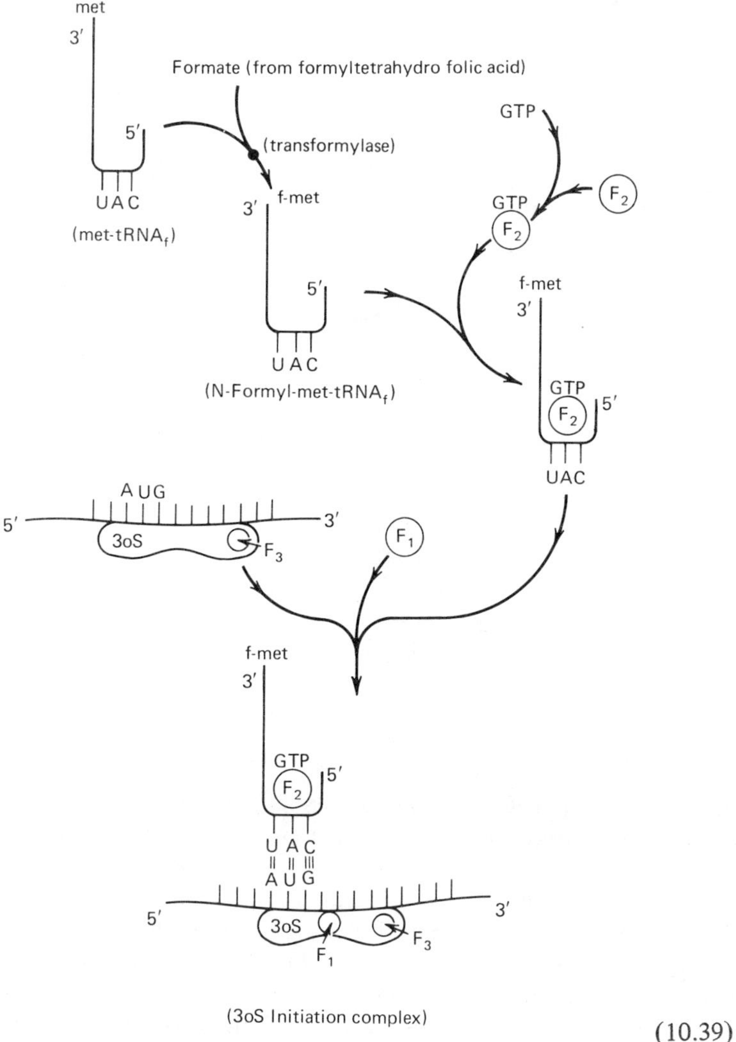

(3oS Initiation complex)

(10.39)

ribosomal subunit and releases all three protein factors F_1, F_2, and F_3, as well as cleaving the terminal phosphate from GTP to yield guanosine diphosphate (GDP) and phosphate (P) [see (10.40)]. This completes the story of initiation and prepares the way for a description of protein elongation.

Elongation of the protein involves another protein factor, the "elongation factor," *F-T*. It also requires another molecule of GTP that binds to *F-T*. This combination then binds to the amino acyl tRNAs, preparing them for incorporation into protein by the ribosomal complex. The 50S subunit appears to have two sites for binding amino acyl tRNA. One of them, the *P* site, is occupied

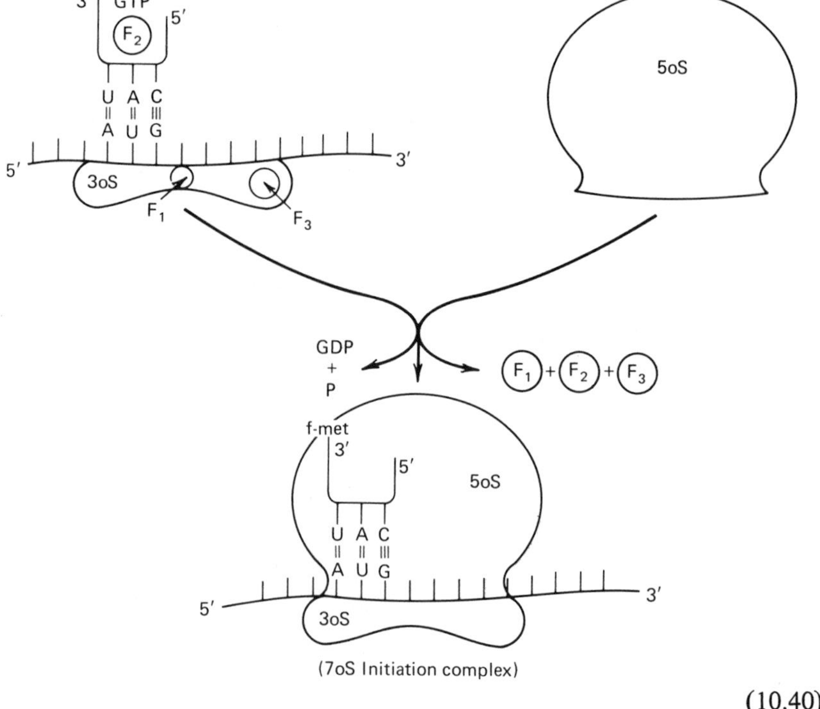

(7oS Initiation complex)

(10.40)

by formyl methionine $tRNA_f$ in the 70S initiation complex; the other site, the *A* site, is the place where the prepared amino acyl tRNA complex will bind, provided that the tRNA anticodon involved can properly base pair with the triplet codon on the mRNA adjacent to AUG on the 3′ side [see (10.41)]. The protein *F-T* is released and GTP is cleaved to GDP and P. At this stage the "peptidyl transferase" activity of a region of the 50S subunit that is located near the amino acyl ends of the bound tRNAs catalyzes the formation of a peptide bond between the carboxyl group of formyl methionine (fmet) and the free amino acid group of the adjacent tRNA-bound amino acid. The methionine $tRNA_f$ is no longer covalently bound to the fmet and leaves the ribosomal *P* site to be recycled for further cellular use. This leaves the complex (10.42). From here, a "translocation" step is required to move the peptidyl tRNA from the *A* site into the *P* site so that another properly prepared amino acyl tRNA can occupy the *A* site for the formation of another peptide bond. This translocation step requires protein factor *F-G* with bound GTP. The *F-G* factor is released and GTP is cleaved to GDP and P after translocation is completed. The result is to end up with (10.43). By repeating steps (10.41) through (10.43),

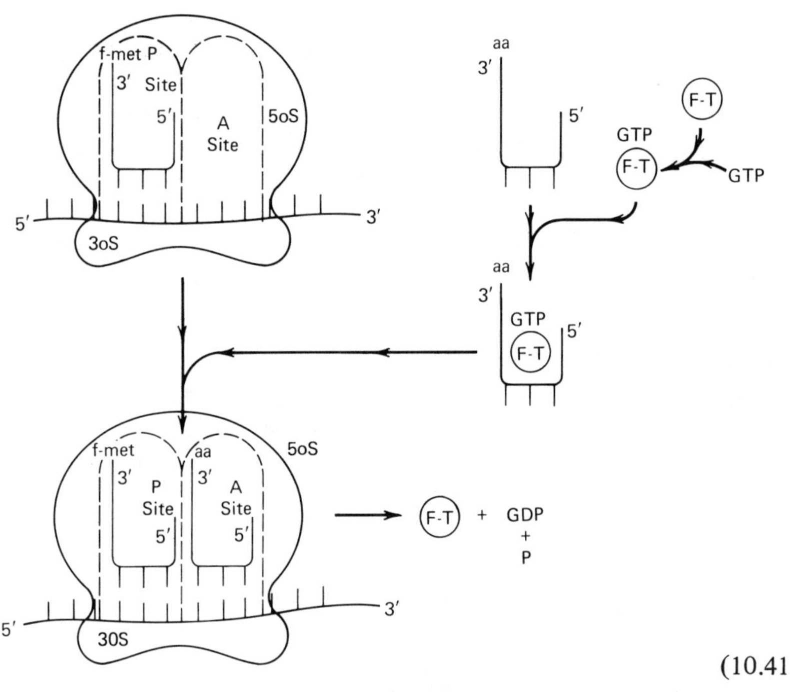

(10.41)

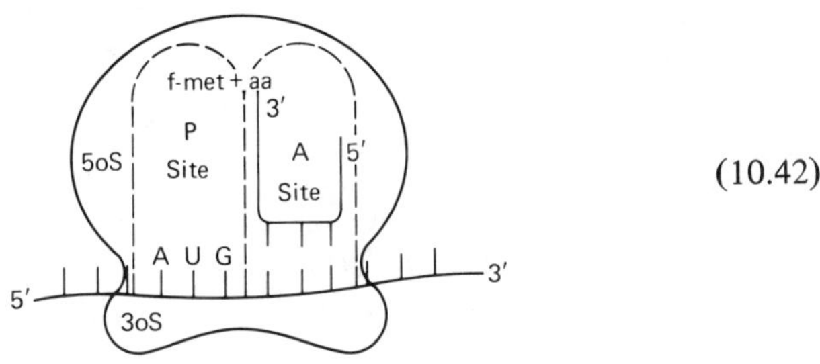

(10.42)

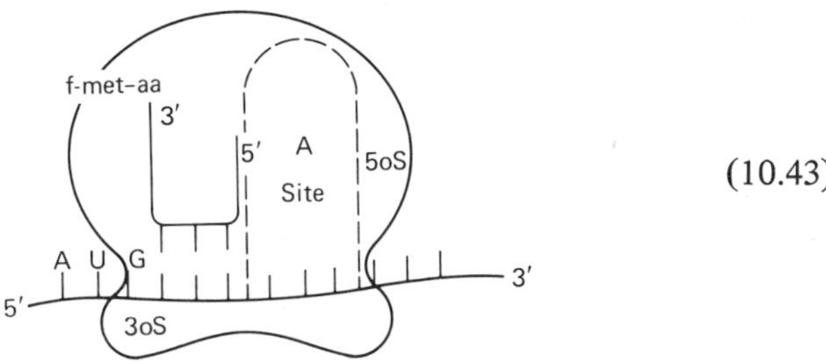

(10.43)

the polypeptide is elongated and the amino acid sequence follows from the sequence of triplets on the mRNA.

Eventually this process comes to the triplets UGA, UAA, or UAG, which do *not* code for amino acids but instead signal termination. These are the analogs of periods as punctuation in written English. Protein-releasing factors, R_i, one specific for each of the three termination codons, bind the 30S subunit and promote the hydrolysis of the polypeptide ester linkage to the last tRNA on the ribosomal complex. With this hydrolysis, the polypeptide (protein) is freed, the last tRNA is freed, the mRNA is released, and the releasing factor is recycled. At this stage the 70S ribosome can again dissociate into its 30S and 50S subunits. Formyl methionine is usually removed from the protein product, after translation.

This concludes the sketch of protein biosynthesis. The complexity of this process is not fully appreciated until a more detailed description of the structure of synthetases, tRNAs, and ribosomes is contemplated. Rapid progress in the determination of these structures has been made, as is demonstrated below by a relatively detailed introduction to these complex molecules. In the process the functional significance of the genetic code will be emphasized, and the "mystery of the second code" will be discussed.

The Genetic Code and the Mystery of the Second Code

The key to DNA replication and DNA transcription into RNAs is *base pairing* of purines and pyrimidines, A=T (A=U), G=C, T=A (U=A), C≡G. In addition, the connection between the nucleotide base sequence and the gene-directed amino acid sequence of biosynthesized proteins has been seen above to be dependent on the base pairing of the anticodon of tRNA to the codon of mRNA. Thus, base pairing plays a key role in the biosynthesis of DNA, RNA, and protein. The lexicon of amino acid codon assignments, alluded to in the discussion of (10.41), is called the "genetic code" and is given in Table 10.1. These assignments were identified in the early 1960s in the laboratories of Nirenberg, Khorana, and Ochoa.

The fact that there is redundancy in the code was attributed to the fact that there are 64 triplet codons and only 20 amino acids plus two "punctuation" signals (initiation and termination) to code for. Nevertheless, while serine has six codons, only five tRNAs are believed to be required to read them, and generally only 30 to 40 tRNAs are needed altogether for all 64 codons. Even after 3 is subtracted for the three termination codons for which there are no cognate tRNAs, 61 is still roughly 50% larger than the number of tRNAs observed. This is a result of tRNA structural properties which cause it to be "wobbly" with regard to its fidelity of base pairing between its base on its 5′ end of the anti-

Table 10.1

	U		C		A		G		Amino Acid	Number of Codons
U	UUU	Phe	UCU	Ser	UAU	Tyr	UGU	Cys		
	UUC	Phe	UCC	Ser	UAC	Tyr	UGC	Cys	Ala	4
									Arg	6
	UUA	Leu	UCA	Ser	UAA	End	UGA	End	Asn	2
	UUG	Leu	UCG	Ser	UAG	End	UGG	Trp	Asp	2
C	CUU	Leu	CCU	Pro	CAU	His	CGU	Arg	Cys	2
	CUC	Leu	CCC	Pro	CAC	His	CGC	Arg	Gln	2
									Glu	2
	CUA	Leu	CCA	Pro	CAA	Gln	CGA	Arg	Gly	4
	CUG	Leu	CCG	Pro	CAG	Gln	CGG	Arg	His	2
									Ile	3
A	AUU	Ile	ACU	Thr	AAU	Asn	AGU	Ser	Leu	6
	AUC	Ile	ACC	Thr	AAC	Asn	AGC	Ser	Lys	2
									Met	1
	AUA	Ile	ACA	Thr	AAA	Lys	AGA	Arg	Phe	2
	AUG	Met	ACG	Thr	AAG	Lys	AGG	Arg	Pro	4
									Ser	6
G	GUU	Val	GCU	Ala	GAU	Asp	GGU	Gly	Thr	4
	GUC	Val	GCC	Ala	GAC	Asp	GGC	Gly	Trp	1
									Tyr	2
	GUA	Val	GCA	Ala	GAA	Glu	GGA	Gly	Val	4
	GUG	Val	GCG	Ala	GAG	Glu	GGG	Gly	End	3

codon and the base at the 3′ end of the codon on the mRNA to which the tRNA bonds. This is depicted in (10.44) [8]. In (10.45), the tRNA structure for phenylalanine tRNA is schematized by its phosphodiester backbone, and its amino acid acceptor arm and anticodon are indicated. This structure is quite general for tRNAs, which have the sequential structure called the "clover-leaf" model; although it clearly shows the linear sequence of the bases in tRNAs as well as the regions of intramolecular base pairing along the "stems," it does not provide the faithful three-dimensional rendering of the structure given by (10.44). The "variable loop" shows the most variation between different tRNAs, being virtually nonexistent in some and much longer than depicted in (10.45) in others.

Steric strain at the location of the 5′ end of the anticodon as well as tRNA–ribosome and mRNA–ribosome interactions result in the infidelity of the base pair in question; the base pairs in Table 10.2 are possible as a result. Inosine (I), a metabolic precursor of purines, is one of the many unusual bases found in

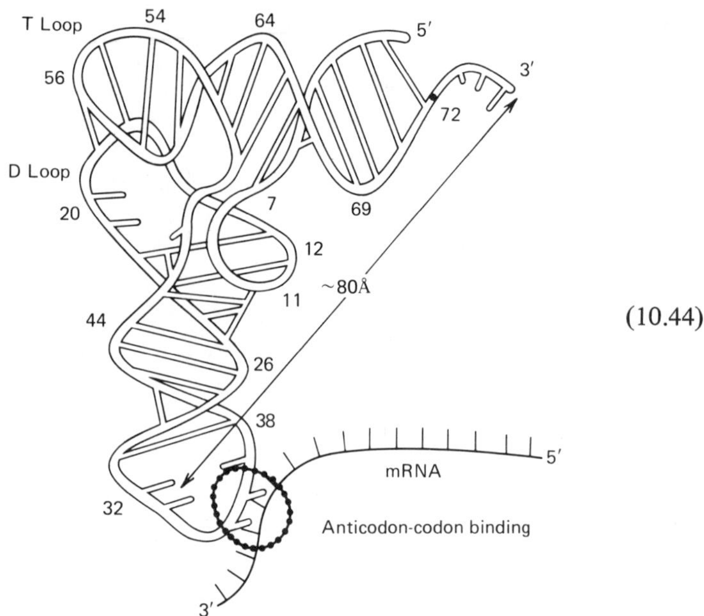

(10.44)

tRNAs. These modified bases are produced post-transcriptionally by cellular enzymes that modify the tRNA-gene transcriptional product, the tRNA precursor, which contains only the four natural ribonucleotides (A, C, G, U).

If most tRNA anticodons have for their 5′ end either G or U, then the U or C and A or G redundancy of Table 10.1 is explained by the wobble (Table 10.2). The distinction between isoleucine and methionine, which have codons differing only at their 3′ ends, is achieved by an isoleucine tRNA with an anticodon with I at its 5′ end and a methionine tRNA with an anticodon with C at its 5′ end. Thus, the isoleucine tRNA anticodon is $^{3'}$UAI$^{5'}$ and the methionine tRNA anticodon is $^{3'}$UAC$^{5'}$. Similarly, cysteine tRNA has the anticodon $^{3'}$ACG$^{5'}$, tryptophan tRNA must have the anticodon $^{3'}$ACC$^{5'}$, and fortunately the codon UGA is a termination codon, for which there is no tRNA, because the wobble (Table 10.2) would not permit a tRNA anticodon to read *only* UGA! In most cases there is more than one way to construct tRNA anticodons that obey Table 10.2 and explain Table 10.1, and nearly all of these possibilities are observed in organisms.

The tRNA structure is responsible for three functions during protein biosynthesis: it carries the activated amino acid on its 3′ end; it possesses the anticodon that reads the codon on the mRNA by base pairing; and it interacts with the ribosome so that it is properly positioned for peptidyl transferase activity on the ribosome during protein elongation. The third function is apparently

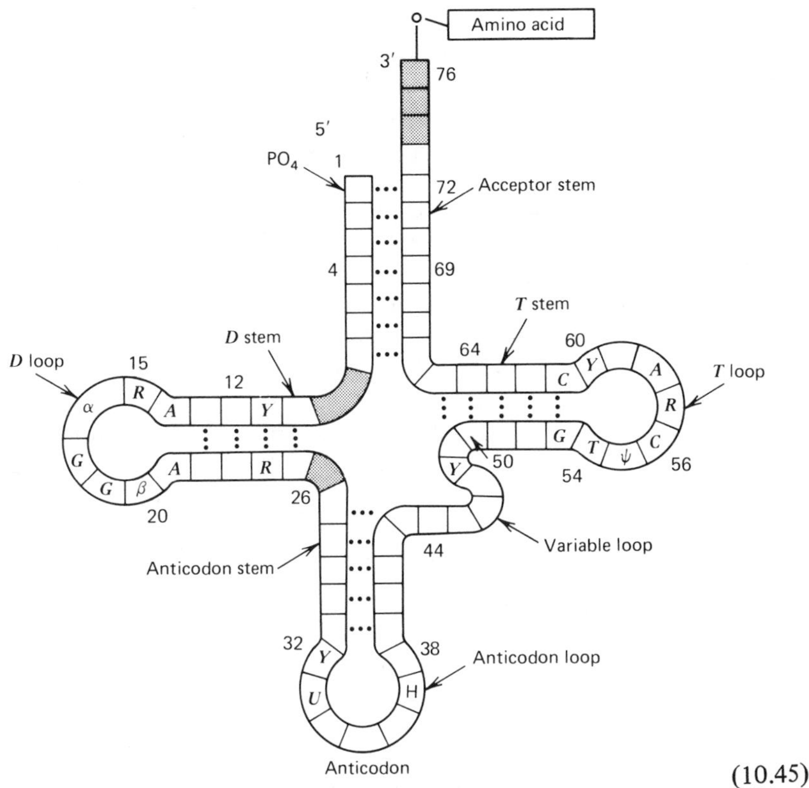

(10.45)

largely the result of an interaction between the D and T loop region of tRNA (10.44) and the 5S rRNA component of the ribosomal 50S subunit. This structure, which in (10.44) looks like an upside down L, is called the L-structure. It is believed to prevail in solutions of free tRNAs, although Ried [9] has argued for a U-conformation as well, which is acquired by bending the tRNA at its D loop-T loop "corner." Thid conformation positions the anticodon region of the

Table 10.2

Base at 5′ End of Anticodon	Base of 3′ End of Codon
A	U
C	G
G	U or C
U	A or G
I	A, C, or U

tRNA in adjacence to the amino acid acceptor, arm's stem region, in particular, the "discriminator" base of the acceptor stem. The discriminator base is located at the fourth base in from the 3′ end of the tRNA. *All* tRNAs begin their 3′ ends with $^{3'}ACC^{5'}$. The next base, the fourth base in, is called the "discriminator" base because its identity is correlated with the identity of the amino acid for which the tRNA is specific. Since the anticodon is also correlated with the amino acid, there may be significance in having a conformation in which the acceptor arm's discriminator is close to the anticodon during amino acid activation. Indeed, Ried [9] suggests that during interactions between tRNA and synthetase the synthetase recognizes the U conformer, and not the L conformer. Synthetases are either single polypeptides or dimers of single chains (a few tetramers have been observed). Not all synthetase subunits have amino acid recognition and binding sites (in particular the tetrameric ones do not), but the subunits that do, appear to have the schematic structure (10.46), where the X

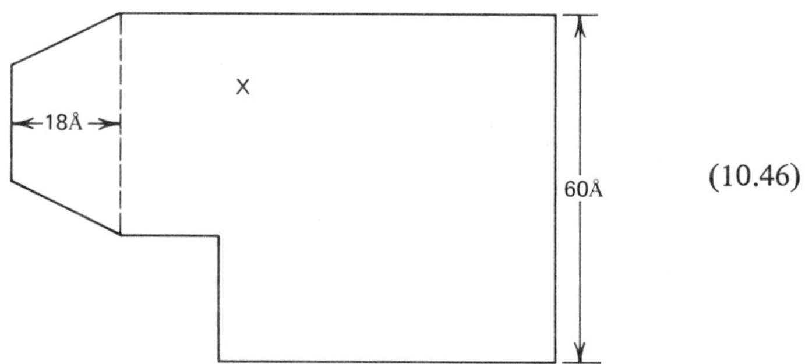

(10.46)

marks the amino acid binding site. The L-conformation of tRNA does not fit very well if the amino acid acceptor arm is placed adjacent to X [see (10.47)].

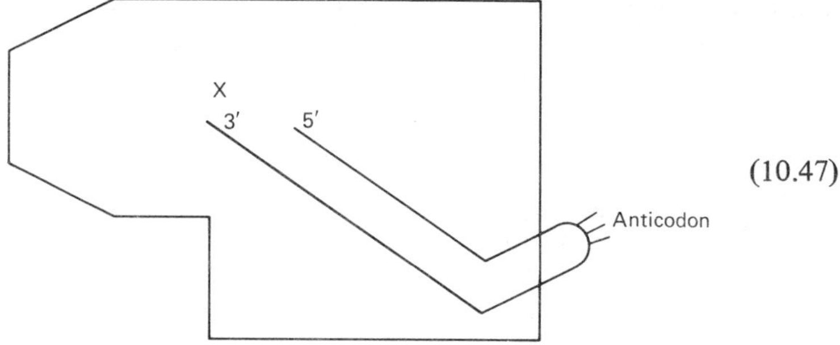

(10.47)

No matter how the synthetase and the tRNA are oriented, they do not fit so that the amino acid can be attached *while* the anticodon region is available for recognition by the synthetase. The dimeric synthetases would allow for over-

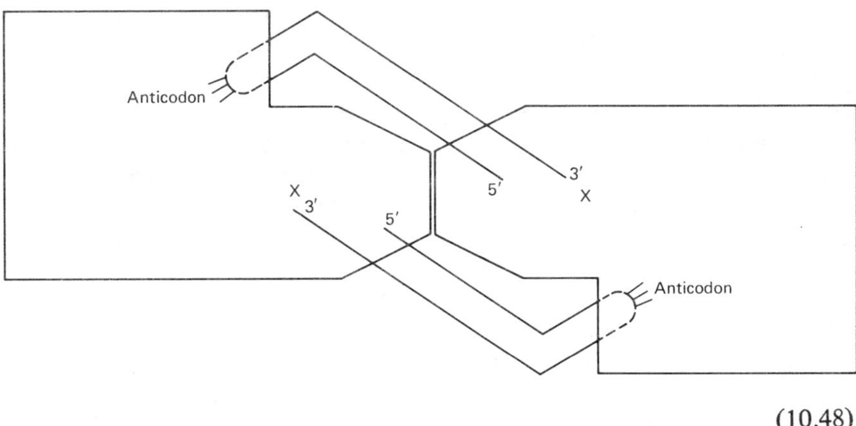

(10.48)

lapping recognition, as depicted in (10.48), in which one synthetase subunit recognizes the amino acid while the other identifies the anticodon. However, this is difficult to rationalize with the observation that monomeric synthetases (single subunits) are fully capable of recognizing both tRNAs and amino acids. In the U-conformation, the tRNA can bind to a single synthetase subunit with ease, as shown in (10.49), and the discriminator base (⫯) is also found close to the anticodon.

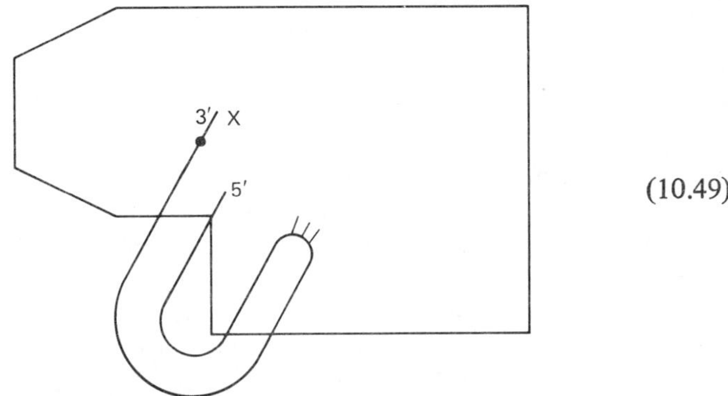

(10.49)

It may be objected that there is no need to have direct anticodon recognition, and repeated observations have shown that enzymatically modified tRNAs whose anticodon regions have been removed are faithfully recognized by their cognate synthetases *and* can even become charged with their cognate amino acid. Thus the U-conformation may not be necessary, and to date it is still controversial.

We can now introduce the fundamental unsolved problem in the elucidation of protein biosynthesis: the mystery of the second code. While the base-pairing code explains much, it does not explain how the synthetases and their cognate tRNAs recognize each other. The obvious guess is that synthetase enzymes contain polynucleotide components whose bases pair with the anticodons of the tRNAs. Each specific amino acid type of synthetase is imagined to possess an RNA, say, which has on it the codon associated with the amino acid and to which the tRNA would bind. This would explain recognition, even if it did leave the question of how this arrangement arose in evolution. However, careful analysis shows that synthetases are *pure* proteins, containing no RNA! Recognition between proteins and polynucleotides is required for mutual recognition between DNA and polymerase, DNA and transcriptase, DNA and repressor, DNA and CAP, DNA and unwinding protein, rRNA and ribosomal protein, and others. Thus, recognition between synthetase and tRNA is a special case of protein–nucleic acid recognition generally. It is this problem that leads to the idea of a second code. It is an especially acute problem for recognition between tRNA and synthetase because this step in protein biosynthesis is critical for the faithful translation of a DNA gene into a protein. The base-pairing code is of absolutely no utility if the wrong amino acid and the wrong tRNA are hooked together by the synthetase! Much work has gone into solving this problem, but very little real progress has yet been made, except in eliminating possibilities.

The problem of the second code will be addressed again in Chapter 14. The context will not be the general question of protein–nucleic acid recognition but rather the crucial special case of recognition between synthetase and tRNA and its evolution.

Ribosome Structure [5] [11]

The most complicated structural problem for protein biosynthesis involves the structure of the ribosome. This results from its impressively large size and the large number of subunits (see Chapter 6). Many techniques have been brought to the study of this simplest of all cellular "organelles." These include bifunctional reagents to cross-link protein pairs that are topologically adjacent; singlet energy transfer between pairs of fluorescence-labeled ribosomal proteins; chemical modification of ribosomal protein and rRNA subunits; affinity labeling; base sequencing of rRNA regions and limited nuclease digestion; neutron scattering of selectively deuterated ribosomes; electron microscopy; immunological labeling and antibody binding; and combinations of these and other methods.

The three-dimensional model of the 30S ribosomal subunit of *E. coli*, (10.50), shows the location of all 21 proteins and the detailed shape, which in (10.37) through (10.43) was schematized. The three-dimensional model of the 50S subunit is shown in (10.51), depicting the location of 19 of its 30 to 35 ribosomal proteins. The structure of the 70S complex is seen in (10.52). This

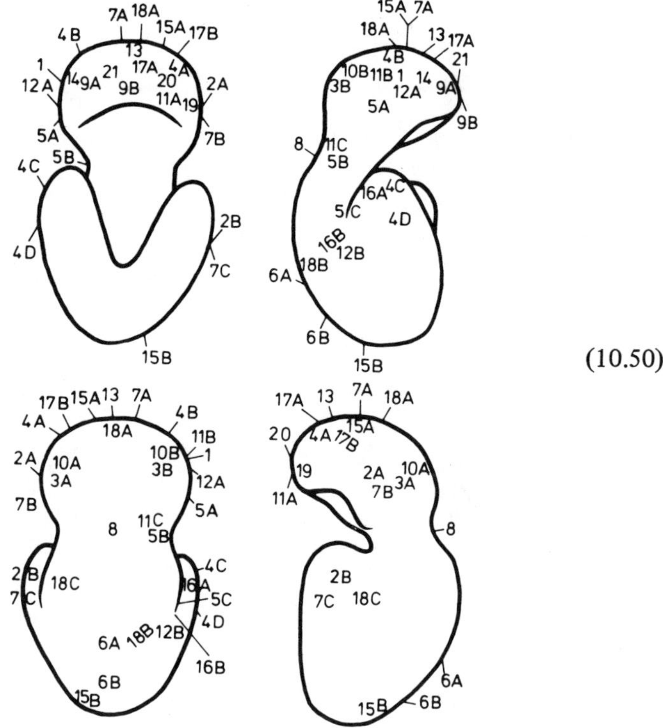

(10.50)

Reproduced, with permission, from Ref. 5.

figure shows the location of the mRNA and the tRNA (dotted line) on the 70S complex, as has been tentatively determined by immuno electron microscopy.

Even the detailed arrangement of proteins in the peptidyl transferase region of the 50S subunit is now known with some certainty according to (10.53). Interactions among these proteins are probably responsible for the full biological activity of this region. In the not too distant future, the details of mRNA ribosome binding and all of the other interactions between macromolecules and ribosomes should become well worked out.

Summary of Polymer Synthesis

The preceding description of polynucleotide and protein synthesis was designed to be an as up to date as possible sketch of current knowledge about these processes. Enormous effort by numerous researchers has compiled a massive amount of information about the details of these processes. Similarly remarkable

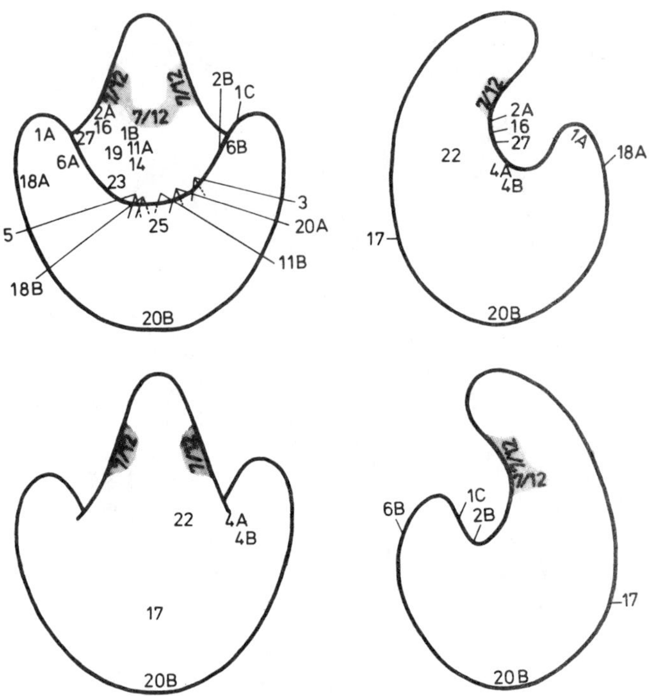

Reproduced, with permission, from Ref. 5.

(10.51)

accounts of polysaccharide synthesis, as well as fatty acid synthesis (about which something will be said in the next chapter), are also possible. The readers for whom the account of protein synthesis given above is essentially a first introduction to the subject are cautioned that the sketch given in no way does justice to the documentation that could be brought to bear on this most remarkable phenomenon. Indeed, whole books (see references) have been written about single components (tRNAs, for example) of the protein biosynthesis machinery.

The organizational structure of polymer biosynthesis manifests the typical biological feedback character: protein biosynthesis requires polynucleotides, whose synthesis requires proteins. Moreover, the energy requirement for ATP, which is absolutely obligatory for polymer synthesis, is filled by an energy metabolism that depends, first of all on proteins for its numerous specific catalytic enzymes, and second of all on polynucleotides to maintain the genetic information for enzyme synthesis as well as for synthesis machinery and regulatory functions.

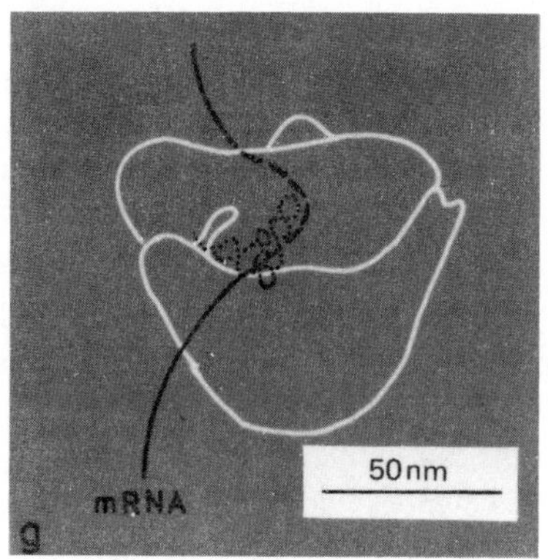

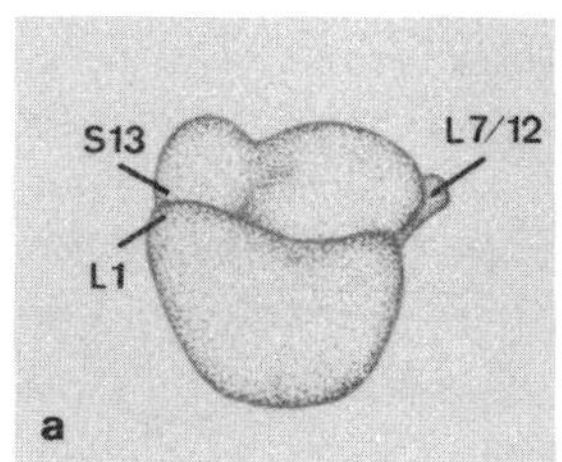

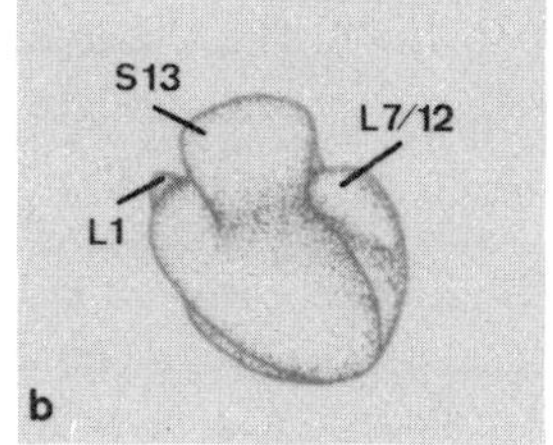

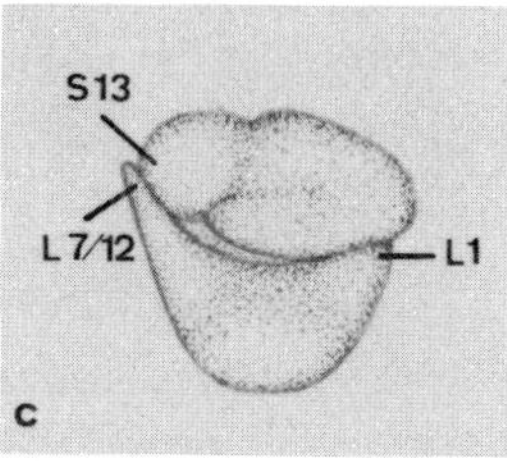

Reproduced, with permission, from Ref. 11.

(10.52)

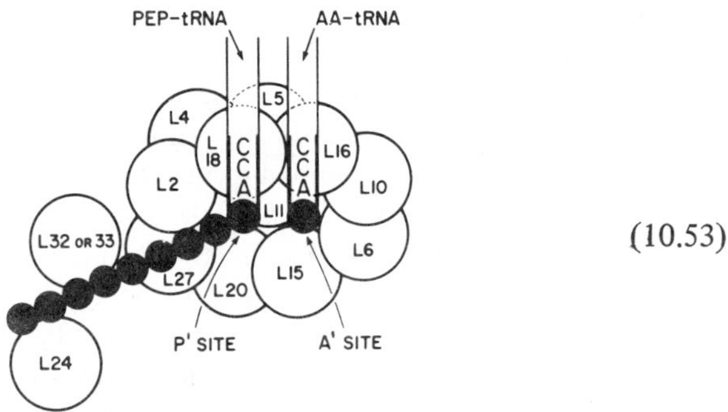

(10.53)

Reproduced, with permission, from Ref. 5.

When one contemplates that the pyruvate dehydrogenase complex (see Chapter 6) has a molecular weight of 4,400,000 and only accounts for a short segment, from pyruvate to acetyl-CoA, of the energy metabolism pathways; *and* that ribosomes, with a molecular weight of 2,700,000, *and* numerous other macromolecular structures, are required *simultaneously* for the proper functioning of the whole polymer biosynthesis machinery, it is difficult to understand how such a complex "integron" (see Chapter 7) could have emerged during evolution. Referring to the discussion of the types of free energy transductions in Chapter 5, the polymer biosynthesis process, including its obligatorily associated energy metabolism, can be classified as an autocatalytic network of great complexity. As shown in Chapter 5 component portions of this integrated whole (e.g., glycolysis and the carbon cycle) are themselves autocatalytic networks. The polymer biosynthesis network, and it must again be explicitly emphasized here that this *includes* all of the associated energy metabolism for making ATP, is so complex by comparison with the other examples of autocatalysis that it deserves a special category to itself. The word "uroboros" (sometimes spelled ouroboros) is henceforth used to refer to this kind of complex polymer synthesis process, including its energy supply mechanisms. This word "uroboros" comes from a phonetically similar Greek word, *ουροβορος*, which denotes the concept of "self-begetting." In western Europe, the word has been used over the centuries in association with alchemical and mystical symbolism, specifically with the image of a dragon, or a serpent, looped in a circle biting its own tail [10]. Similar serpent images also occur in relics found in the New World and in Africa and India. The spelling "uroboros" has been chosen over "ouroboros" in order to emphasize the conundrum posed by the question of the *origin*, or *evolutionary emergence*, of biological uroboroses because "ur" is an English prefix meaning "original" and "primitive." It is curious that the ancient city of the Sumerians on the Euphrates River was named Ur. One could think of Ur as one of the earliest "societal" *uroboroses* manifesting large-scale cooperative and collective human behavior.

The biological polymer uroboros is schematized in (10.54). This diagram exhibits many facts simultaneously. All processes occur at constant pressure in H_2O and are isothermal or "thermally buffered." The Gibbs free energy is consequently the governing thermodynamic function. Each of the five major component processes depicted by a circle involves only individual molecular reaction steps that obey the second law of thermodynamics for thermally buffered constant-pressure systems, $\Delta G < 0$. Energy ultimately is harvested from sunlight, the prime source for all contemporary biological energy requirements: It is transduced into chemical energy, primarily as ATP and NADPH, which is used in metabolism, depicted in the lower circle, to build up monomers from simple precursors, indicated by CO_2, NH_3, CH_4, H_2O, and others. The left-most circle not only depicts the light reaction production of ATP but also

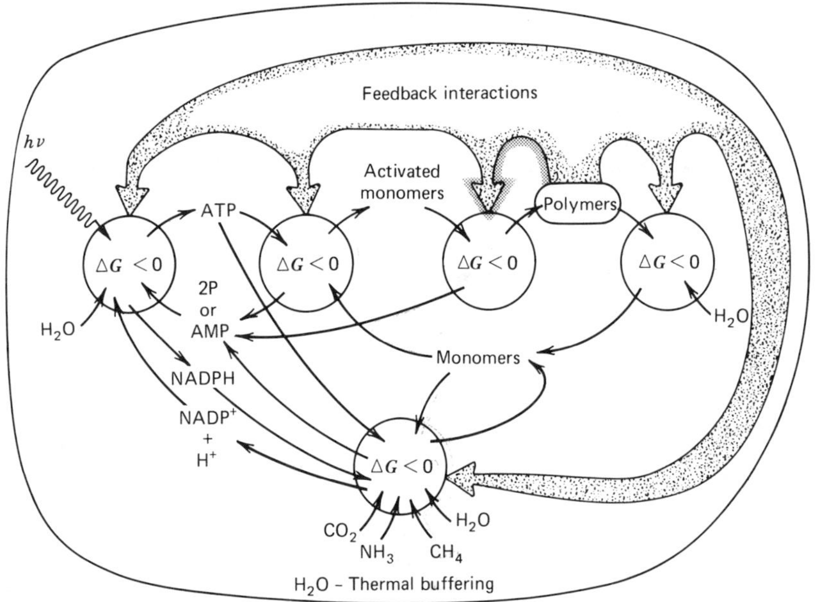

Free Energy Polymer Biosynthesis Uroboros.

(10.54)

serves to depict the production of ATP by the other energy metabolism pathways. The monomers and the ATP are used to make activated monomers that will release energy coupling degradation products such as AMP or 2P, depending on the specific polymer and its particular mechanism of ATP utilization for activation (2P for amino acids and AMP for nucleosides). The activated monomers are condensed into polymers by complex processes dependent on many specific polymeric products, as depicted in the polymer-producing circle. The polymers produced are either proteins (stippled areas), which have obligatory feedback effects on *all* steps composing this uroboros, or polynucleotides (shaded areas), which have obligatory feedback effects on only polymer biosynthesis itself. The right-most circle depicts the inevitable hydrolysis of these polymeric, *metastable* macromolecules, which recycles the monomers.

The stippled and shaded polymer feedback loop in (10.54) can be exhibited in greater detail by a diagram in which the source of energy is ATP and in which the complex of steps leading to the regeneration of ATP is not explicitly depicted in order to isolate the polymer loop for scrutiny. In (10.55) the large spots depict the protein components that are obligatory. For example, the DNA replication process is known to involve at least six different obligatory protein components. The transcription process also involves a large polymerase, etc. The

1) 2ATP + NMP → NTP + 2ADP
   2ATP + dNMP → dNTP + 2ADP $\Delta G < 0$

2) DNA + {dNTP} → DNA + $\overline{\text{DNA}}$ + {P~P} $\Delta G < 0$

3) DNA + {NTP} → DNA + $\overline{\text{RNA}}$ + {P~P} $\Delta G < 0$

4) tRNA^{AA} + ATP + (AA) → (AA) ~tRNA^{AA} + AMP + P~P $\Delta G < 0$
   {$\text{synthetase}^{\text{AA}}$}

5) rRNA + rProteins → Ribosome {ribosome} $\Delta G < 0$

6)

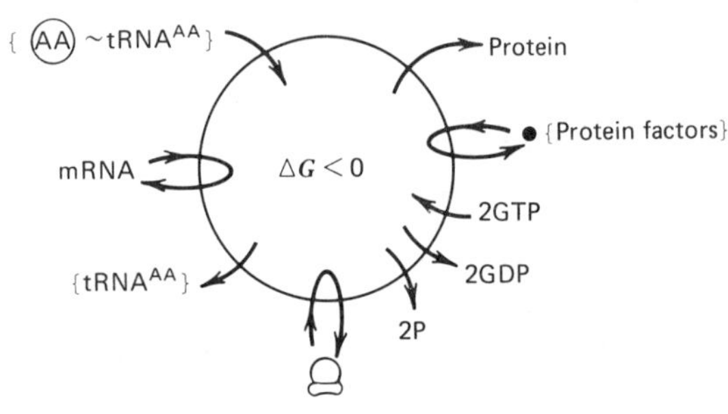

7) Hydrolysis of DNA, RNA, and Protein. $\Delta G < 0$

Contemporary Polymer Biosynthesis Uroboros.

(10.55)

notation $\overline{\text{RNA}}$ is used to signify that the RNA transcript is *complementary* in base sequence to the transcribed DNA gene, and $\overline{\text{DNA}} \equiv \text{DNA}$ for double-stranded DNA. Amino acids are designated AA, and amino acid specificity of tRNAs and synthetases is indicated by the superscript AA. Once again, the validity of the second law of thermodynamics for isothermal processes, that is, $\Delta G < 0$ for each step, is explicitly emphasized.

The story of how ATP is produced by oxidative or light reaction energy metabolism has been broached in Chapter 8, but no detailed mechanism has yet been presented. Such detail will occupy Chapters 11 and 12, especially Chapter 12. By now the reader should be impatient to get to this remarkable part of the whole story. As will be seen, it involves phenomena and mechanisms every bit as amazing and complex as those that have already been described for polymer biosynthesis.

The conundrum posed by the existence of the uroboros structure is the question of its origin and evolution. The contemporary genetic polymer biosynthesis uroboros in (10.55) represents the mechanism used by *E. coli* for polymer synthesis. Modern molecular and microbial genetics has clearly and compellingly explained how this mechanism, through genetic "mutations," can evolve into a more sophisticated mechanism such as would be required by eukaryotic bacteria or multicellular eukaryotes. It provides a detailed macromolecular realization of Darwin's concept of natural selection. The mechanism is not able to explain its own origin, however, and the origin of any uroboros, with its incredible requirement for enough components to be self-propagating, is a *mysterium tremendum*!

Part III of this book is devoted to exploring the question of the origin and evolution of an uroboros, specifically the origin and evolution of the uroboroses shown by (10.54) and (10.55).

The concepts of the uroboros and the integron (see Chapter 7) are closely related but not equivalent. Integrons, according to Jacob (see Chapter 7), occur at every morphological level of organomolecular manifestation. Looking upward from the atomic level of viewpoint, the first integron to be truly uroboric is the cell. Thus, an uroboros is an integron, but not all integrons are uroboroses. Moreover, the definition of integron is on a general, and consequently a somewhat vague, plane; the definition of uroboros, however, requires very specific properties and functions, specialized for the uroboric integrons only, and not characteristic of integrons generally.

Appendix 10A

In contrast with the calculation of the entropy change for self-assembly given in Chapter 6, the entropy change contemplated here for the polymerization process is a calculation of the *standard-state change* discussed in Chapter 4; that is, here the initial state could be described as *unit molarity* for each monomer and the final state could be described as *unit molarity* for polymers. In the self-assembly entropy change calculation, the final state was the *equilibrium* state. The equilibrium state for the polymerization process contains virtually no polymer at all.

In the self-assembly entropy change calculation in Chapter 6, the entropy of mixing change was actually positive, and the overall negative value for the entropy change was created by the translational, or volume-dependent, part of the entropy. This was because during self-assembly the translational degrees of freedom possessed by the subunits are lost as they aggregate. In contrast, during polymerization, which at first sight appears to behave the same way with regard to translational freedom, the release of H_2O molecules maintains these degrees of translational freedom, so that in (10.12) there is no volume dependence and only the residual mass dependence remains. Because nearly all the mass does end up in a *single molecule*, the polymer, and the rest remains with a *single species* of molecule, H_2O, the entropy change is negative. The term for the entropy of mixing is likewise negative.

During polymerization, some rotational entropy is probably lost because the monomers are unable to rotate as freely in the polymer as when they are still free monomers. Water is released, but its moment of inertia is small [see (2A-15)], so it does not compensate for the loss of monomer rotational entropy completely. On the other hand, the polymer possesses vibrational levels, as well as electronic levels, which are nonexistent in the free monomeric state. A long polymer possesses many low-frequency vibrational modes that extend over large regions of the polymer molecule, affecting much larger masses than occur in individual monomers. Therefore the vibrational entropy will *increase* on polymerization, but insufficiently to reverse the overall *decreasing* trend.

References

1 B. Lewin, *Gene Expression 1* (Wiley, London, 1974).

2 D. E. Metzler, *Biochemistry* (Academic Press, New York, 1977).

3 R. Losick and M. Chamberlin, editors, *RNA Polymerase* (Cold Spring Harbor Laboratory, 1976).

4 S. H. Wickner, "DNA Replication Proteins of *E. Coli*," *Ann. Rev. Biochem.*, **47**, 1163–1191 (1978).

This review provides details regarding DNA replication and RNA polymerase requirements, as well as descriptions of viral systems in which other priming proteins are required.

5 H. Weissbach and S. Pestka, editors, *Molecular Mechanisms of Protein Biosynthesis* (Academic Press, New York, 1977).

This book is an outstanding update on what has been learned over 25 years of intense study. Chapter 3 contains amazing, detailed pictures of ribosomes.

6 Sidney Altman, editor, *Transfer RNA* (MIT Press, Cambridge, Massachusetts, 1978).

See especially the chapters by Crothers and Cole, and by Kim.

7 P. R. Shimmel and D. Söll, "Aminoacyl-tRNA Synthetases: General Features and Recognition of Transfer RNAs," *Ann. Rev. Biochem.*, **48**, 601 (1979).

An up-to-date account of the "lack of progress" in answering the difficult question of recognition between tRNA and synthetase.

8 S. H. Kim, "The Three-Dimensional Structure of Transfer RNA," *Scientific American*, **238**, 52 (1978).

9 H. J. Vogel, editor, *Nucleic Acid–Protein Recognition* (Academic Press, New York, 1977).

See Ried's chapter for the story of the tRNA U-conformation.

10 Erich Neumann, *The Origins and History of Consciousness* (Princeton University Press, Princeton, N.J., 1954).

Although this book is directed at a Jungian study of the phenomenon of human consciousness, Chapter 1, entitled "The Uroboros," provides an illuminating account of the historical origins of the concept of the uroboros and the symbolism associated with it.

11 G. Stöffler and M. Stöffler-Meilicke, "Structural Organization of the Escherichia Coli Ribosomes and Localization of Functional Domains" in *Cell Biology*, H. G. Schweiger, editor (Springer-Verlag, 1981).

CHAPTER

11

Complexes and Membranes [1, 2]

In Chapter 6 the phenomenon of self-assembly was described and partially explained. It was observed that enzyme complexes are integral components of energy metabolism, and pyruvate dehydrogenase was described as an example. Additionally, the structure of the ribosome was described, and in Chapter 10 its significance in protein biosynthesis was elucidated. In Chapter 12, it is shown how enzyme complexes, particularly membrane-associated complexes, solve the problem of ATP synthesis, as well as explain a variety of other energy-requiring processes, such as the transport of a specific solute across membranes. In this chapter, the structural features of membrane-associated energy metabolism pathways will be described. These structures self-assemble from their precursor macromolecular constituents (nearly always proteins) after these components have been translated from their cognate genes. It is difficult to overemphasize the importance of self-assembly in making these structures; so much so, that before describing the membrane-associated complexes, a description of three more complexes which are simply extraordinary examples of self-assembly is given first; and then the fact of the existence of the membrane-associated complexes will seem less miraculous, or at least a *biological* "commonplace."

The three complexes to be described share with pyruvate dehydrogenase the property that they contain a "rotor," a freely rotating molecular component that is the key to their function. Two of them, fatty acid synthetase and tyrocidine synthetase, possess long rotating side chains terminating with a sulfhydryl group to which ester intermediates are attached. The third example, the bacterial flagellum, is a membrane-anchored rotating complex. For comparison, the description of the pyruvate dehydrogenase complex in Chapter 6 should be consulted.

The fatty acid synthetase complex [1] is comprised of seven distinct protein subunits. It is responsible for storing excess energy in the cell. The ultimate precursor of fatty acids is acetyl-CoA (see Chapters 5 and 8). Most of the acetyl-CoA is carboxylated by acetyl-CoA carboxylase to malonyl-CoA, which is then directly used by the fatty acid synthetase complex. The complex incorporates the two carbon atoms of malonate derived from the acetate carbons of acetyl-CoA into the fatty acid hydrocarbon chain while releasing as CO_2 the carbon incorporated during the carboxylase reaction. The explanation for this seemingly unnecessary incorporation and then elimination of CO_2 is that the direct utilization of acetyl-CoA by the fatty acid synthetase complex in the steps that actually use malonyl-CoA would involve steps for which $\Delta G > 0$, in violation of the second law for isothermal processes; with malonyl-CoA these same steps occur with $\Delta G < 0$. The carboxylation of acetyl-CoA to malonyl-CoA requires the input of Gibbs free energy, which accounts for the difference just mentioned between $\Delta G > 0$ and $\Delta G < 0$. This input of free energy is supplied by ATP in (11.1). [In (11.1) it is actually carbonic acid, rather than CO_2, that supplies the

$$\underset{\left(\begin{array}{c}\text{H}\\ \text{HCH}\\ |\\ \text{O=C—S—CoA}\end{array}\right)}{\text{acetyl - CoA}} + HCO_3^- + H^+ + ATP \rightleftharpoons \underset{\left(\begin{array}{c}\text{O}\\ \|\\ \text{C—OH}\\ |\\ \text{HCH}\\ |\\ \text{O=C—S—CoA}\end{array}\right)}{\text{malonyl - CoA}} + ADP + P \tag{11.1}$$

extra carbon, which eventually comes out as released CO_2.] In parallel with polynucleotide and protein synthesis, ATP is used to *activate* a monomeric constituent from which a polymeric molecule is to be made. The synthesis of the hydrocarbon "polymer" chain of fatty acids requires the activation of acetate to malonate in order to get the carbon atoms into the chain. Carbon atoms are added to the chain until there are 16 carbons in the fatty acid end product, palmitic acid. Palmitic acid is the precursor of other saturated and unsaturated fatty acids.

The structure of fatty acid synthetase involves a central protein, acyl carrier protein (ACP), surrounded hexagonally in a specific order by six distinct enzyme subunits: ACP-acyl transferase, β-ketoacyl-ACP synthase, ACP-malonyl transferase, β-ketoacyl-ACP reductase, enoyl-ACP hydratase, and enoyl-ACP reductase. This complex is depicted in (11.2). Attached to ACP is a freely rotating, 20-Å-long 4′-phosphopantetheine "arm" that terminates with a sulfhydryl (SH) group. This arm is attached covalently to serine 36 of ACP as shown in (11.3). There is also a sulfhydryl group on a specific cysteine residue of β-ketoacyl-ACP synthase, as depicted in (11.2).

The sequence of steps is as follows:

1 Acetyl-CoA (the only one *directly* used) reacts with the ACP carrier arm in a reaction catalyzed by ACP acyl-transferase, which transfers acetate from acetyl-CoA to the ACP carrier arm and releases CoA-SH [see (11.4)].

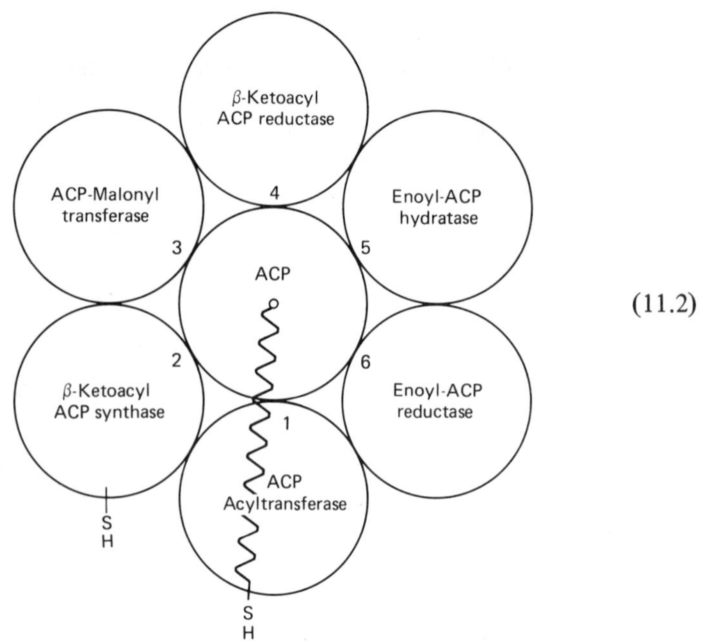

(11.2)

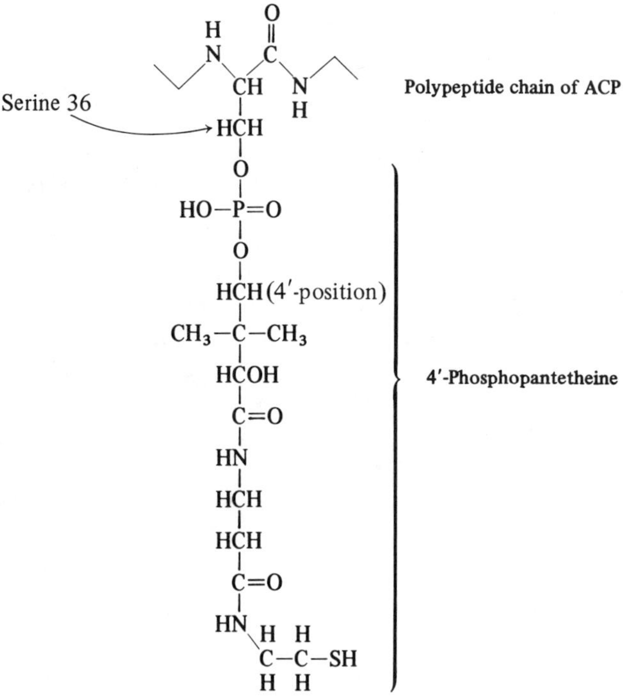

Serine 36

Polypeptide chain of ACP

```
         H    O
         N    ‖
          \  C
           CH  N
           |   H
  Serine 36 → HCH
           |
           O
           |
      HO—P=O
           |
           O
           |
          HCH (4′-position)
           |
     CH3—C—CH3
           |
          HCOH
           |
           C=O
           |
          HN
           |
          HCH
           |
          HCH
           |
           C=O
           |
          HN  H  H
            \ C—C—SH
              H  H
```

4′-Phosphopantetheine

(11.3)

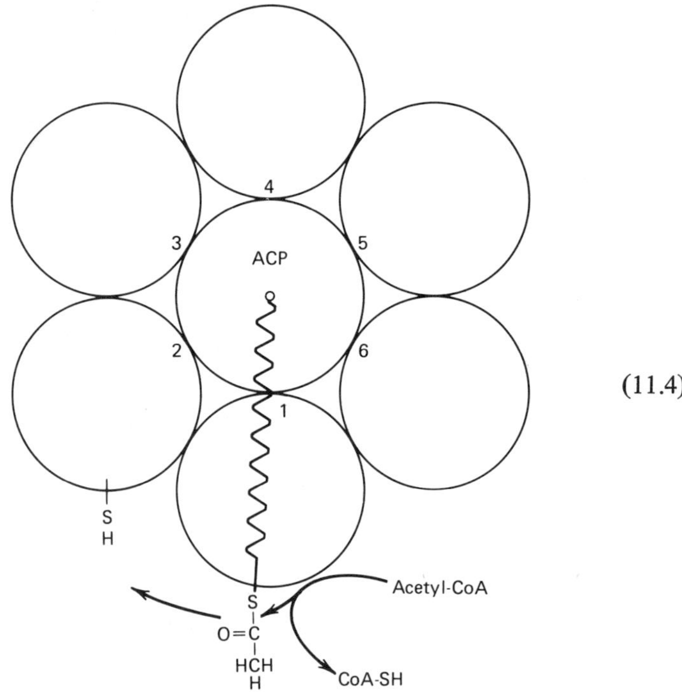

(11.4)

2 The carrier arm moves by virtue of thermal agitation (rotational diffusion) to over the SH group of the synthase component and a "transthiolation" takes place in which the acetate ends up on the synthase and the carrier arm again terminates with an SH group [see (11.5)].

3 Malonyl transferase catalyzes the transfer of malonate from malonyl-CoA to the SH arm of ACP and releases another molecule of CoA-SH [see (11.6)].

4 Thermal motion is again responsible for the movement of the carrier arm, which now lines up over the "thio ester" of acetate on the synthase subunit, and the acetate is transferred to the second carbon of the malonate moiety with the concommitant release of CO_2 in (11.7). At this stage, acetoacetate is esterified to the carrier arm, and it is this compound that could have been achieved chemically with two molecules of acetyl-CoA rather than with one molecule each of acetyl-CoA and malonyl-CoA, *except* for the energy requirement mentioned above.

5 From this point, the carrier arm swings around to the enzyme components numbered 4, 5, and 6 and is successively reduced, dehydrated, and again reduced. The reductions are achieved by the oxidation of NADPH, which is a coenzyme for components 4 and 6 [see (11.8)].

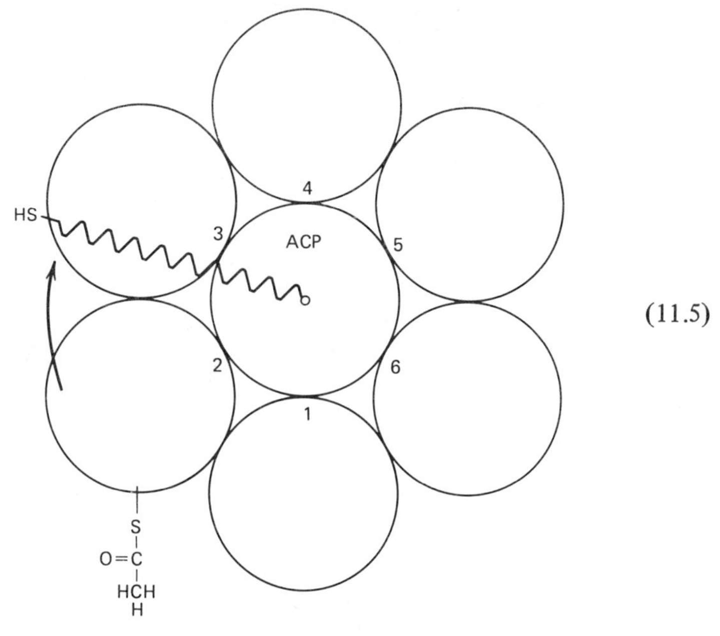

(11.5)

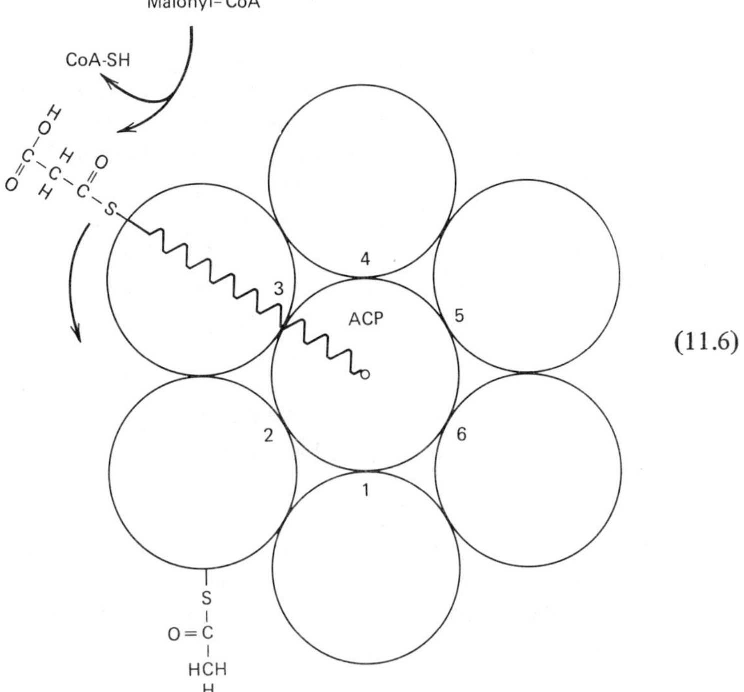

(11.6)

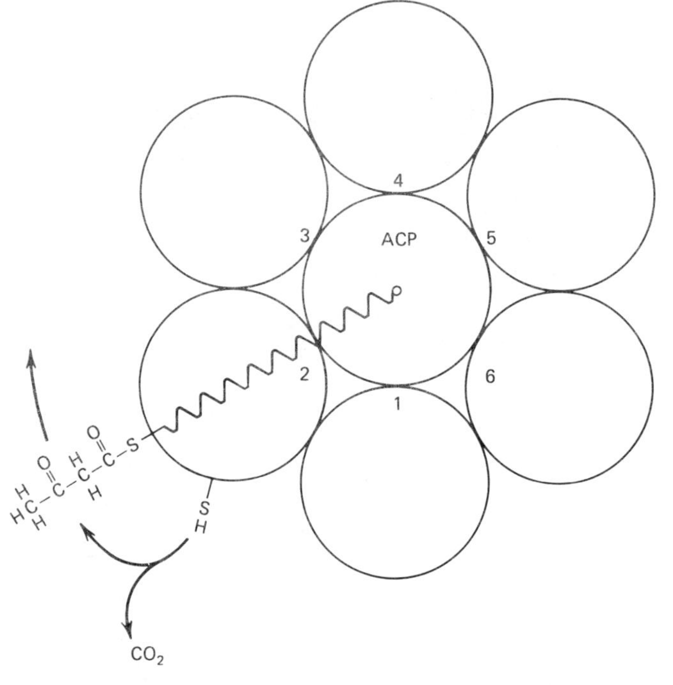

(11.7)

6 Now, the nascent hydrocarbon chain is "transthiolated" to the SH group of the synthase component so that the carrier arm can pick up another malonate moiety at the malonyl transferase subunit. Thus the carrier arm swings over component 2, deposits the nascent hydrocarbon chain, swings up to component 3, is recharged with malonate as CoA-SH is released, and returns to the synthase (2) to pick up the nascent chain [see (11.9)].

7 The transfer of the nascent chain to the terminus of the malonyl carrier arm of ACP involves the release of another CO_2 molecule and sets up the acylated complex for another round of reductions [see (11.10)].

8 This cycle is repeated seven times until a hydrocarbon chain with seven $(CH_2)_2$-units is assembled. Apparently, the synthase (2) will not accept a longer chain from the carrier so that palmitic acid is released and the complex reverts to the form depicted in (11.2).

Each step involves $\Delta G < 0$, as explained above, and thermal agitation is solely responsible for the movement of the carrier arm. Together these forces create a rotary behavior that involves changes in direction during each cycle. A most remarkable self-assembled molecular machine is fatty acid synthetase!

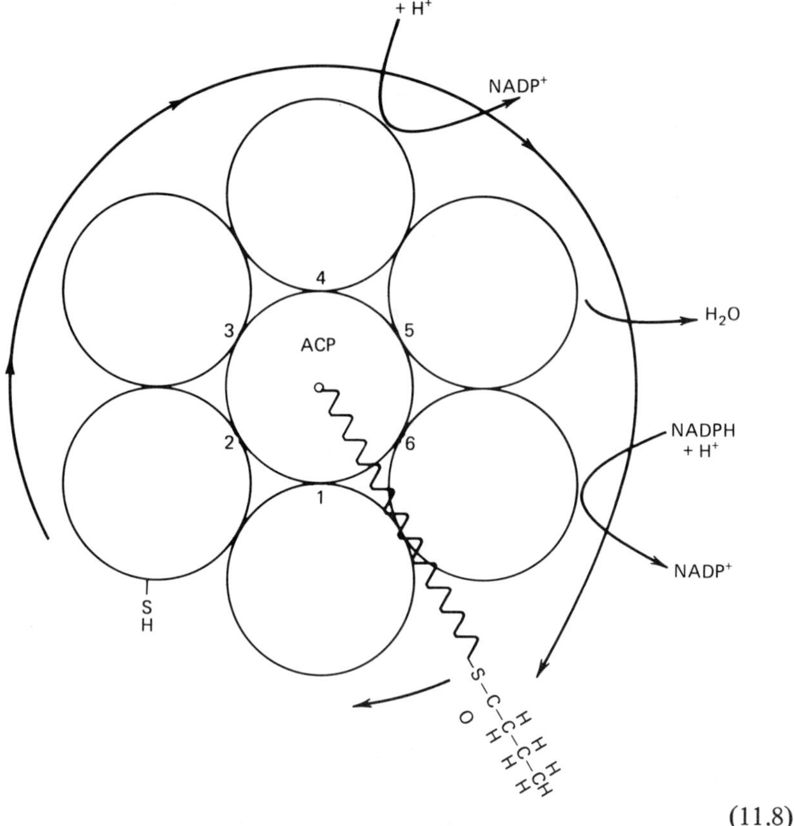

(11.8)

The tyrocidine synthetase complex [3] is just as marvelous a molecular machine. Tyrocidine is a cyclic decapeptide antibiotic synthesized by the bacterium *Bacillus brevis*. A decapeptide is really a very small protein, but tyrocidine is not synthesized on ribosomes like other proteins. Instead, it is constructed on a polyenzyme complex that uses a pantetheine arm in much the same fashion such an arm is used by fatty acid synthetase.

The tyrocidine synthetase complex consists of four protein subunits, each of which is itself the product of a gene-directed ribosomal translation typical of normal protein synthesis. These subunits self-assemble as shown in (11.11). The central core of the complex is a protein to which is attached the pantetheine arm, which terminates with an SH group. This protein has a molecular weight of about 20,000 and is tightly bound to the polyenzyme containing the six sites numbered 5, 6, 7, 8, 9, and 10, which has molecular weight 440,000. A smaller polyenzyme containing three sites numbered 2, 3, and 4 is complexed to this

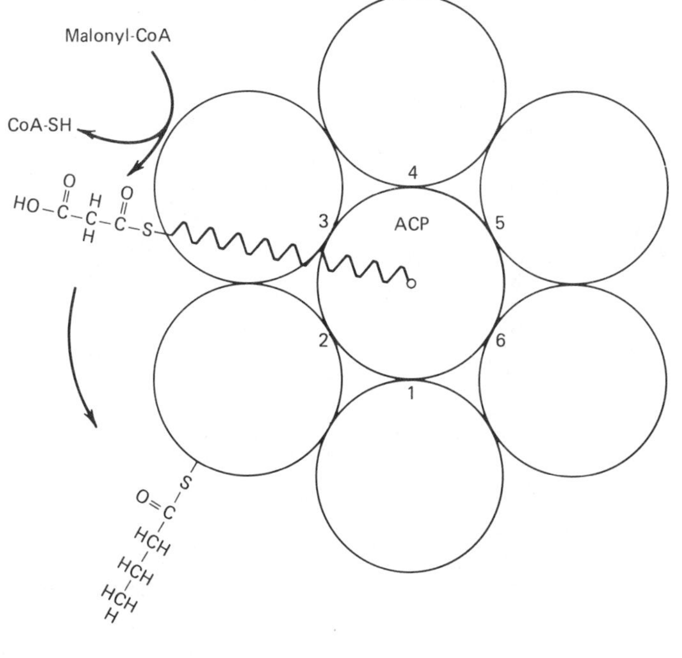

(11.9)

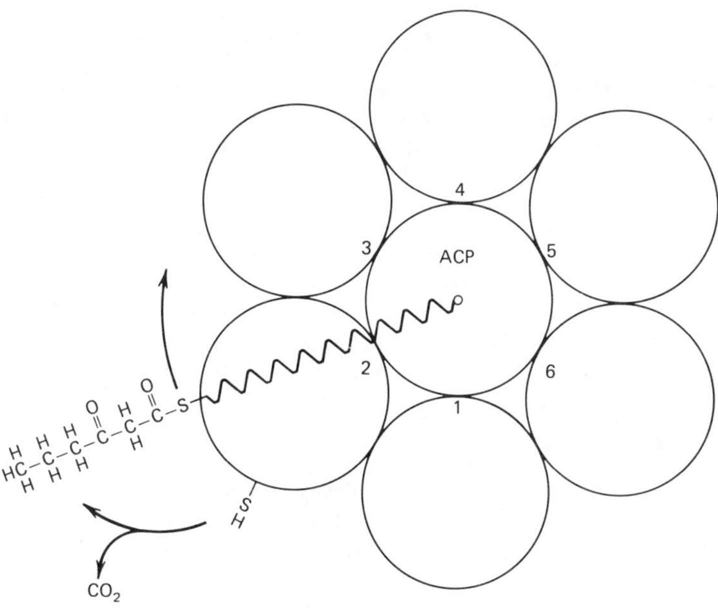

(11.10)

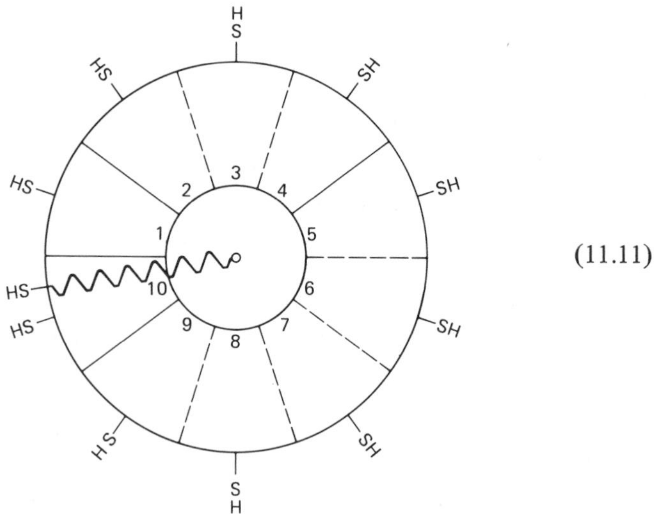

(11.11)

larger component. It has a molecular weight of 230,000. Finally, a smallest enzyme containing only the site numbered 1 and with molecular weight 100,000 completes the structure. The numbered sites indicate the location of SH groups, presumably cysteine residues, which are the specific sites for the attachment of the amino acids that ultimately comprise the decapeptide product, tyrocidine.

Each site is specific for a specific amino acid. This specificity is listed in Table 11.1. Two features of Table 11.1 are peculiar. First of all, D-phenylalanine appears twice, as well as the usual L-amino acids. Actually, L-phenylalanine is in fact used initially and is racemized to the D-form by the enzyme complex. Second, the amino acid ornithine, not found in gene-directed proteins, occurs.

Table 11.1

Site	Amino Acid
1	D-Phenylalanine
2	Proline
3	Phenylalanine
4	D-Phenylalanine
5	Asparagine
6	Glutamine
7	Phenylalanine
8	Valine
9	Ornithine
10	Leucine

The attachment of the amino acids requires prior activation with ATP to form the activated amino acyl adenylates that react with the SH group to form thiocarboxyl esters:

$$\text{aa} + \text{ATP} \longrightarrow \text{aa-AMP} + \text{PP}$$

$$\text{aa-AMP} + \text{enzyme-SH} \longrightarrow \text{enzyme-S}{-}\overset{\overset{\displaystyle O}{\|}}{C}{-}\underset{\underset{\displaystyle H}{|}}{\overset{\overset{\displaystyle R}{|}}{C}}{-}\underset{\underset{\displaystyle H}{|}}{\overset{\overset{\displaystyle H}{|}}{N^+}}{-}H + \text{AMP} \tag{11.12}$$

Thio ester

These thio esters maintain enough of the free energy of activation in the amino acyl adenylates to promote peptide formation.

Tyrocidine synthesis begins after all 10 sites are esterified with their specific amino acids. The D-phe on site 1 transfers to the proline at site 2 to form a thioesterified dipeptide, N-terminal D-phe-pro-S-enzyme. When this dipeptide transfers to the phe at site 3, the rest of the polymerization is triggered until the decapeptide thio ester, N-terminal D-phe-pro-phe-D-phe-asn-glu-phe-val-orn-leu-S-enzyme is formed. At this point, cyclization occurs as a peptide bond forms between the N-terminal amino group of D-phe and the carboxyl group of leucine, which is attached to the enzyme complex through a carboxyl thio ester at site 10.

The fascinating mechanism of elongation from site 3 through site 10 involves the panthetheine arm. What happens resembles the acetate activation step in fatty acid synthesis. The peptide at any one of these sites gets to the next site by first being "transthiolated" to the SH group of the swinging pantetheine arm. The arm swings to the next adjacent site, and the peptide "transpeptidates" onto the amino group of the esterified amino acid beneath it at that site. Then this elongated peptide transthiolates back to the pantetheine arm, which carries it one more site along for a repetition of transpeptidation and transthiolation. This is depicted in (11.13).

As in fatty acid synthesis, $\Delta G < 0$ for each step, and the pantetheine arm movement is caused by thermal agitation.

The third complex to be described before a detailed treatment of membrane-associated complexes is the bacterial flagellum [4]. The flagellum of a bacterium is a self-assembled complex of subunit protein constituents of molecular weight 40,000, called flagellin. The flagellum usually has the structure of three intertwined strands of flagellin subunits, all twisted helically. This structure is rigid and permanently twisted like a corkscrew [see (11.14)].

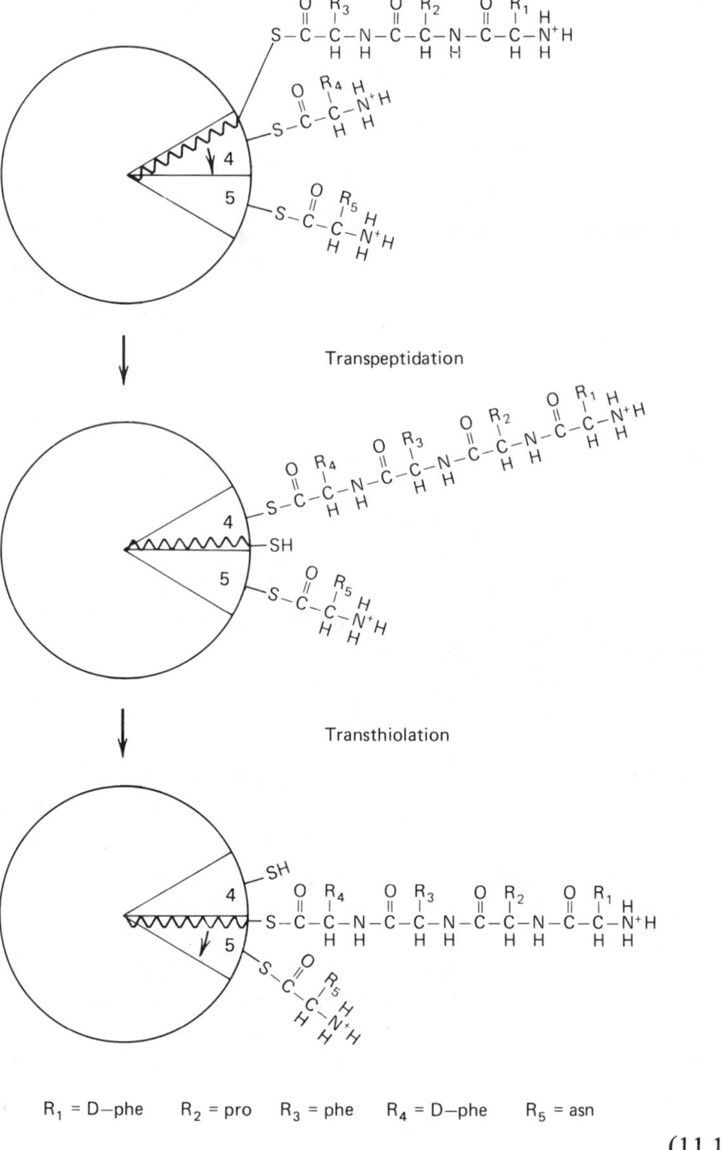

(11.13)

Salmonella (Two flagella) (11.14)

Propulsion is achieved for the bacterium by its ability to *rotate* the flagellum up to 40 times per second. Speeds of 500,000 Å/sec are possible.

Typically, bacterial flagella are from 100 to 400 Å in diameter and can be over 100,000 Å in length. This is several times longer than the bacterium, which is only about 2 μm long. The flagellum is anchored in the bacterial membrane, and the mechanism of rotation generation is also in the membrane. Two impressions of this structure are given in (11.15) and (11.16).

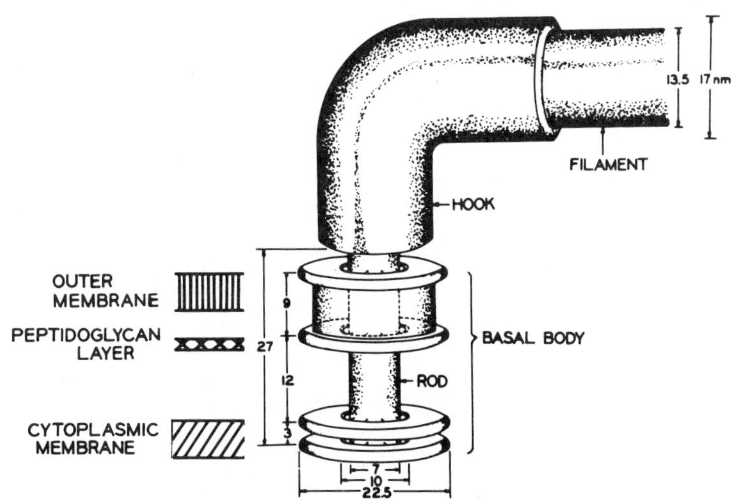

Reproduced, with permission, from Ref. 4.

(11.15)

In (11.15), taken from Adler's paper [4], special self-assembled regions of the "basal body" of the flagellum fit into the bacterial outer membrane (a lipid bilayer structure) and also into the cytoplasmic membrane (another lipid bilayer). These regions, which appear as disks and rods in the figure, are made of protein subunits, as depicted schematically in (11.16) [5]. In (11.16) the outer membrane is called the cell wall and the inner membrane is depicted explicitly as a lipid bilayer. The protein subunits of the basal plates, or disks, are depicted by spheres. The picture also suggests that the motive power for rotation is supplied by protons (H^+'s) traversing the inner membrane at 256 protons per revolution. It is this particular energy transduction, the conversion of "protonic transmembrane electrochemical energy" into the work of flagellum propulsion, that shares something with the still mysterious transduction of electron transport chain energy to ATP energy. As is shown in the description of bacterial, mitochondrial, and chloroplast membrane-associated energy transductions, the *transmembrane electrochemical energy content of protons* is the unifying principle in energy metabolism and transduction.

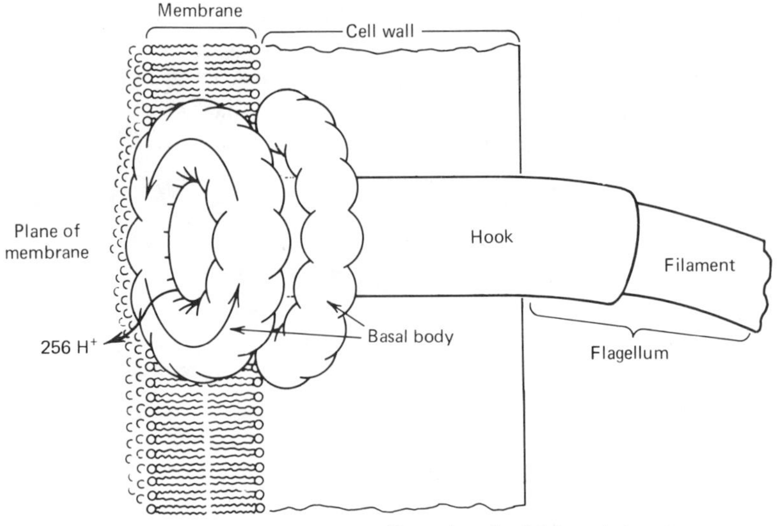

Reproduced, with permission, from Ref. 5.

(11.16)

An explanation of how Gibbs free energy can be transduced into the transmembrane electrochemical free energy of protons is provided in the next chapter. The transduction mechanism was given the name "chemiosmosis" by its most recent and compelling advocate, Peter Mitchell. Ever since Lipmann enunciated the "ATP energy principle" in 1941, researchers have tried to elucidate the detailed molecular steps involved in converting metabolically generated energy into ATP energy. In glycolysis (see Chapters 5 and 8) ATP is synthesized from ADP and phosphate by direct chemical coupling to either 1,3-diphosphoglycerate or phosphoenolpyruvate. By 1950, Lehninger and Kennedy had shown that energy metabolism is organized inside the cellular organelles *mitochondria*. The citric acid cycle, the electron transport chain, and fatty acid oxidation were all found to be confined to these energy organelles. It was hoped by many researchers that the transduction of the energy, harvested during electron transport from NADH to oxygen, into ATP energy would occur by a similar chemical coupling scheme. The mitochondrial membrane was ruptured, separated away, and chemical coupling intermediates, usually called X, Y, or Z, were pursued. They were never found! Today, it is known that they do not exist. An entirely different transduction mechanism has been discovered, and it is based on the general principle of "chemiosmosis," championed most vigorously since 1961 by Mitchell. As already indicated, this mechanism requires an *intact* membrane, the key being the electrochemical potential of protons across this intact membrane. No amount of breaking up and fractionating the membrane and its constituents and contents will ever exhibit this type of energy. Only the intact membrane is capable of manifesting the membrane potential.

The structural properties of membrane-associated energy metabolism and transduction, to be described below, make the new chemiosmotic view easy to accept. This knowledge is rather recent, however, so it is not surprising that many researchers, until very recently, had a difficult time understanding the resolution of the mystery of ATP synthesis. As will be emphasized in the next chapter, not all the details are yet completely clear, but the bulk of the evidence is enormously compelling. Perhaps the most significant methodological breakthrough that is responsible for creating a rapid acceptance of the new view was the discovery of lipid vesicles, the liposomes, that could be self-assembled in the laboratory. These vesicles are small lipid-bilayer spheres ranging in radius from 100 Å to more than 1 μm. They can be prepared with specific contents and immersed in media of specific composition. Into their lipid membrane phase, lipophilic proteins can be incorporated in a selective way. In fact, Racker and colleagues in the middle 1960s were able to reconstitute oxidative phosphorylation in mitochondrial vesicles by such techniques. This provided great impetus for more diverse and detailed studies, some of which are described in the next chapter. The result has been an explosion of successful experimental studies, all of which underscore the basic significance of the transmembrane electrochemical potential.

Bacterial Membrane [2,5]

In (8.9) bacterial electron transport was depicted as a succession of oxidation-reduction reactions. Chapter 8 concluded by stating that the energy harvested by these oxidation-reduction sequences is transduced into the energy of ATP. No mechanism for this transduction was presented in Chapter 8, and the mechanism that does explain this fundamental transduction is presented for the first time in this book in the next chapter. However, a diagram exhibiting how the components of the electron transport chains are actually organized in the bacterial membrane provides the basis for this explanation. The discovery that electron transport was in fact organized in the membranes of organisms was made by Kennedy and Lehninger in the late 1940s. They showed that such membrane-associated complexes for electron transport were observable in the mitochondrial membranes of rat livers. In (11.17) the organization of electron transport in bacteria is shown.

The curved lines represent the outline of the lipid-bilayer membrane of a bacterium. Near the top and bottom of the figure, schematic lipid phosphoglyceride molecules are drawn. They fill in all the space except for that portion of the membrane that is occupied by proteins. The sequence of redox reactions

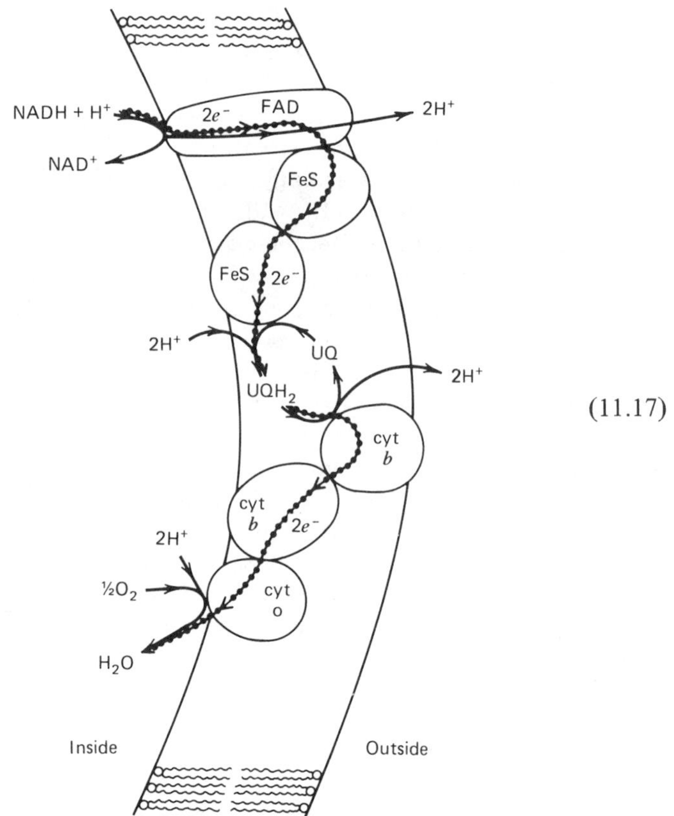

(11.17)

given in (8.9) appears in (11.17). An FAD-binding protein followed by two iron-sulfur proteins (FeS) connects NADH to ubiquinone (UQ), which freely diffuses cyclically between FeS protein and cytochrome *b*. Two molecules of cytochrome *b* and one molecule of cytochrome *o* complete the sequence. Because some redox reactions involve whole hydrogen atoms and others involve only the electrons, as was explained in Chapter 4, protons are taken up and released at specific steps in the sequence. The electrons engage in a continuous flow from NADH to H_2O. Two electrons are removed from NADH, and two electrons are used to make H_2O. These electrons are a dotted line in (11.17). A net transfer of four protons from inside to outside the bacterium takes place.

The next chapter gives an explanation of how the transfer of protons from inside to outside the bacterium provides a transmembrane electrochemical potential.

While there is good evidence for the topological arrangement of components in (11.17), the diagram is still partially tentative. As shown in Chapter 12, there is almost certainly a necessity for additional protein components associated with the terminal cytochrome sequence ($b \rightarrow b \rightarrow o$) of the chain. This is seen to follow from the need to harvest more energy from electron flow than is yielded by the four transferred protons depicted in (11.17).

The bacterial membrane also contains a variety of other proteins for the transport of ions and small molecules. An "ATPase" complex is present in addition, and it is responsible for transducing the *electron-transport generated transmembrane proton electrochemical potential* into ATP synthesis. These components are depicted in (11.18). In (11.18) phosphate ($H_2PO_4^- \equiv P^-$), proline,

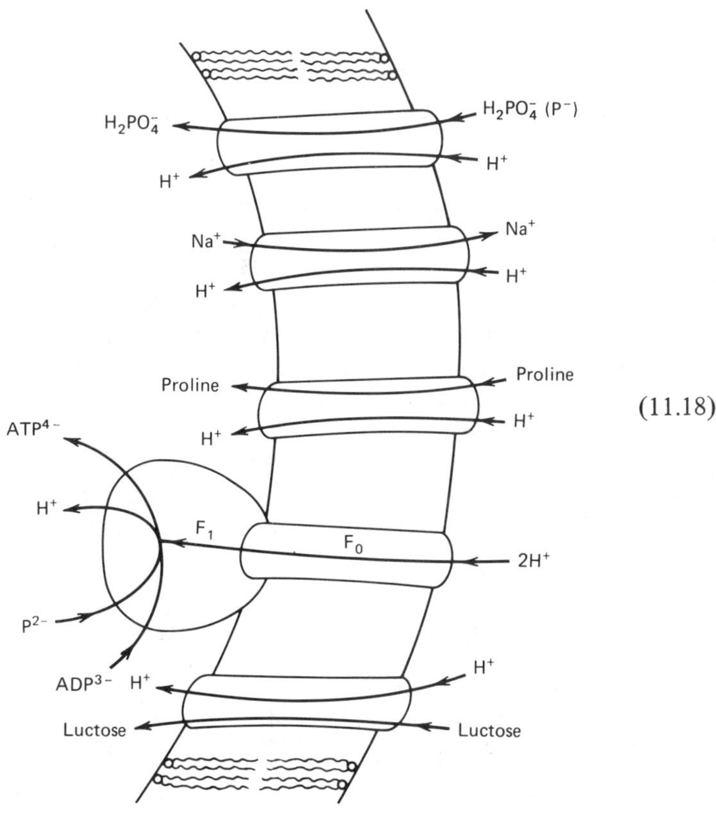

(11.18)

and lactose "symporters" are shown. They transport specific substrates in the same direction as protons. The sodium "antiporter" transports Na^+ and H^+ in opposite directions. How protons are able to drive these transporters is explained in Chapter 12. The large protein complex labeled F_1 and F_0 is the ATPase complex. The F_0 subunit transports two protons inwardly for each ATP synthesized on the active catalytic site located on the F_1 multisubunit portion of the ATPase. Once again, the explanation for this process is given in Chapter 12. Nevertheless, between (11.17) and (11.18) it is already seen that electron transport generates across the membrane a proton gradient that is used to drive ATP synthesis at the ATPase when protons come back through the membrane via a channel in the F_0-F_1 complex. The electron transport chain complex and the ATPase complex are physically separated in the membrane. Their coupling is achieved by an imbalance of protons across the membrane that is communicated from its locus of generation at an electron transport chain complex to its locus of action, which can be *anywhere* on the membrane. The only constraint is an intact membrane. No chemical intermediates are involved.

Mitochondrial Membranes [2,5]

The organization of the electron transport chain in mitochondria, the energy organelles of eukaryotic bacteria and higher organisms, is similar to that for prokaryotic bacteria. It involves a different and more complex cytochrome sequence, as well as flavinmononucleotide (FMN) as the flavin coenzyme instead of FAD. As with the prokaryotic bacteria case, the diagram for mitochondria is tentative, and almost certainly requires additional components in the cytochrome sequence if sufficient energy transduction is to be achieved for ATP synthesis. Diagram (11.19) depicts a whole mitochondrion, and (11.20) exhibits a portion of its inner membrane, showing electron transport, the ATPase complex, and an ATP^{4-}-ADP^{3-} antiporter. The portion of the inner membrane circled in stippling is shown in (11.20). As in (11.17), the electron flow is shown

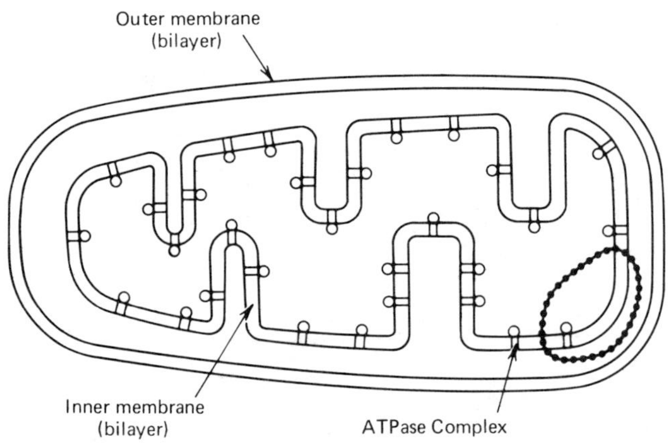

(11.19)

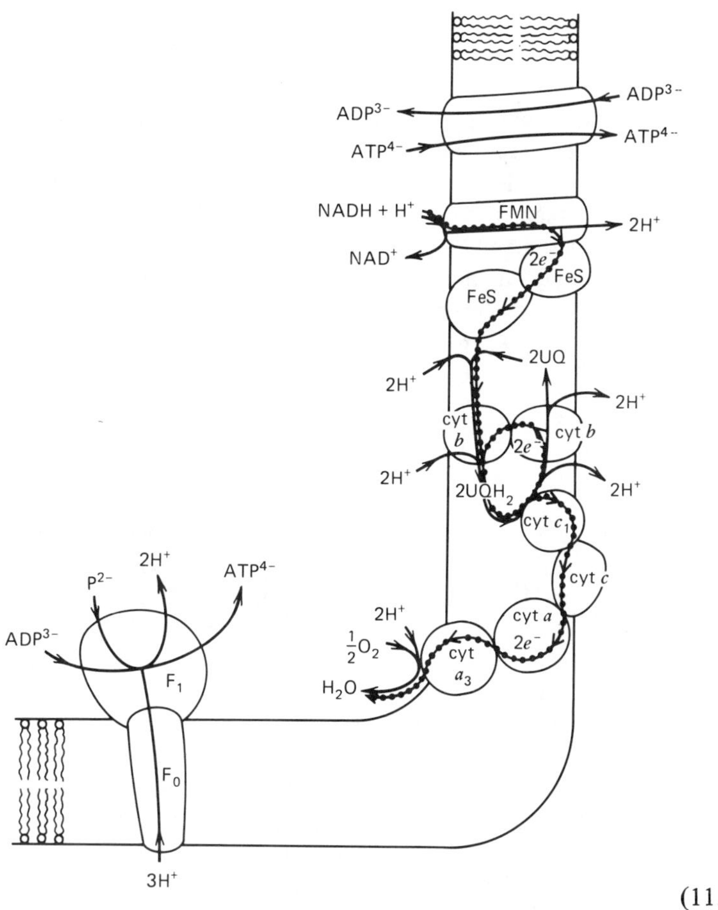

(11.20)

as a dotted line from NADH to H_2O. The line represents the transport of two electrons. A more complicated ubiquinone cycle is shown in (11.20) than in (11.17). In (11.17) two electrons are transported to the second FeS protein from NADH. Both electrons are then used to reduce UQ to UQH_2 while two protons are taken up from the medium. Both protons are released on the other side of the membrane as the two electrons are passed along the cytochrome ($b \rightarrow b \rightarrow o$) chain, one electron at a time, until both electrons are used, along with two more protons from the media, to reduce oxygen to H_2O. In all, only four protons are extruded by the mechanism in (11.17). In (11.20), however, a total of six protons is extruded. This is achieved by the more complex ubiquinone cycle as follows: The two electrons that arrive at the second FeS protein are used, along with a proton each from the medium, to partially reduce each of two UQ molecules to a pair of "semiquinones" UQH. Each of these two semiquinones borrows an electron from the first cytochrome *b* and takes up a proton from the medium to form UQH_2, the dihydroquinone species. Each of these two molecules of UQH_2 gives up one electron to cytochrome *c*, and one proton to the other side of the membrane, *and* then gives up its other electron to the second cytochrome *b* while releasing to the outside the second proton. The second cytochrome *b* returns the electron to the first cytochrome *b* from which an electron was borrowed, completing the electron-borrowing minicycle. The previous step also completes the ubiquinone cycle by regenerating UQ. This clever, still hypothetical "double" quinone cycle was suggested by Mitchell in order to get more protons across the membrane for one pair of electrons transported from NADH to H_2O. In Chapter 12 much more is said about the energy budgeting of this mechanism.

Chloroplast Membrane [2,5]

Photosynthetic organisms have developed a variety of mechanisms and structures for transducing sunlight energy into the free energy contained in ATP and NADPH. In Chapter 12, a halobacterium will be described that can collect sunlight and transduce it into a transmembrane proton potential, suitable for driving the synthesis of ATP on an ATPase complex, through the intermediation of a *single* protein, bacteriorhodopsin. Most photosynthetic organisms use chlorophyll and cytochromes for this purpose and possess organized organelles in which these constituents are located in orderly complexes. Photosynthetic bacteria may contain "chlorobium vesicles," "chromatophores," or loosely arranged "cytoplasmic lamellae" as their chlorophyll-associated organelles. Examples include green sulfur bacteria, purple bacteria, and blue-green algae, respectively. All higher photosynthetic organisms contain chloroplasts as their sunlight-utilizing organelles. The energy metabolism pathway for the light reactions in (8.1) is typical of the pathways that are organized in chloroplasts. Diagram (11.21) depicts an entire chloroplast. The portion of the inner membrane circled with stippling is exhibited in (11.22). Several features of this diagram require emphasis. The ATPase complex is different from that found in bacteria in (11.18). Three protons instead of only two are apparently required, as was also the case for mitochondria in (11.20). The migration of excitation in the pigment complexes from the point of sunlight absorption to the points where electrons are finally excited by this energy is depicted by dashed arrows. The actual electron transport pathway is shown by a dotted line. Two photons are required at photosystems I *and* II in order to excite two electrons from H_2O to NADPH. If only photosystem I is functioning, then the cytochrome b_6 loop for electron transport is used for *cyclic* electron transport. Otherwise, however, this shunt is not used, and electrons flow from H_2O to NADPH along the rest of the path-

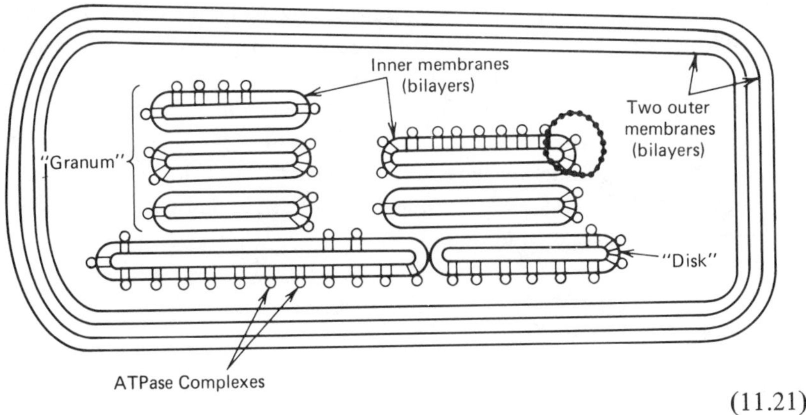

(11.21)

way, excluding the Z → cyt b_6 → cyt b_3 shunt. Plastoquinone (PQ) is responsible for proton translocation.

Again, it must be emphasized that this diagram is still partially tentative. In Chapter 12 it is shown whether or not sufficient energy transduction is achieved by this mechanism to account for the ATP synthesis energy requirements.

Each of the three membrane systems just described is self-assembled from

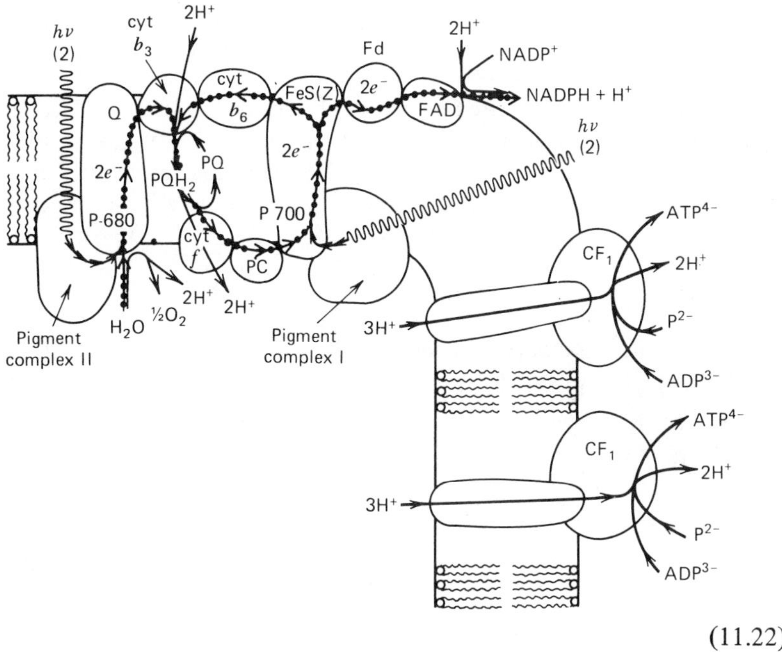

(11.22)

lipids and proteins. Usually, the biological perspective emphasizes how *function* follows from *structure*, particularly with respect to proteins and polynucleotides. This relationship was especially apparent in protein biosynthesis. Here, however, it is difficult to avoid the impression that the *structure* of the membrane-associated electron transport chain is the *result* of its *function*; that is, that energy flow and transduction have created, through evolution, the structure of the mechanism [6]. This view is expanded in Part 3, and exhibits the energy flow ordering idea discussed in Chapter 3.

References

1 A. Lehninger, *Biochemistry*, 2nd ed. (Worth, New York, 1975).

2 E. J. DuPraw, *Cell and Molecular Biology* (Academic Press, New York, 1968).

This book contains several superb pictures of cellular organelles, both as diagrams and as photomicrographs.

3 F. Lipmann, "A Mechanism of Polypeptide Synthesis on a Protein Template," in *Molecular Evolution*, edited by D. L. Rohlfing and A. I. Oparin (Plenum, New York, 1972).

4 J. Adler, "Chemotaxis in Bacteria," in *Ann. Rev. Biochem.*, **44**, 341 (1975).

5 P. C. Hinkle and R. E. McCarty, "How Cells Make ATP," *Scientific American*, **238**, 104 (1978).

The diagrams in this paper are *updated* in (11.20) and (11.22). "Stoichiometry of Vectorial H^+ Movements Coupled to Electron Transport and to ATP Synthesis in Mitochondria," by A. Alexandre, B. Reynafarje, and A. L. Lehninger, *Proc. U.S. Natl. Acad. Sci.*, **75**, 5296-5300 (1978), is a closely related paper.

6 A. Szent-Györgyi, *Introduction to Submolecular Biology* (Academic Press, New York, 1960).

One of the earlier exponents of the idea that life is driven by an "electric current." Szent-Györgyi provides a provocative treatment of redox reactions in this book.

CHAPTER

12

The Mechanism of ATP Synthesis: "Chemiosmosis"

In a timely review [1] entitled "Membranes and Energy Transduction in Bacteria," Harold began his account with a section titled: "The Revolution in Membrane Biology." He wrote:

> The study of bacterial bioenergetics, long overshadowed by molecular genetics, has flowered prodigiously in the past decade. The impetus came partly from the introduction of new and powerful tools—mutants defective in energy coupling, ionophores, and membrane vesicles. Equally significant has been the emergence of a unifying conceptual framework that links bacterial bioenergetics to that of mitochondria, chloroplasts, muscle and nerve. Broadly stated, it is increasingly recognized that biochemical reactions may be so organized within membranes as to bring about the translocation of molecules, ions, or chemical groups across the membrane; that some of these reactions lead to the separation of electrical charges within and across the membrane; and that the recombination of charges underlies the performance of osmotic, chemical, and mechanical work.
>
> The roots of these ideas reach back more than half a century [2] (Robertson, 1968), but they found little welcome among biochemists or microbiologists until quite recently. The decisive event was the formulation by Peter Mitchell of his chemiosmotic hypothesis (1961) [3], which sharpened the contrast between the traditional biochemistry of soluable enzymes and metabolic intermediates, and a new order of vectorial pathways linked by topology and ion gradients. This provocative proposal instigated a furious debate, now in its second decade, which has many of the hallmarks that Thomas Kuhn (1970) [4] found characteristic of scientific revolutions. It has generated a quite excessive volume of print, some personal animosities, at least six more or less distinct models of energy coupling, but also a very respectable body of sound experimental work. Predictably, the controversy failed to produce consensus but is gradually being transcended by a new generation of investigators and problems.

In the three short years since Harold wrote these paragraphs, real consensus has emerged. The experimental evidence is diverse, plentiful, and overwhelming. In 1978, Peter Mitchell was awarded the Nobel prize in chemistry for his contributions to this revolution. Nevertheless, numerous points of detail are not yet fully settled. Many of Mitchell's specific mechanisms [5] are known not to be correct, even though the general theme of chemiosmosis has become more firmly established. Serious questions still remain regarding the stoichiometry and efficiency of energy transduction by the electron transport chain. Lehninger, a codiscoverer of the organized nature of electron transport in mitochondria and for many years an eloquent spokesman for the "chemical coupling" mechanism of transduction, has done much during the last three years [6–9] to sharpen, clarify, and elucidate the detailed molecular mechanisms of chemiosmosis. While the work of Mitchell has provided the general setting for the material to be presented in this chapter, it is the very recent work of Lehninger and his colleagues that has provided most of the specific details regarding the bioenergetics of ATP synthesis.

Before confronting the question of how cellular membranes are used to generate ATP, the general principles of chemiosmosis are presented. Numerous special cases are presented in order to demonstrate the variety of experimental evidence that has been accumulated to support chemiosmotic mechanisms. Following this material is a presentation of the problem of ATP synthesis and its solution, as it is understood at present.

In the preceding chapter it was shown that electron transport is organized in the bacterial membrane, or in the inner membranes of mitochondria and chloroplasts, in such a way that protons are translocated across the membrane as electrons move through the electron transport chain. In bacteria, this process results in the establishment of a transmembrane electrical potential *and* a transmembrane pH differential. Because the protons carry a positive electrical charge, the interior of the bacterium becomes electrically negative as protons are extruded, and the pH inside rises relative to the outside as well. This is depicted

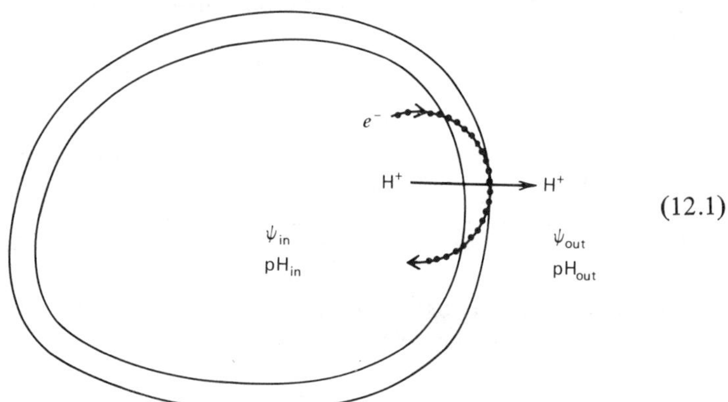

(12.1)

schematically in (12.1), where the stoichiometry of H^+ extrusion for e^- transported is presently ignored. Using the convention $\Delta\psi = \psi_{out} - \psi_{in}$ and $\Delta pH = pH_{out} - pH_{in}$ leads to

$$\begin{aligned} \Delta\psi &> 0 \\ \Delta pH &< 0 \end{aligned} \tag{12.2}$$

In Chapter 4 it was shown how to compute the Gibbs free energy for a molecular species partitioned into two regions by a membrane. For the protons inside and outside the bacterial membrane depicted in (12.1), the Gibbs free energy can be written in parallel with (4.81) and (4.82), except that here we again ignore the effects resulting from γ terms.

$$G_{in} = N_{H^+}^{in}(U_{H^+}^0 + |e|\psi_{in}) + N_{H^+}^{in}PV_{H^+}^0 - N_{H^+}^{in}T\bar{S}_{H^+}^{0\,in} + N_{H^+}^{in}k_BT\ln\left(\frac{N_{H^+}^{in}}{e}\right) \tag{12.3}$$

and

$$G_{out} = N_{H^+}^{out}(U_{H^+}^0 + |e|\psi_{out}) + N_{H^+}^{out}PV_{H^+}^0 - N_{H^+}^{out}T\bar{S}_{H^+}^{0\,out} + N_{H^+}^{out}k_BT\ln\left(\frac{N_{H^+}^{out}}{e}\right)$$

where the entropy terms $\bar{S}_{H^+}^{0\,in}$ and $\bar{S}_{H^+}^{0\,out}$ are different because of their explicit volume dependences:

$$\bar{S}_{H^+}^{0\,in} = S_{H^+}^0 + k_B\ln\left[\left(\frac{2\pi M_{H^+}k_BT}{h^2}\right)^{3/2} V^{in}\right] \tag{12.4}$$

and

$$\bar{S}_{H^+}^{0\,out} = S_{H^+}^0 + k_B\ln\left[\left(\frac{2\pi M_{H^+}k_BT}{h^2}\right)^{3/2} V^{out}\right]$$

Unlike the analysis in Chapter 4, in which the *equilibrium state* was determined by equating the Gibbs free energies per molecule

$$\mu_{in} \equiv \frac{G_{in}}{N_{H^+}^{in}} = \mu_{out} \equiv \frac{G_{out}}{N_{H^+}^{out}}$$

the analysis here corresponds with a *nonequilibrium* situation because a continuous energy flux, manifested by electron flow down the electron transport chain, continuously drives protons out of the bacterium, thereby establishing an *inequality* of the chemical potentials μ_{in} and μ_{out}. From (12.3) and (12.4), it follows that

$$\mu_{out} - \mu_{in} \equiv \Delta\mu = |e|\Delta\psi + k_B T \ln \left(\frac{N_{H^+}^{out}/V^{out}}{N_{H^+}^{in}/V^{in}} \right) \tag{12.5}$$

This expression refers to a single proton in units of energy per proton. It can be converted to its equivalent in units of energy per mole by multiplying through by Avogadro's number. In addition, the definition of pH given by (4.14) and (4.15) can be used to convert the logarithm term into a ΔpH term. Finally, the notation $F\Delta p$ is used for $6 \times 10^{23}\ \Delta\mu$ because this leads to the expression for the proton electrochemical potential most frequently found in the bioenergetics literature [10–12]:

$$F\Delta p = F\Delta\psi - 2.3RT\,\Delta\text{pH} \tag{12.6}$$

or

$$\Delta p = \Delta\psi - 2.3\,\frac{RT}{F}\,\Delta\text{pH}$$

where Δp is in units of volts. Because protons carry a *positive* charge, they tend to move toward more negative values of volts, just the reverse of what electrons do. From (12.2) it is seen that $\Delta p > 0$ for the situation depicted in (12.1). Therefore, there is a tendency for protons to move back across the membrane into the interior of the bacterium *if they can find a pathway*. Part of this tendency is purely electrical, the $\Delta\psi$ term, and part is purely "osmotic" or "chemical," the $-2.3(RT/F)\Delta$pH term. Mitchell uses the term "chemiosmosis" for this phenomenon and refers to Δp as the "protonmotive force." Equation (12.6) is the fundamental equation for the chemiosmotic hypothesis and for all further considerations of membrane energetics in this chapter.

How big are Δp, $\Delta\psi$, and ΔpH for bacteria, mitochondria, and chloroplasts? First of all, note that the bacterial membrane and the mitochondrial inner membrane have equivalent topologies with respect to inside and outside. The chloroplast inner membrane is inside-out relative to these other two cases, as can be seen in (11.17) through (11.22). Therefore, after working out the details for the bacterial membrane, the mitochondrial case can be done in parallel, whereas the chloroplast case requires reversing all signs on Δp, $\Delta\psi$, and ΔpH in order to do it explicitly. The size of $\Delta\psi$ is determined by the membrane's capacitance C and the steady-state number of protons $N_{H^+}^*$, extruded by electron transport. According to (7.7), this is

$$\Delta\psi = \frac{N_{H^+}^*|e|}{C} \tag{12.7}$$

This equation requires careful interpretation. Electron transport drives protons out of the bacterium, but eventually a steady state develops because pathways for the return of protons to the interior are operating and exactly balance the outflow. This steady state is characterized by an amount of negative counter-anions left inside the bacterium as the protons go outside. According to the capacitance analysis in Chapter 7, the negative anions are electrically bound to the inner surface of the membrane, each being associated with a proton bound on the outside. This amount is given by $N^*_{H^+}$. Protons are continuously ejected from the interior and are continuously coming back inside along various pathways during the steady-state flow. Consequently, the steady-state amounts of counterions on the inside and protons on the membrane's outside surface are constants. For bacteria grown at pH 7, it is seen below that their interiors are at a pH of about 9. For a bacterium 1 μm in radius, the volume is 4.2×10^{-12} cm^3 = 4.2×10^{-15} liter. At a pH of 9 the H_3^+O concentration is 10^{-9} molar, or 6×10^{14} protons per liter. Therefore the bacterium contains at any instant only on the average of 2.5 protons! It must be emphasized here that pH = 9 is used because in many bacteria there is a difference of 2 pH units across the membrane, interior alkaline. Chemiosmosis requires this, and it has been repeatedly verified experimentally. Nevertheless, most books and paper implicitly suggest that the pH of the medium is the pH of the cellular interior. Chemiosmosis has changed all that, as indicated here. Even at pH = 7 *for the interior*, there would be only about 250 protons in a bacterium with a radius of 1 μm. In Chapter 7 it was shown that the electrical potential generated by the transport of one unit charge across the membrane of a cell 1 μm in radius is only 2.5 microvolts. Measurements give values of $\Delta\psi$ of about 100 millivolts. In parallel with the analysis in Chapter 7, this means here that $N^*_{H^+}$ is about 4×10^4. How does a bacterium containing only 2.5 protons (pH = 9) manage to extrude enough protons so that it will leave behind, inside, as many as 4×10^4 negative counterions? The answer is that the cytoplasm possesses considerable pH "buffering" capacity. As protons are removed to the exterior by electron transport, ionizable groups on numerous cytoplasmic molecules with pK near pH = 9 release additional protons into the cytoplasm and become the required additional negative counterions. Thus, at respiratory steady state (i.e., the electron transporting steady state of energy metabolism), a bacterium 1 μm in radius with internal pH = 9 maintains a population of an average 2.5 free protons (H_3^+O actually) and builds up a steady-state value of about 4×10^4 negative counterions on the inner surface of the membrane, for a $\Delta\psi$ value of about 100 mV. The number of protons $N^*_{H^+}$ is 4×10^4, *not* 2.5 (or 250 for interior pH = 7).

Many microbiologists and biochemists will be surprised to read that for bacteria grown in a medium of pH 7 the interior pH is as high as 9. Some studies of DNA polymerase activity have shown a pH optimum, in vitro, of about 8 to 8.5, whereas for the growth of the bacteria whose polymerase is being studied the optimum pH for the medium is 7. Until the advent of the chemiosmotic theory, this observation was viewed as an anomaly, not understood. If looked for, other comparable "anomalies" may be found.

Because $N^*_{H^+} \sim 4 \times 10^4$, some strain is placed on the cytoplasmic pH-buffering capacity, so that, in fact, the pH inside the bacterium does increase some during respiration. This leads to a negative ΔpH on the order of 1 to 2 pH units in bacteria and mitochondria. In chloroplasts, which are "inside out," the protons are accumulated inside the chloroplast's inner membrane space (thylakoid *disk* interior). This is a relatively small volume compared with bacterial or mitochondrial volume, and the accumulation of protons in this case leads to a larger pH change, becoming as low as pH = 3.5 inside the disks.

In summary, the following values are representative of $\Delta\psi$ and ΔpH for bacteria, mitochondria, and chloroplasts [10]:

Membrane	$\Delta\psi$ (mV)	ΔpH
Bacteria	70	-2
Mitochondria	140	-1.4
Chloroplasts	~0	+3.5

(12.8)

The value of ~0 for $\Delta\psi$ in chloroplasts is a result of an intrinsically high permeability of chloroplast membrane to the Cl^- anion, which leaks across to kill the $\Delta\psi$. As explained above, chloroplasts, by being "inside out," are ideally suited to compensate for this loss through a larger ΔpH. Using $(RT_{300}/F) = 25$ mV in (12.6) leads to the conversion of (12.8) into

Membrane	Δp (mV)
Bacteria	185
Mitochondria	220
Chloroplasts	-200

(12.9)

These values of Δp are steady-state values for actively respiring cells or organelles. The energy required to move the unit charge of one proton through an electrochemical potential Δp of approximately 200 mV is $1.6 \times 10^{-19} \times 0.2$ J = 3.2×10^{-20} J = 3.2×10^{-13} erg. This is $8k_BT_{300}$. The required energy must be supplied by electron transport energy. The feasibility of this requirement is discussed later in this chapter.

The decomposition of Δp into $\Delta\psi$ and $-2.3(RT/F)\,\Delta$pH permits experimental tests for chemiosmosis that selectively check the $\Delta\psi$ or the $-2.3(RT/F)$ ΔpH term. Two distinct K^+ transport systems are ideally suited for this purpose, valinomycin and nigericin. The valinomycin-induced transport of K^+ was discussed in Chapter 7. It was shown there that the Nernst equation, (4.84) or

(7.13), determines the ratio of K^+ concentration inside and outside a membrane vesicle as a result of a transmembrane electrical potential, $\Delta\psi$, by (12.10), in which the γ terms are usually ignored for simplicity.

$$\Delta\psi = \frac{RT}{F} \ln\left(\frac{\gamma[K^+]_{in}}{\gamma[K^+]_{out}}\right) \tag{12.10}$$

In this equation, the $\Delta\psi$ is generated by electron transport driven proton extrusion that maintains a steady-state value for $\Delta\psi$ as long as respiration proceeds. The K^+ concentrations come to *equilibrium* with this $\Delta\psi$, which is itself a *nonequilibrium* steady-state value. Since the valinomycin transport mechanism has no H^+ dependence, except indirectly through $\Delta\psi$, the $-2.3(RT/F)\,\Delta pH$ term in (12.6) is of no importance in determining the K^+ concentration ratio. The ionophore nigericin, on the other hand, is a membrane-associated antiporter of K^+ and H^+. This ionophore catalyzes the *electroneutral* exchange of K^+ for H^+:

$$K^+_{in} + H^+_{out} \rightleftharpoons K^+_{out} + H^+_{in} \tag{12.11}$$

With steady-state values of $\Delta\psi$ and $-2.3(RT/F)\,\Delta pH$ imposed by electron transport, the overall chemical potential change associated with the exchange can be written

$$\Delta E_{e^-} + \Delta\mu_{H^+} + \Delta\mu_{K^+} = -\,|e|\Delta\psi + |e|\Delta\psi + 2.3k_BT\,\Delta pH + k_BT\ln\left(\frac{[K^+]_{out}}{[K^+]_{in}}\right) \tag{12.12}$$

where the first and third terms on the right-hand side correspond with the negative of (12.5) because the proton is going in the opposite direction here as compared with (12.5); and the second and fourth terms correspond with the K^+ ion going from the interior to the exterior. The negative value of $\Delta\mu_{H^+} = -|e|\Delta\psi + 2.3k_BT\,\Delta pH$ corresponds with the fact that the inward flux of a proton, at the steady state, delivers an amount of energy, $|\Delta\mu_{H^+}|$, for the K^+ transport system. The maintenance of the *steady-state* values of $\Delta\psi$ and $-2.3(RT/F)\,\Delta pH$ costs the electron transport system an amount of energy equal to ΔE_{e^-}, which is just $-\Delta\mu_{H^+} = |\Delta\mu_{H^+}|$. Therefore, $\Delta E_{e^-} + \Delta\mu_{H^+} = 0$ in (12.12) *at steady state*. The K^+ concentration ratio will come to *equilibrium* with the proton steady state when $\Delta\mu_{K^+} = 0$. From (12.12), these conditions imply that

$$\ln\left(\frac{[K^+]_{out}}{[K^+]_{in}}\right) = -2.3\,\Delta pH \tag{12.13}$$

The electroneutrality of this exchange of H^+ for K^+ shows up in (12.12) as the cancellation of $-|e|\Delta\psi$ by $+|e|\Delta\psi$. As long as electron transport maintains $\Delta pH < 0$ and nigericin is available to contribute to the reverse flow of protons, which is essential for the development of a steady state, potassium ions will be transported outwardly until (12.13) is achieved.

Note that the electrical case, exhibited by (12.10), results in a K^+ concentration ratio $[K^+]_{in}/[K^+]_{out} > 1$, whereas the osmotic case, exhibited by (12.13), results in a K^+ concentration ratio $[K^+]_{out}/[K^+]_{in} > 1$.

Both valinomycin and nigericin function by freely diffusing across the membrane. Nigericin does this in two forms: its anionic form will bind K^+ and the electroneutral complex then traverses the membrane; its protonated form is also electroneutral but does not bind K^+. Consequently, it goes one way with K^+ and the other way with H^+. Thermal energy alone makes it move, and the transmembrane protonic electrochemical potential causes the asymmetric accumulation of K^+.

Each K^+ transport system has been repeatedly tested experimentally, and good quantitative results have been achieved. In fact, the K^+-valinomycin system works so well that it has become an important tool for measuring the proton efflux generated by electron transport. Lehninger [8] has recently used this tool to reevaluate the stoichiometry of H^+ extrusion, originally measured using this tool with a different method by Mitchell and Moyle [13] during the late 1960s. These new measurements play a critical role in the analysis of ATP generation that appears later in this chapter.

The mechanism of action of so-called "uncoupling" agents serves to support the chemiosmotic viewpoint as well [1]. Carbonyl cyanide *m*-chlorophenylhydrazone (CCCP) uncouples electron transport from the phosphorylation of ADP. That is, electron transport continues to take place when CCCP is present, but no energy suitable for the generation of ATP, by the phosphorylation of ADP, remains available. This phenomenon has been known for a long time, but only after it was realized that CCCP is an H^+ carrier, and the significance of this fact from the chemiosmosis point of view, was it possible to understand its action. By virtue of its being an ionophoric proton carrier, it causes Δp to leak away across the membrane. Electron transport continues and drives protons across the membrane, but CCCP brings them right back, so that $\Delta p = 0$. This means that there is no Δp to drive the ATPase complex discussed in Chapter 11.

The ATPase complex provides another test of chemiosmosis because it can be dissociated into its F_1 and F_0 portions. The F_1 portion can be removed, leaving the F_0 embedded inside the membrane just as it is in (11.18), (11.20), and (11.22). The F_0 component serves as the proton channel for functional, intact ATPase complexes as depicted in these diagrams. When the F_1 portion is removed, the F_0 becomes a proton leak. Its effect is to dissipate Δp just as CCCP does. Furthermore, dicyclohexyl-carbodiimide (DCCD) has been known for a long time to uncouple electron transport and ATP synthesis. Its action is to block the flow of protons through the F_0 portion of the ATPase complex without effecting electron transport. Consequently, preparations in which the F_1 portion of the ATPase has been removed *and* DCCD is present are found *not* to leak protons.

Phosphate transport across membrane is achieved by a phosphate-H^+ symporter, as depicted in (11.18). Phosphate has three different ionizable hydroxyl groups [see (12.14)]. Therefore, at pH 9, most phosphate is P^{2-} according to

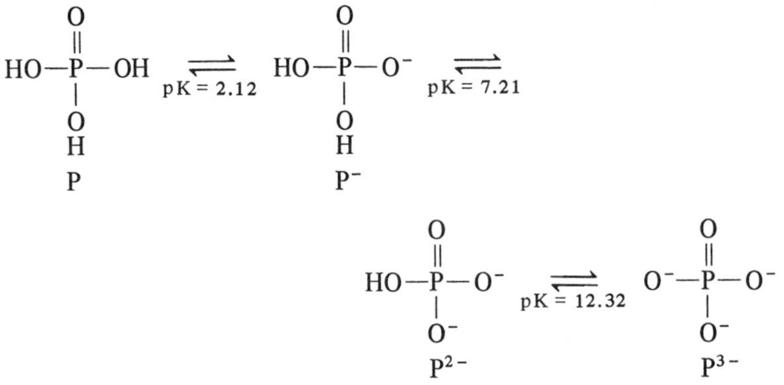

$$\underset{\mathrm{P}}{\mathrm{HO{-}P(=O)(OH){-}OH}} \underset{\mathrm{pK}=2.12}{\rightleftharpoons} \underset{\mathrm{P^-}}{\mathrm{HO{-}P(=O)(OH){-}O^-}} \underset{\mathrm{pK}=7.21}{\rightleftharpoons}$$

$$\underset{\mathrm{P^{2-}}}{\mathrm{HO{-}P(=O)(O^-){-}O^-}} \underset{\mathrm{pK}=12.32}{\rightleftharpoons} \underset{\mathrm{P^{3-}}}{\mathrm{O^-{-}P(=O)(O^-){-}O^-}} \qquad (12.14)$$

the Henderson-Hasselbalch equation (4.32). In the culture medium (pH 7), the dominant ion is P^-. Virtually no P nor P^{3-} are present at pH 9. One known phosphate-H^+ symporter appears to transport P^- and H^+ in an electroneutral process driven by only the $-2.3(RT/F)\,\Delta\mathrm{pH}$ term in Δp. This symporter provides another pathway for the return of protons, which is necessary for a steady state to be achieved. The process can be written, in parallel with (12.11),

$$P^-_{out} + H^+_{out} \rightleftharpoons P^-_{in} + H^+_{in} \qquad (12.15)$$

with an overall chemical potential change associated with this process at steady state given by

$$\Delta E_{e^-} + \Delta\mu_{H^+} + \Delta\mu_{P^-} = -|e|\Delta\psi + |e|\Delta\psi + 2.3k_BT\,\Delta\mathrm{pH} + k_BT\ln\left(\frac{\gamma[P^-]_{in}}{\gamma[P^-]_{out}}\right) \qquad (12.16)$$

This is the parallel of (12.12) *except* that the direction of P^- transport here and K^+ transport there is reversed. The $\Delta\psi$ terms again cancel, this time because P^- carriers a negative charge, and it is again necessary that $\Delta E_{e^-} + \Delta\mu_{H^+} = 0$ for maintaining the steady state. Phosphate equilibrium is achieved when $\Delta\mu_{P^-} = 0$, or

$$\ln\left(\frac{\gamma[P^-]_{in}}{\gamma[P^-]_{out}}\right) = -2.3\Delta pH \tag{12.17}$$

This parallels (12.13) except that P^- accumulates inside, whereas in (12.13) K^+ accumulates outside, again because of the opposite charges in these two cases. Once inside the bacterium or mitochondrial inner membrane (*outside* the chloroplast inner membrane), P^- can serve as one of the components required for pH buffering and give up a proton for proton extrusion, as was discussed above, becoming P^{2-}, which is the precursor for ATP^{4-} generation on the ATPase. The ΔpH across the membrane is such that, if the exterior medium has pH 7, then the interior will be around pH 8 to 9, which favors the P^{2-} state of phosphate according to the pK values in (12.14). The phosphate symporter is inhibited by N-ethylmaleimide (NEM), which does not significantly impair electron transport.

In (11.20) an ATP^{4-}-ADP^{3-} antiporter system is shown. This transport system is driven solely by the $\Delta\psi$ contribution to Δp. As shown, it enables the product of ATP synthesis, ATP^{4-}, to leave the region of synthesis in exchange for the precursor ADP^{3-}. This exchange carries a net charge of $-|e|$ outside the mitochondrial inner membrane and into the cellular cytoplasm. Equilibrium with the steady-state value of $\Delta\psi$ is achieved when

$$0 = -4|e|\Delta\psi + 3|e|\Delta\psi + k_B T \ln\left(\frac{\gamma[ATP^{4-}]_{out}\,\gamma[ADP^{3-}]_{in}}{\gamma[ATP^{4-}]_{in}\,\gamma[ADP^{3-}]_{out}}\right) \tag{12.18}$$

Bacteria have no need for this antiporter since they use ATP where they make it, inside; but mitochondria, as energy organelles, need to transport ATP into the cellular cytoplasm, where it is also needed, and they must transport ADP from the cytoplasm back inside themselves. This mechanism is also needed by the chloroplasts' outer membranes.

Diagram (11.18) indicates that amino acid transport and sugar transport are driven by proton symporters. In *E. coli*, this has been clearly demonstrated for β-galactosides (lactose), galactose, arabinose, glucose-6-phosphate, alanine, glycine, serine, threonine, proline, phenylalanine, cysteine, lysine, and gluconic acid; other molecular species are being added each year. Lysine is exceptional in that it is transported by a "uniporter" that is driven by only the $\Delta\psi$ in Δp and does not involve an H^+ ion. Lysine carries a net positive charge because of its basic residue (see Chapter 1), and this positive charge responds directly to $\Delta\psi$, just like K^+ does when valinomycin is the ionophore. Isoleucine, however, has a net charge of zero but is transported as a cation by a H^+ symporter that is therefore sensitive to the full Δp. Glutamic acid, on the other hand, has a net negative charge and is also transported by a H^+ symporter. The combination is neutral, so that glutamate transport is responsive to only the $-2.3(RT/F)\ \Delta\text{pH}$ term in Δp. Thus, amino acid transport reflects the entire variety of transport specializations possible within the general chemiosmotic mechanism.

In addition to demonstrating a variety of transport processes in intact bacteria, mitochondria, and chloroplasts, the technique of vesicle preparation has permitted experimentation of a more precisely controlled nature. Some vesicles are prepared directly from bacterial or mitochrondrial membranes, whereas other experimenters use wholely artificial lipid vesicles. It is possible to prepare vesicles that are loaded with a potassium salt and into the lipid phase of which may be incorporated specific, purified transport proteins and complexes. By adding valinomycin to these preparations, K^+ efflux is generated and as the K^+ ions come out of the vesicles, they leave behind their counteranions, thereby rendering the vesicle interior electrically negative relative to its exterior. Transport systems incorporated into the vesicle lipid phase which are responsive to $\Delta\psi$ can be driven in this way, whereas the transport systems depending exclusively on $-2.3(RT/F)\ \Delta\text{pH}$ are not. A considerable number of variations on this theme have been tried and have corroborated the general ideas about chemiosmosis.

Perhaps one of the most striking demonstrations of chemiosmosis in synthetic vesicular preparations is the work of Stoeckenius and Racker [14] on reconstituted "purple" membranes. Racker and his associates were the first to isolate the F_1 portion of the ATPase complex in 1960. Stoeckenius has made extensive studies of the bacterium *Halobacter halobium*, a halophilic organism that is happy at *4 molar* salt concentrations. This organism can be either aerobic or "photosynthetic," depending on the availability of oxygen. When oxygen is unavailable, it synthesizes a single polypeptide, called bacteriorhodopsin, that is very similar to the visual pigment rhodopsin. Stoeckenius and his associates showed during the mid-1970s that this protein acts as a light-driven proton pump that on illumination cyclically absorbs light and carries protons from one side of the cellular membrane to the other. Together, Racker and Stoeckenius prepared lipid vesicles into which they incorporated previously isolated bacteriorhodopsin and previously isolated ATPase complex. On illumination, these preparations generated a transmembrane Δp that drove protons back through the ATPase on which the synthesis of ATP from ADP and P was observed. Other bacteriorhodopsin vesicle preparations have been used to demonstrate that light-driven proton transport will also drive amino acid and rubidium accumulation when the appropriate transport systems are incorporated into the vesicles. These systems are schematized in (12.19).

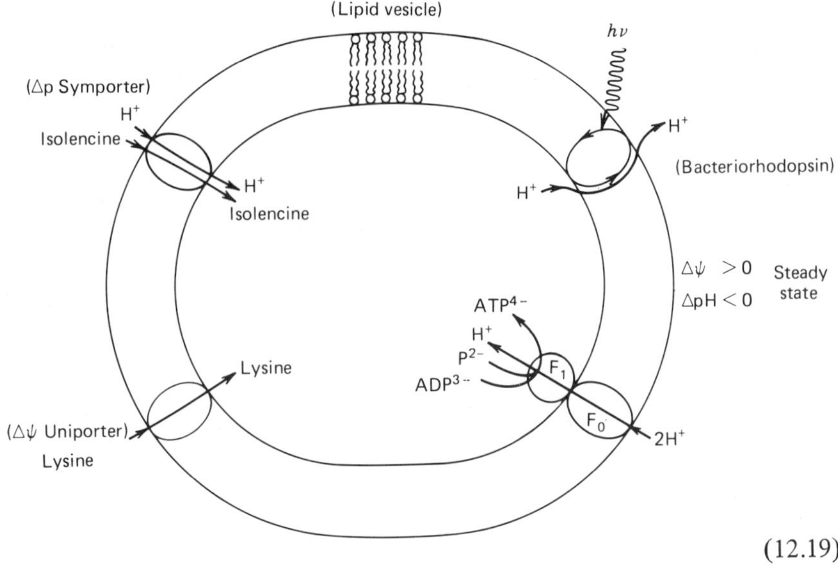

(12.19)

ATP Synthesis

A quantitative assessment of the efficiency of energy transduction from the metabolic energy provided by the electron transport chain to the energy carried by ATP requires great care. There are several considerations that affect substantially the quantitative values obtained and have not been taken into account in books and articles written without appropriate concern for the implications of chemiosmosis. In Chapter 4 some of these issues were already addressed, specifically corrections to standard-state results caused by physiological pH and H_2O concentrations. Even these particular points have to be reassessed in the light of the topological picture of the arrangement of electron transport chain components in the membrane. Other points have not been addressed explicitly so far. These include corrections to the redox potential for $(\frac{1}{2}O_2/H_2O)$ resulting from nonstandard-state cellular oxygen partial pressure; pK values for phosphate, ADP, and ATP as a function of "ionic strength"; and corrections resulting from non-standard-state concentration values at steady state for each redox pair (oxidized/reduced) in the electron transport chain.

Because the most extensive experimental work has been done with mitochondria (from rat liver), it is this case that is discussed in detail. The topological arrangement [15] of the electron transport chain in the mitochondrial inner membrane was given in (11.20). It must be remembered that this picture is still somewhat tentative and may become modified by more studies. As drawn, it exhibits Mitchell's quinone cycle, which has not been firmly established as the correct mechanism. Nevertheless, all of the important considerations required for a careful, quantitative accounting of energy budgeting are necessitated by this scheme.

In (12.8) the typical ΔpH for the mitochondrial inner membrane is listed as -1.4 pH units. This means that, for an exterior pH of 7, the inside pH is 8.4. Therefore, it is *not* correct to use $E^{0\prime}$ values for the redox potentials of the various redox pairs that have been determined exclusively at pH 7. Specifically, the reduction of FMN by NADH, the two reductions of UQ to UQH_2, and the reduction of $\frac{1}{2}O_2$ to H_2O by cytochrome a_3 occur at pH 8.4. On the other hand, the oxidation of $FMNH_2$ by FeS protein and the two oxidations of UQH_2 back to UQ occur at pH 7. All of the other redox reactions involve either FeS proteins or cytochromes, which both contain iron. In Chapter 4, following (4.73), it was pointed out that, for redox reactions *not* involving H^+, there is *no* pH correction. This means that even though the redox potentials for iron-containing redox pairs are given in Chapter 4 in the same tables as other redox pairs that do depend on pH, the $E^{0\prime}$ values for the iron-containing redox pairs are pH-independent. This is most fortunate from the point of view of (11.20) because it is not at all clear how one would determine the pH value *inside* the lipid phase of the membrane where some of these redox pairs react. Fortunately, the redox pairs that are pH-dependent appear to take up and release their protons at H_2O-membrane interfaces, for which pH values can be assigned without difficulty.

While the authors of most modern books and articles do account for pH effects, although not quite properly as was just explained, they rarely correct for H_2O concentration effects. This was also discussed in Chapter 4 following the description of (NAD^+/NADH) in (4.76). While it is a smaller effect than the pH correction, it is nevertheless important to include. This effect is also important for quinone redox reactions, as was also discussed in Chapter 4 in (4.77) through (4.79). In (11.20) this H_2O concentration correction will be required for (NAD^+/NADH), (UQ/UQH_2), and ($\frac{1}{2}O_2/H_2O$) as well. Finally, the redox potential for ($\frac{1}{2}O_2/H_2O$) will also have to be corrected for non-standard-state O_2 partial pressure. The standard-state value is 1 atm, but in cells it may be less than 0.05 atm.

Each of these corrections is worked out below and tabulated in Table 12.1. From (4.74) and (4.75) it follows that for (NAD^+/NADH)

$$E = E^0 - 2.3\frac{RT}{2F}\,\mathrm{pH} - \frac{RT}{2F}\ln\left(\frac{[H_2O]}{1\ \text{molar}}\right) - \frac{RT}{2F}\ln\left(\frac{\gamma[\mathrm{NADH}]}{\gamma[\mathrm{NAD}^+]}\right)$$

$$(E^0 = -0.110\ \mathrm{V}) \quad (12.20)$$

At standard state for γ[NADH], γ[NAD^+], and [H_2O], and with pH = 7, this yields $E = E^0 - 0.210\ \mathrm{V} \equiv E^{0\prime} = -0.320$ V, because of (4.54). This is the value

Table 12.1

Redox Pair	E^0 (volts)	$E^{0'}$ (volts)	$E^{0''}$ (volts)	$E^{0'''}$ (volts)
$(NAD^+/NADH)$	-0.110	-0.320	-0.414	–
$(FMN/FMNH_2)$	+0.298	-0.122	-0.311	-0.227
$(Fe^{3+}\text{-}S/Fe^{2+}\text{-}S)$				
(UQ/UQH_2)	+0.530	+0.110	-0.079	+0.005
(Cyt b^{3+}/Cyt b^{2+})	+0.050	+0.050	+0.050	+0.050
(Cyt c_1^{3+}/Cyt c_1^{2+})	+0.220	+0.220	+0.220	+0.220
(Cyt c^{3+}/Cyt c^{2+})	+0.254	+0.254	+0.254	+0.254
(Cyt a^{3+}/Cyt a^{2+})	+0.290	+0.290	+0.290	+0.290
(Cyt a_3^{3+}/Cyt a_3^{2+})	+0.550	+0.550	+0.550	+0.550
$(\frac{1}{2}O_2/H_2O)$	+1.235	+0.815	+0.535	–

for $E^{0'}$ found in Tables 4.1 and 4.2. For standard-state values of $\gamma[NADH]$ and $\gamma[NAD^+]$, but with $[H_2O] = 55.55$ molar and pH = 8.4, $E = E^0 - 0.252\text{ V} \equiv E^{0''} = -0.414$ V, which defines $E^{0''}$ for use below.

From (4.77) and (4.78) it follows that for (UQ/UQH_2)

$$E = E^0 - 2.3\frac{RT}{F}\text{pH} - \frac{RT}{F}\ln\left(\frac{[H_2O]}{1\text{ molar}}\right) - \frac{RT}{2F}\ln\left(\frac{\gamma[UQH_2]}{\gamma[UQ]}\right) \quad (E^0 = 0.530\text{ V}) \quad (12.21)$$

Notice the factors of 2 in the denominators in (12.20) and (12.21). No error has been made here, as the discussion of (4.79) reveals. At standard state for $\gamma[UQH_2]$, $\gamma[UQ]$, and $[H_2O]$, and with pH = 7, this yields $E = E^0 - 0.420\text{ V} \equiv E^{0'} = 0.110$ V. This is the value of $E^{0'}$ in Tables 4.1 and 4.2. For standard-state values of $\gamma[UQH_2]$ and $\gamma[UQ]$, but with $[H_2O] = 55.55$ molar and pH = 8.4, $E = E^0 - 0.504\text{ V} - 0.105\text{ V} \equiv E^{0''} = -0.079$ V. This applies to the inside of the mitochondrial inner membrane. On the outside, pH = 7 and $E = E^0 - 0.420\text{ V} - 0.105\text{ V} \equiv E^{0'''} = 0.005$ V. An argument very similar to the one just given for (UQ/UQH_2) applies to $(FMN/FMNH_2)$:

$$E = E^0 - 2.3\frac{RT}{F}\text{pH} - \frac{RT}{F}\ln\left(\frac{[H_2O]}{1\text{ molar}}\right) - \frac{RT}{2F}\ln\left(\frac{\gamma[FMNH_2]}{\gamma[FMN]}\right) \quad (E^0 = 0.298\text{ V}) \quad (12.22)$$

This leads to $E^{0'} = -0.122$ V, in agreement with Table 4.1, to $E^{0''} = -0.311$ V for the inside of the mitochondrial inner membrane where pH = 8.4, and to $E^{0'''} = -0.227$ V for the outside of the inner mitochondrial membrane. For $(\frac{1}{2}O_2/H_2O)$, the half-reaction is written $\frac{1}{2}O_2 + 2H^+ + 2e^- = H_2O$, which can be rewritten according to the methods in Chapter 4 as $\frac{1}{2}O_2 + 2H_3^+O + 2C^- = 3H_2O + 2C$. This leads to

$$E = E^0 - 2.3 \frac{RT}{F} \text{pH} - \frac{RT}{F} \ln \left(\frac{[H_2O]}{1 \text{ molar}}\right) - \frac{RT}{2F} \ln \left(\frac{[H_2O]}{[\frac{1}{2}O_2]}\right)$$

$$(E^0 = 1.235 \text{ V}) \quad (12.23)$$

This leads to $E^{0\prime} = 0.815$ V, in agreement with Table 4.1 and Table 4.2, for pH = 7, $[H_2O]$ = 1 molar, and $[\frac{1}{2}O_2]$ = 1 molar. Actually, O_2 concentrations are measured as partial pressure for which the standard-state value is 1 atm. Here a correction to this value is required because mitochondrial partial pressures for oxygen are as low as 0.05 atm. Therefore, instead of the standard-state values in the ln $([H_2O]/[\frac{1}{2}O_2])$ term, the values in ln (55.55/0.05) will be used. For the inside of the mitochondrial inner membrane, this leads to $E^{0\prime\prime} = E^0 -$ 0.504 V - 0.105 V - 0.091 V = 0.535 V.

As was already stated, the iron-containing redox pairs, FeS protein and the cytochromes, are unaffected by either pH or H_2O concentration. One qualification to this statement is required. The protein components of these redox pairs are pH sensitive with respect to their conformations, and this could indirectly produce a pH variation of the redox potential. To the extent that this indirect effect can be ignored, $E^0 = E^{0\prime} = E^{0\prime\prime} = E^{0\prime\prime\prime}$ for each of these iron-containing redox pairs (see Table 4.2) for $E^{0\prime}$ values. All of these results are tabulated in Table 12.1, where the values are compiled from Table 4.1, Table 4.2, and Section J-27 of the *Handbook of Biochemistry* [16]. A reliable value for the FeS protein was unobtainable, although a tentative value of about +0.005 V does fit in nicely.

These different kinds of E^0 ($E^{0\prime}$, $E^{0\prime\prime}$, and $E^{0\prime\prime\prime}$) values reflect pH, $[H_2O]$, and $[\frac{1}{2}O_2]$ values inside and outside the mitochondrial inner membrane. In addition, the steady-state electron transport chain values for each member of each redox pair (e.g., $\gamma[NAD^+]$, $\gamma[NADH]$, $\gamma[UQH_2]$, $\gamma[\text{Cyt } c^{3+}]$) must be determined in order to obtain E values, as distinct from E^0 ($E^{0\prime}$, etc.) values.

Each redox pair in the electron transport chain sequence of redox pairs acts *catalytically*. Only ($NAD^+/NADH$) and ($\frac{1}{2}O_2/H_2O$) are substrate level components. Both $[H_2O]$ and the partial pressure of O_2 are environmentally fixed. The condition $[H_2O]$ = 55.55 molar is not changed by the production of H_2O at the end of the electron transport chain. The partial pressure of O_2 is an ambient condition imposed from a large environment containing the exterior of the membraneous microspherule. On the other end, the ratio $\gamma[NADH]/\gamma[NAD^+]$ inside the bacterium or mitochondrion is determined by the metabolic activity of the energy pathways from glucose to NADH. When glycolysis and the citric acid cycle are functioning at a high level of activity, the ratio $\gamma[NADH]/\gamma[NAD^+] > 1$. Call this ratio R_{NAD}. For $R_{NAD} > 1$, there is a strong reducing potential that drives the electron transport chain. The overall free energy change for the entire pathway is determined solely by its end points, that is, ($NAD^+/NADH$) and ($\frac{1}{2}O_2/H_2O$):

$$H_3^+O + NADH + \tfrac{1}{2}O_2 \rightleftharpoons H_2O + NAD^+ + H_2O$$

$$\Delta E = \Delta E^0 - \frac{RT}{2F} \ln\left(\frac{\gamma[NAD^+][H_2O]^2}{\gamma[NADH][H_3^+O][\frac{1}{2}O_2]}\right) \qquad (12.24)$$

On the basis of the analysis used to obtain Table 12.1, this redox potential change per electron given by (12.24) can be written

$$\Delta E = \Delta E^{0''} + \frac{RT}{2F} \ln R_{NAD} \geqq \Delta E^{0''} \qquad (R_{NAD} > 1) \qquad (12.25)$$

Therefore, for actively metabolizing cells, $\Delta E \geqq \Delta E^{0''}$, where

$$\Delta E^{0''} = 0.535 - (-0.414) = 0.949 \text{ V} \qquad (12.26)$$

This corresponds to a decrease in the Gibbs free energy of $-F\Delta E^{0''}$ per electron, or -21.8 kcal/mole. When the common error of using $\Delta E^{0'}$ instead of $\Delta E^{0''}$ is made, the corresponding value for ΔG is -26.03 kcal/mole, which is about 25% greater. The quantity ΔE is greater than $\Delta E^{0''}$ by $(RT/2F) \ln R_{NAD}$, where $(RT/2F) = 0.013$ V for $T = 298°$K. For $R_{NAD} = 10^3$, $(RT/2F) \ln R_{NAD} = 0.089$ V.

The energy yielded per electron by electron transport from NADH to O_2 is -21.8 kcal/mole. For the complete oxidation of glucose, as was seen in Chapter 8, a total of 24 electrons is transferred, thereby releasing a total of 24 × 21.8 kcal/mole = 523 kcal. This is nearly 100 kcal/mole less than the value in (8.14), which was obtained by using -26 kcal/mole for each electron for a total of 624 kcal/mole of glucose released. [The value in (8.14) is 611 and not 624 because four of the electrons travel from $FADH_2$ instead of NADH, and this yields a little less energy.] The major difference in these values, 624 and 523, is the *non-standard-state values* of [H_2O] and [$\frac{1}{2}O_2$].

The amount of Gibbs free energy required to make ATP from ADP and P depends on pH and ionic strength as well as on the "phosphorylation state ratio," $R_P \equiv (\gamma[ATP])/(\gamma[ADP]\,\gamma[P])$. Ionic strength affects [17] the pK of the ionizable groups on P, ADP, and ATP. For ions of charge $Z_i|e|$, at concentrations C_i (in moles per liter), the ionic strength μ is defined by

$$\mu = \frac{1}{2} \sum_i C_i Z_i^2 \qquad (12.27)$$

The pK is known to vary with μ according to

$$\mathrm{p}K' = \mathrm{p}K - a\sqrt{\mu} + b\mu \tag{12.28}$$

where a and b are characteristics of the specific molecular species involved. For phosphate, the second pK given in (12.14) is 7.21 when $\mu = 0$, but for $\mu = 0.2$, the typical biological value for the ionic strength, with $a = 1.52$ and $b = 1.96$, yields

$$\mathrm{p}K' = 7.21 - 1.52 \times 0.447 + 1.96 \times 0.2 = 6.92 \tag{12.29}$$

Similar effects account for pK' values for ADP and ATP:

$$\mathrm{ATP^{3-} + H_2O \rightleftharpoons ATP^{4-} + H_3^+O}$$
$$\mathrm{p}K = 7.65$$
$$\mu = 0$$
$$\mathrm{p}K' = 7.07$$
$$\mu = 0.2$$

(12.30)

$$\mathrm{ADP^{2-} + H_2O \rightleftharpoons ADP^{3-} + H_3^+O}$$
$$\mathrm{p}K = 7.18$$
$$\mu = 0$$
$$\mathrm{p}K' = 6.80$$
$$\mu = 0.2$$

The three values for pK' with $\mu = 0.2$ are 6.92, 6.80, and 7.07 for P, ADP, and ATP, respectively. Therefore, inside the mitochondrial inner membrane, or inside a bacterium, where ATP is to be made, the proper charge form of the reaction is

$$\mathrm{ADP^{3-} + P^{2-} \rightleftharpoons ATP^{4-} + O^{-}H} \tag{12.31}$$

because the pH is 8 to 9, so that these groups are mostly ionized according to (4.32). If charged states are ignored, the reaction is simply

$$\mathrm{ADP + P \rightleftharpoons ATP + H_2O} \tag{12.32}$$

which is an ordinary dehydration condensation. The correct reaction can also be expressed as

$$ADP^{3-} + P^{2-} + H_3^+O \rightleftharpoons ATP^{4-} + 2H_2O \tag{12.33}$$

which is the summation of (12.31), and the reaction

$$H_3^+O + O^-H \rightleftharpoons 2H_2O \tag{12.34}$$

which was studied in Chapter 4. The ΔG^0 for (12.33) is -1.29 kcal/mole at pH = 0, and at pH = 7 $\Delta G^{0\prime} = +8.24$ kcal/mole. The $\Delta G^{0\prime}$ at pH = 7 for (12.34) is exactly 0 since the *equilibrium* for this reaction is exactly pH = 7. Therefore, the $\Delta G^{0\prime}$ for (12.31) is $+8.24 - 0 = +8.24$ kcal/mole. For (12.33) at pH = 8.4 in the inside of the mitochondrial inner membrane, $\Delta G^{0\prime\prime} = \Delta G^0 + 2.3RT\Delta\text{pH} = -1.29 + 8.4 \times 1.362 = 10.15$ kcal/mole. The $\Delta G^{0\prime\prime}$ for (12.34) at pH = 8.4 is still 0 because an equilibrium for H_3^+O and O^-H is achieved when pH + pOH = 14. Therefore, $\Delta G^{0\prime\prime}$ for (12.31) at pH = 8.4 is +10.15 kcal/mole. Consequently, under non-standard-state conditions at pH = 8.4 and $\mu = 0.2$,

$$\Delta G = \Delta G^{0\prime\prime} + RT_{298} \ln (R_P \times 1 \text{ molar}) \tag{12.35}$$

with

$$\Delta G^{0\prime\prime} = 10.15 \text{ kcal/mole}$$

The value of R_P for mitochondria is determined by a combination of the activity densities of ATP, ADP, and P and by the transport processes shown in (12.17) and (12.18). The two transport processes result in lower values of R_P because they decrease the ratio $\gamma[ATP^{4-}]_{in}/\gamma[ADP^{3-}]_{in}$ and increase $\gamma[P^{2-}]_{in}$. The absolute concentrations matter because for values of 10^{-2} molar for each of $\gamma[ATP^{4-}]$, $\gamma[ADP^{3-}]$, and $\gamma[P^{2-}]$, $R_P = 10^2\ M^{-1}$, whereas for values of 10^{-3} molar, $R_P = 10^3\ M^{-1}$. These two values of R_P provide $RT_{298} \ln (R_P \times 1\ M)$ values of +2.72 and +4.08 kcal/mole, respectively. It would be very useful to know an accurate value for R_P for *steady-state* conditions, so that the cost of ATP could be accurately determined. Probably, $R_P > 1$ under nearly all biological conditions, so that $\Delta G^{0\prime\prime}$ measures the *least* cost for ATP. A cost of 10 kcal/mole corresponds with an electrical potential of 0.422 V for a unit charge.

From (12.26) and (12.35) it can be concluded that, for each *pair* of electrons that traverses the electron transport chain, at least $1.898\ \text{V} \times 1.6 \times 10^{-19}\ \text{C} = 3 \times 10^{-12}$ erg of energy is generated; and at least 10 kcal/mole, or 7×10^{-13} erg, is required for each molecule of ATP made. With R_P as large as $10^3\ M^{-1}$, the cost for a molecule of ATP is 9.93×10^{-13} erg. Thus, each pair of electrons that goes down the electron transport chain could supply enough energy for three ATP molecules to be made. Since it takes two electrons to reduce $\frac{1}{2}O_2$ to H_2O, this implies a P/O ratio of 3 for mitochondria. This is indeed the observed value. However, the electron transport energy must first be converted into the transmembrane proton electrochemical potential, which is then used to drive ATP synthesis on the ATPase complex.

In the early days of research on ATP synthesis (1950–1970) *chemical coupling* (see Chapter 8), which explains the phosphorylation of ADP to ATP during glycolysis, was widely believed to be the mechanism for energy transduction from the electron transport chain to ATP [18]. It was natural then to divide up the electron transport chain into three segments, each of which would be sufficiently exergonic to drive the generation of ATP. Referring to (11.20) and to Table 12.1, the three segments suggested, and also found as subcomplexes during isolation procedures, were

$$\begin{gathered} \text{NADH} + \text{H}^+ \longrightarrow \text{UQH}_2 \text{ (inside)} \\ 2 \times \Delta E^{0''} = 2 \times 0.335 \text{ V} = 0.670 \text{ V} \\ \text{UQH}_2 \text{ (inside)} \longrightarrow 2 \text{ cytochrome } c_1 \text{ (outside)} \\ 2 \times \Delta E^{0''} = 2 \times 0.299 \text{ V} = 0.598 \text{ V} \\ 2 \text{ cytochrome } c_1 \text{ (outside)} \longrightarrow \text{H}_2\text{O (inside)} \\ 2 \times \Delta E^{0''} = 2 \times 0.315 \text{ V} = 0.630 \text{ V} \end{gathered} \tag{12.36}$$

(Remember that for cytochromes $E^{0'''} = E^{0''}$.) Clearly, each of these segments readily accounts for the minimum requirement for ATP synthesis, 0.422 V (~10 kcal/mole) and would also account for the energy requirements for ATP synthesis for R_P values as high as $10^3\ M^{-1}$.

In spite of this nice fit in energy values and requirements, the mechanism of ATP synthesis does not involve chemical coupling, but instead is a consequence of *chemiosmotic coupling*. In (11.20) the Mitchell model for chemiosmotic coupling is represented. It attempts to account for the indirect transduction of electron transport energy into ATP energy through the intermediary transmembrane electrochemical potential for protons. In (12.8) typical mitochondrial values for $\Delta\psi$ and ΔpH are given, and together they provide a $\Delta p = 0.220$ V according to (12.9). Therefore, during ***steady-state*** energy metabolism, it costs $0.220 \text{ V} \times 1.6 \times 10^{-19} \text{ C} = 3.52 \times 10^{-13}$ erg to get one proton across the membrane. Since electrons and protons carry charges of identical absolute magnitude, it is sufficient to match up *volts* to account for *energy*. In (11.20) two protons cross the membrane between NADH and FeS protein. From Table 12.1, the potential change for two electrons going from NADH (inside) to $FMNH_2$ (outside) is $2[-0.227 - (-0.414)] = 0.374$ V. This does not quite provide enough energy for two protons, that is, 0.440 V. However, electrons will be drawn on through the FeS proteins to UQH_2, and this tendency of the electrons will pull protons across via $FMNH_2$. As was seen in (12.36), 0.670 V is available from NADH to UQH_2. From UQH_2 to two molecules of cytochrome c_1, the change in potential given in (12.36) is 0.598 V. In (11.20), however, the clever UQ cycle of Mitchell drives four protons across. This corresponds with a potential of $4 \times 0.220 = 0.880$ V. Once again, it is the tendency of the electrons to keep on going that drives the protons across. Finally, the electrons release additional energy between two molecules of cytochrome c_1 and H_2O, corresponding to a potential change of 0.630 V according to (12.36). In (11.20) Mitchell does not imagine that any protons cross the membrane in this segment of the electron transport chain. Instead, this segment's energy is used to pull electrons through in order to bias the tendency of protons to come across the membrane via the UQ cycle, which was just seen to be insufficient by itself to get four protons across. Consequently, it makes more sense to see whether the total energy released by a pair of electrons traversing the electron transport chain altogether will provide the energy needed to get across the membrane all of the protons that the model requires. In this case the energy available corresponds to 1.898 V for a pair of electrons, and the energy needed for six protons corresponds to 1.320 V. This appears to fit nicely, with excess energy to spare.

Everything is *not* self-consistent, however. For every pair of electrons going from NADH to H_2O, three molecules of ATP are synthesized according to measurements with mitochondria. The six protons driven across the membrane would have to account for this, two at a time. That is, it would be necessary to make one ATP molecule using the energy provided by just two protons coming back inside through the ATPase complex. At steady state, this provides only 0.440 V, whereas ATP costs at least 0.422 V and probably as much as 0.600 V when $R_P \sim 10^3 M^{-1}$. Moreover, it is now known, as depicted in (11.20), that three protons are required by the ATPase complex for each ATP made.

There are several explanations for these discrepancies. The main reason is that more than six protons are driven across the membrane during the transport of two electrons from NADH to H_2O. By the clever use of inhibitors and special electron donors, it is possible to measure the number of protons driven across the membrane for each pair of electrons that is transported along a specific segment of the electron transport chain. For the three segments described in (12.36), Mitchell originally measured values of two protons per segment, although his model (11.20) distributes two of them in the first segment and four in the second segment, with none in the third segment. Lehninger discovered recently systematic errors in Mitchell's technique and has reported values of four protons *per* segment, distributed as four for *each* segment of the chain. Now, the first problem with these new values is that a total of 12 protons will be driven across the membrane for each pair of electrons going from NADH to H_2O. At Δp = 0.220 V, this requires 2.640 V from the pair of electrons. The electrons provide only 1.898 V, however! Two facts mitigate this particular difficulty. For the system used by Lehninger in his measurements, he reports a Δp of 0.150 - 0.190 V, not 0.220 V. Moreover, he believes that only about 20% of this is $-2.3(RT/F)\ \Delta\text{pH}$, as compared with the 36% for $-2.3(RT/F)\ \Delta\text{pH}$ in (12.8) and (12.9). If the value $\Delta p = 0.180$ V with $-2.3(RT/F)\ \Delta\text{pH} = 0.036$ V is assumed, then $\Delta\text{pH} = 0.6$ rather than the 1.4 used throughout the analysis leading to Table 12.1. Therefore 12 protons would cost only 2.16 V, while the energy released between NADH and H_2O for a pair of electrons would *increase* to 1.946 V because the pH of the mitochondrial inner membrane interior is 7.6, which changes $E^{0''}$ for (NAD^+/NADH) to -0.390 V and $E^{0''}$ for ($\frac{1}{2}O_2/H_2O$) to 0.583 V, so that 2[0.583 - (-0.390)] = 1.946 V. This still leaves a discrepancy of 0.214 V, but this is a great improvement over the discrepancy of 0.742 V above. An increase in partial pressure of O_2 over the 0.05 atm assumed for Table 12.1 would easily account for this residual difference by increasing $E^{0''}$ for ($\frac{1}{2}O_2/H_2O$) well beyond 0.600 V. A second problem with Lehninger's new values of four protons for each segment is that there is no simple mechanism like that depicted in (11.20) to account for more protons getting across the membrane. In fact, a comparison of (11.17) with (11.20) shows that the quinone intermediate UQH_2 is most easily envisaged to carry only two protons for each pair of electrons. The UQ cycle in (11.20) is really rather contrived and has not been verified experimentally.

Besides chemical coupling and chemiosmosis, a third mechanism [18] has been studied over the years. This is the *conformational coupling* model, and it is modeled after the Bohr effect in hemoglobin. The basic idea is that, as electrons flow down the electron transport chain, particularly along the cytochrome segment, the alternating oxidation and reduction steps create alternating conformational changes that are communicated to *other proteins* that do not directly transport electrons but do transport protons from inside to outside as they alternate from one conformation to the other. This is pictured in (12.37). By associating enough of these conformationally driven proteins with the electron transport chain, enough protons could be driven across the membrane. Some such mechanism appears obligatory!

It may be objected that, because the ATPase only uses three protons to make ATP, only nine protons need to be extruded for each pair of electrons and then three molecules of ATP could be made, as observed. The three protons would bring 0.540 V (3 × 0.180 V) of energy for ATP synthesis, which is about right. The fourth proton is needed, however, for the P^- symporter in (11.18), which is also a component of the mitochondrial inner membrane [not depicted

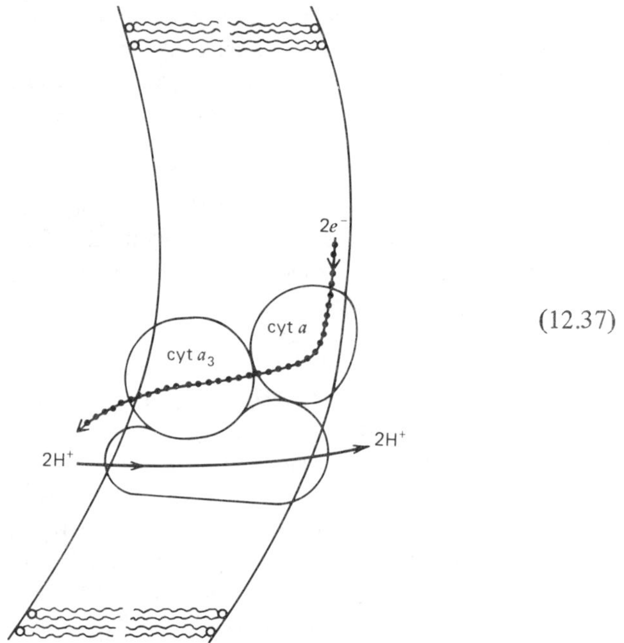

(12.37)

in (11.20)]. Note that the ATP^{4-}-ADP^{3-} antiporter depends only on $\Delta\psi$ and does not put additional demands on protons for transport. Therefore 12 protons really are required for the overall process, as depicted in (12.38).

It is instructive to consider in some detail the reasons why Lehninger [7–9] measured more protons per pair of electrons for each segment of the transport chain than did Mitchell. The tool they used, described earlier, involved valinomycin and potassium. Mitochondria are preincubated anaerobically in a KCl medium with a respiratory substrate to serve as an electron donor for the transport chain. Valinomycin is added, and a pulse of oxygen is injected into the system. The oxygen stimulates electron transport and the extrusion of protons. The electrical potential that begins to build up draws K^+ ions in via the valinomycin carrier complex. In this way all of the Δp for protons is converted into the $-2.3(RT/F)\ \Delta pH$ term. The oxygen pulse is brief, so that after the initial, rapid efflux of protons, there follows a relatively slow, passive reentry of ejected protons back inside the mitochondria with a return to the original pH, both inside and outside. By measuring this backflow relaxation, it is possible to extrapolate the data to zero time—that is, the time at which oxygen was added—and thereby determine how many protons come out during the oxygen pulse. Typical data are shown in (12.39), with ΔH_0^+ denoting the initial amount of protons ejected. Lehninger had no difficulty repeating Mitchell's original experi-

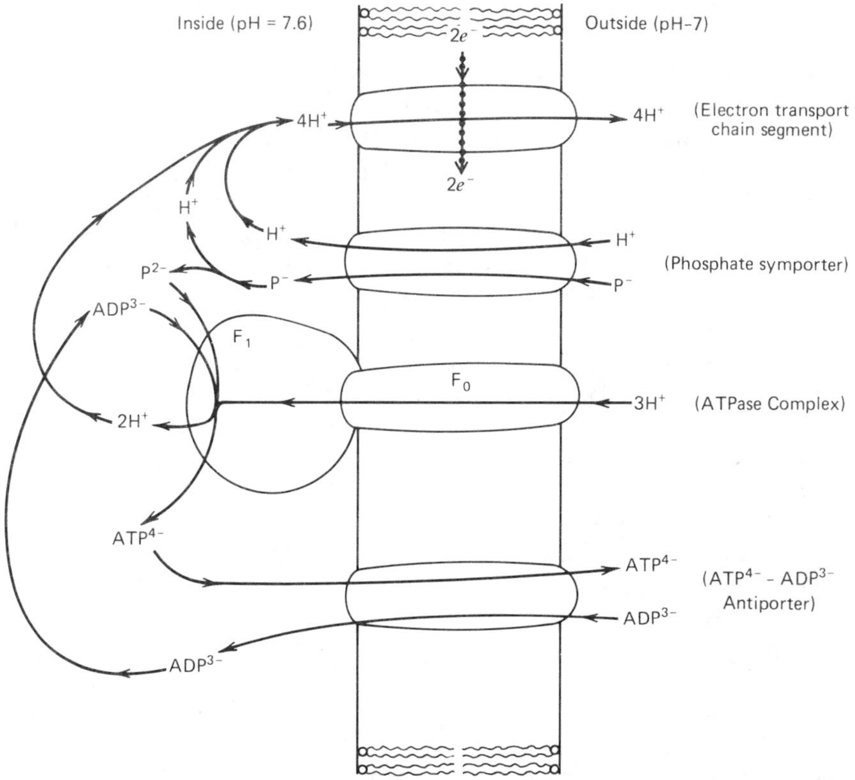

Inside (pH = 7.6)
Outside (pH-7)
2e⁻
4H⁺
4H⁺
(Electron transport chain segment)
2e⁻
H⁺
H⁺
H⁺
(Phosphate symporter)
P²⁻
P⁻
P⁻
ADP³⁻
F₁
F₀
3H⁺
(ATPase Complex)
2H⁺
ATP⁴⁻
ATP⁴⁻
(ATP⁴⁻ - ADP³⁻ Antiporter)
ADP³⁻
ADP³⁻

(12.38)

$t_{1/2}$
ΔH_c^+
ΔH_0^+
30
60
O_2
T, sec
Intercept = log ΔH_0^+
log ΔH_c^+
30
60
T, sec

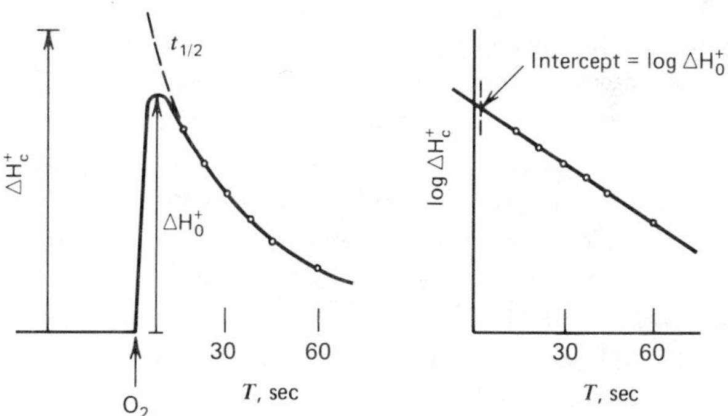

Reproduced, with permission, from Ref. 6.

(12.39)

ments and obtained identical results. $2H^+$ per segment for each pair of electrons. However, Lehninger also noticed and proved that the method employed led to an underestimation of the correct value, subsequently shown to be more nearly $4H^+$ per segment for each pair of electrons. Two systematic errors are involved. As protons initially come out, as a result of the oxygen pulse, there is a *very rapid* partial reuptake of protons through the phosphate-proton (P^--H^+) symporter, which was discussed earlier. This causes the apparent value of ΔH_0^+ to be lower than it really is. By using the specific P^--H^+ symporter inhibitor N-ethylmaleimide (NEM) and then repeating the measurements, this backflow of H^+ was eliminated, and Lehninger got a value of $3H^+$ per segment for each pair of electrons. The second source of systematic error has to do with an extremely rapid and irreversible increase in the volume of mitochondria on injection of the oxygen pulse. This is largely due to chloride ion permeability, which increases as protons are extruded and potassium is drawn inside the mitochondria. With this change, the intrinsic permeability to protons themselves increases, causing a rapid reentry of protons. By doing the measurements in media containing relatively little potassium salt and an abundance of either LiCl or sucrose instead, the swelling of mitochondria and the concommitant H^+ permeability are greatly reduced. Values of 3.4 to 3.7 H^+ per segment for each pair of electrons were then obtained. Ultimately, Lehninger developed a new "steady-state" method that does not use an oxygen pulse. The mitochondria are kept in an aerobic state all of the time. Electron flow is initiated by adding the electron donor. Simultaneous recordings with a glass electrode for H^+ and an oxygen electrode for O_2 are made. N-Ethylmaleimide is used as well as a LiCl medium to prevent the systematic errors discussed above. Values very close to $4H^+$ per segment for each pair of electrons were obtained.

It should be noted that the value $4H^+$ per segment is in agreement with much earlier work, on K^+ and Ca^{2+} transport, done at about the same time as Mitchell's original measurements which at that time was in conflict with Mitchell's values of $2H^+$ per segment.

A detailed analysis of energy budgeting in bacteria and chloroplasts [19] paralleling that just presented at length above for mitochondria cannot be given here. These other two systems have not yet been as thoroughly studied. Nevertheless, it is clear that pH corrections other than the usual pH = 7 correction and related considerations will be needed to balance the energy yielded by electron transport against the energy required for ATP synthesis. At a qualitative level of analysis, it is clear from (11.22) that chloroplasts can generate only enough protons for one ATP made for each pair of electrons traversing the electron transport chain. Furthermore, the transmembrane proton electrochemical potential of 200 mV is consistent with the requirement for three protons activating the ATPase, as indicated in (11.22). Measurements are consistent with these observations. In bacteria, (11.17) implies that two molecules of ATP can be made for each pair of electrons. Measurements are not inconsistent with this.

There remains one final question regarding the synthesis of ATP, for which very little solid experimental evidence exists. How do the protons that activate the ATPase complex perform this function? The structure of the F_1 complex in the ATPase is being vigorously studied, but it may be several years before enough is known about it to be able to elucidate the mechanism of ATP synthesis on this complex. So far, it is known that the F_1 complex has a molecular weight of about 285,000 and consists of at least five different types of polypeptide chains, some of which occur in the complex more than once. Not much more is yet known. In (11.18), (11.20), (11.22), and (12.38), the charge disparity on going from P^{2-} and ADP^{3-} to ATP^{4-} has been depicted as accomplished by using one of the protons, which comes through the F_0 and stimulates ATP synthesis at the active site, to reduce the net charge by one unit. It is possible, though entirely hypothetical at present, that the other two protons (one proton in the bacteria case) are used to momentarily neutralize the negative charges on either P^{2-} or ADP^{3-} so that these two species experience less coulombic repulsion at the active site, where they combine to form ATP^{4-}. Several subtle points are associated with this idea. At an ionic strength of $\mu = 0.2$, the Debye-Hückel shielding length is only a few angstroms. This means that coulombic effects are very short in range. However, for bacteria and mitochondria, the elevated interior pH means that many molecular species are negatively charged, as are P^{2-} and ADP^{3-}. There are positive countercations around, such as Mg^{2+} and Na^+, which will associate with P^{2-} and ADP^{3-} to partially neutralize them. In addition, amino acid residues on lysine and arginine are positive even at pH = 9, and if they are in the active site they can help to neutralize P^{2-} and ADP^{3-}. In order to react, however, P^{2-} and ADP^{3-} must get as close to each other as about 1 Å, a distance at which the Debye-Hückel shielding is quite small, and the high dielectric constant of H_2O is not really appropriate because no H_2O molecules could fit in between P^{2-} and ADP^{3-} at a distance of 1 Å apart. At 1 Å the electrostatic *repulsion* of one minus charge from another minus charge is 23×10^{-12} erg. This is about $600 k_B T_{300}$. Therefore thermal energy alone is very unlikely to bring two unit charges that close together, not to mention that here the charges are 2- and 3- instead. If, however, the protons that come through the ATPase complex momentarily neutralize the P^{2-} (bacteria) or the ADP^{3-} (mitochondria and chloroplasts), then thermal motions can bring the other reaction partner as close as necessary because one of the partners is now neutral. After reaction to form ATP, one proton remains to yield ATP^{4-} while the others (one for bacteria and two for mitochondria and chloroplasts) are released into the medium. If this hypothetical mechanism is correct, then it can be said that the protons that come through the ATPase complex drive ATP synthesis by overcoming an "activation barrier" resulting from electrostatic repulsion. The problem of this mechanism deserves considerable experimental attention, which presently is largely devoted to determining the structure of the F_1 active site.

The model picture of ATP synthesis presented in this chapter suggests that, for each pair of electrons transported down the redox chain, three molecules of ATP can by synthesized in mitochondria. In Chapter 8 it was shown in (8.14) that 12 pairs of electrons are generated for electron transport by each molecule of glucose that begins glycolysis. Thus 36 ATP molecules can be made under optimal conditions. In addition, it was seen in (8.11) and (8.13) that two molecules of ATP and two molecules of GTP, respectively, are also made by glycolysis and the citric acid cycle, respectively, for each glucose molecule. Therefore a total of 40 "high-energy" phosphate bonds is generated for each glucose oxidized to completion. At roughly 10 kcal per mole of phosphate bond and roughly 600 kcal per mole for the oxidation of glucose [this value is 600 and not 700 for reasons given in the discussion following (12.26)], the efficiency of conversion is 66%. If the transmembrane proton electrochemical potential is not exclusively used for ATP synthesis and is instead used for the transport of solutes and ions, then the ATP efficiency will be less.

The field of bioenergetics has made great strides in the last 50 years [20, 21]. Even the progress to date [21], compared with the state of development a little over a decade ago [20], is amazing. As with other scientific revolutions, we may expect this one to continue to expand rapidly for another decade or two.

References

1 F. M. Harold, "Membranes and Energy Transduction in Bacteria," in *Current Topics in Bioenergetics*, Vol. 6, D. R. Sanadi, editor (Academic Press, New York, 1977).

2 R. N. Robertson, *Protons, Electrons, Phosphorylation and Active Transport* (Cambridge University Press, London, 1968).

3 P. Mitchell, "Coupling of Phosphorylation to Electron and Hydrogen Transfer by a Chemiosmotic Type of Mechanism," *Nature*, **191**, 144–148 (1961).

4 T. S. Kuhn, *The Structure of Scientific Revolutions*, 2nd ed. (University of Chicago Press, Chicago, 1970).

5 P. Mitchell, "Possible Molecular Mechanisms of the Protonmotive Function of Cytochrome Systems," *J. Theor. Biol.*, **62**, 327–367 (1976).

6 A. L. Lehninger, "Proton Transport and Charge Separation Across the Mitochondrial Membrane Coupled to Electron Flow," in *Protons and Ions Involved in Fast Dynamic Phenomena* (Elsevier Scientific Publishing Company, Amsterdam, 1978).

7 A. Vercesi, B. Reynafarje, and A. L. Lehninger, "Stoichiometry of H^+ Ejection and Ca^{2+} Uptake Coupled to Electron Transport in Rat Heart Mitochondria," *J. Biol. Chem.*, **253**, 6379–6385 (1978).

8 B. Reynafarje and A. L. Lehninger, "The K^+/Site and H^+/Site Stoichiometry of Mitochondrial Electron Transport," *J. Biol. Chem.*, **253**, 6331–6334 (1978).

9 T. Pozzan, F. DiVirgilio, M. Bragadin, V. Miconi, and G. F. Azzone, "H^+/Site, Charge/Site, and ATP/Site Ratios in Mitochondrial Electron Transport," *Proc. U.S. Natl. Acad. Sci.*, **76**, 2123–2127 (1979).

10 F. M. Harold, "Conservation and Transformation of Energy by Bacterial Membranes," *Bacteriol. Rev.*, **36**, 172–230 (1972).

11 G. Hauska and A. Trebst, "Proton Translocation in Chloroplasts," in *Current Topics in Bioenergetics*, Vol. 6, D. R. Sanadi, editor (Academic Press, New York, 1977).

12 R. D. Simoni and P. W. Postma, "The Energetics of Bacterial Active Transport," *Ann. Rev. Biochem.*, **44**, 523–554 (1975).

13 P. Mitchell and J. Moyle, "Estimation of Membrane Potential and pH Difference Across the Cristae Membrane of Rat-liver Mitochondria," *Eur. J. Biochem.*, **7**, 471–489 (1969).

14 E. Racker and W. Stoeckenius, "Reconstitution of Purple Membrane Vesicles Catalyzing Light-Driven Proton Uptake and Adenosine Triphosphate Formation," *J. Biol. Chem.*, **249**, 662–663 (1974).

15 J. W. DePierre and L. Ernster, "Enzyme Topology of Intracellular Membranes," *Ann. Rev. Biochem.*, **46**, 201–262 (1977).

16 *Handbook of Biochemistry* (Chemical Rubber Company, Cleveland, Ohio, 1968).

17 D. E. Metzler, *Biochemistry* (Academic Press, New York, 1977).

Chapter 3 has excellent coverage of ionic strength effects on pK' values for ADP, ATP, and P as well as reliable values of the Gibbs free energy of ATP hydrolysis.

18 P. D. Boyer, B. Chance, L. Ernster, P. Mitchell, and E. Racker, and E. C. Slater, "Oxidative Phosphorylation and Photophosphorylation," *Ann. Rev. Biochem.*, **46**, 955–1026 (1977).

An illuminating account of the intellectual antagonism between the leaders in the field of bioenergetics is to be found in these short essays.

19 M. Avron, "Energy Transduction in Chloroplasts," in *Ann. Rev. Biochem.*, **46**, 143–155 (1977).

20 H. M. Kalckar, *Biological Phosphorylations: Development of Concepts* (Prentice-Hall, Englewood Cliffs, N.J., 1969).

This volume contains reprints of many of the classic papers in bioenergetics.

21 P. Mitchell, "Keilin's Respiratory Chain Concept and Its Chemiosmotic Consequences," *Science*, **206**, 1148–1159 (1979).

This is Mitchell's Nobel Prize lecture, delivered on December 8, 1978. In it Mitchell explicitly rejects the stoichiometry values obtained by Lehninger. I believe Mitchell is wrong in doing so. The introduction of the lecture contains an important historical account of the evolution of respiratory chain and chemiosmotic ideas.

PART

3

Origins and Evolution

"The Fox knows many things but the Hedgehog knows one great thing."

Archilochus

Part 3 of this book is a speculative discussion of the origin and evolution of an uroboros. In contrast with Parts 1 and 2, which contained generally accepted facts, Part 3 is highly speculative at the molecular level and represents the author's personal view. Most of the speculations presented have no experimental support at this time. Nevertheless, the principles and concepts are derived from the material presented in Parts 1 and 2. It is argued that the conclusions reached are to a large extent inevitable within the conceptual framework already established. The objective of Part 3 is to demonstrate that the physical and biological principles established in the first two parts of the book provide a *complete* basis for a biophysical, molecular "logic" that can be used deductively.

Chapter 13 concerns the problem of the origin of an uroboros and particularly the origin of energy metabolism. Chapter 14 concerns the mystery of the second code, that is, the origin and evolution of protein biosynthesis. Chapter 15 discusses the origin of membranes and cells. It is the author's conviction that the importance of the interface between the physical and the biological sciences (from the point of view of fundamental laws–see Reference [1] in Chapter 1) is not so much with respect to experimental techniques (x-ray diffraction, centrifugation, electron microscopy, chromatography, spectroscopy, nuclear magnetic resonance, etc.), but with respect to questions of origins and evolution. There is no question that modern molecular biology is deeply dependent on numerous physical techniques, originally invented by physicists. These are, however, applications and refinements of established principles. That biology may give rise to profoundly novel physical insights is most readily seen when problems of origins and evolution are broached. These problems involve *self-assembly*, *energy flow ordering*, and the *initiation of uroboric feedbacks*.

The major problem regarding the origin of the "contemporary" uroboros described in Chapters 10, 11, and 12 concerns its complexity. The machinery for protein biosynthesis, including DNA replication and transcription, involves many macromolecular components. Similarly, the energy metabolism pathways, including the membrane-associated components, involve many macromolecular components. The problem is to find a much simpler protein biosynthesis machinery *and* a much simpler energy metabolism for a precursor uroboros that, nevertheless, is richly enough endowed to be able to evolve into the contemporary uroboros. A hypothetical primitive uroboros is presented in the chapters of Part 3.

CHAPTER 13

Energy Metabolism

In Parts 1 and 2 it has been seen that molecular biology can be described in all its details in terms that make use of chemistry and physics without the introduction of new laws. To be sure, polynucleotides and proteins are characteristically biological macromolecules, but their polymeric structure is of a type not unknown today in nonbiological, industrial chemistry. There is no direct parallel to electron transport and proton translocation through membranes in the solid-state microelectronics industry, but nevertheless, electrons and protons are well recognized in physics, and transistors and semiconductors perform functions not all that dissimilar. What sets biology apart from the rest of the physicochemical sciences are the multicomponent complexions of processes with their "lifelike" properties, and not novel physical laws. The biological novelties begin to be discernible when examples of energy flow and self-assembly are contemplated. In this chapter, several such examples are discussed, starting from clearly nonbiological cases and building up toward a truly biological uroboros.

A commonly used example of an energy flow feedback process is the "water cycle" of the earth. In this example, the energy of sunlight is transduced by the earth's atmosphere, hydrosphere, and lithosphere into heat energy by a cyclic process that gives rise to an orderly flow of matter, in this case H_2O. Sunlight is absorbed by the surface of the lithosphere and hydrosphere which convert it into warmth. Water molecules absorb the heat, and some of them enter the H_2O vapor phase of the atmosphere. At elevated strata of the atmosphere, the H_2O vapor condenses into clouds, droplets form, and finally rain falls back to the surface of the earth, completing the water cycle. The condensation of vapor releases the heat energy as the latent heat of condensation, and this heat partly radiates into space. A steady state is eventually achieved in which there is a continuous transformation of sunlight into heat via the cyclic flux of huge amounts of water from ocean vapor into clouds, rain, and then back down rivers into the oceans. Admittedly, the earth's water cycle is quite irregular and complicated as a result of the earth's rotation and irregular distribution of oceans and land masses. It is possible to couple to this cycle by building hydroelectric dynamos on rivers and thereby drive a variety of otherwise totally unrelated energy-requiring processes. The important feature about the water cycle for our purposes is that the mere presence of sunlight inevitably generates H_2O vapor and et sequelae, the entire water cycle. In this case, there is no problem with *initiating* the cycle.

As a second example, consider a wax candle. Faraday [1] explained how the burning of a candle proceeds, once started, as an *autocatalytic* process. The heat of the flame melts the wax, which then vaporizes, rises up into the flame, and keeps the flame going. The flame requires oxygen from the ambient air, and the air currents generated by the flame draw in oxygen, along with the vaporized wax, and drive off (up) the inhibitory byproducts of burning, especially H_2O. Once started, the flame burns brightly until the entire candle is consumed. Before its wick is lighted, the candle and the oxygen of the ambient air comprise a metastable, latent, autocatalytic system. This system is poised for burning, but does not of itself burn. In the water cycle, the mere presence of the energy source, sunlight, automatically drives the cycle. With the candle, and its surrounding oxygen, an energy source is present in the latent energy of combustion of wax with oxygen, *but* the autocatalytic burning cycle does *not* proceed without the specific intervention of an outside agency of ignition, usually a lighted match. It is, of course, possible, though highly unlikely, that spontaneous combustion might occur, igniting the candle. Thus, with a candle, there is a problem in *initiating* the burning cycle, though the problem is not very difficult if matches are available.

Another example of autocatalytic energy flow was discussed in Chapters 5 and 8, where glycolysis was presented. Diagram (5.12) exhibits this system in which the energy latent in glucose is transduced into ATP energy while glucose is converted into lactic acid. Although clearly drawn from biology, it is not an example of an uroboric feedback, but merely a convenient example of autocatalysis, as explained in Chapter 5. It fits into this chapter as an unusually good example of the initiation problem. If all the enzymes of this pathway are prepared in a sol-gel system in a test tube to which glucose, ADP, and phosphate (P) are added, the latent free energy to drive the transduction of energy from glucose to adenosine triphosphate (ATP) is present, but no ATP is present to prime the pathway at its first and third steps. If two molecules of ATP could be introduced into these steps, the pathway would become initiated and the autocatalytic production of ATP would ensue. This is analogous to using a match to light a candle.

Without ATP, however, only a chance event–the spontaneous formation of ATP from ADP and P–could lead to initiation. This possibility is, in fact, not that remote. The equilibrium of an aqueous mixture of ADP and P *requires* that some ATP be formed, albeit only a small amount because $\Delta G^{0\prime\prime} = +10$ kcal/mole. Therefore some ATP will spontaneously form, and the pathway will be initiated–in this example by a simple fluctuation. The fluctuation involves the coming together of only two molecues, ADP and P, and thermodynamic equilibrium favors the formation of some ATP. (This thermodynamic equilibrium involves a relatively short time scale in which the reaction ADP + P $\rightleftharpoons$ ATP + H_2O comes to equilibrium. On a *longer time scale*, there is the ultimate equilibrium of the complete hydrolysis of all ADP and ATP to ribose, adenine, and phosphate.)

By way of contrast, the carbon cycle, also described in Chapters 5 and 8, serves as another example of an autocatalytic energy transduction with special initiation requirements. The details of this pathway are given in (5.13). In this case, the source of energy is ATP and NADPH. This energy is transduced by the cycle into carbohydrate energy in the form of sugars with three to seven carbon atoms. Their precursors are CO_2 and H_2O. If all of the enzymes are in a test tube to which ATP, NADPH, CO_2, and H_2O are added, the system is poised for energy transduction. Its autocatalytic nature, however, requires at least three molecules of glyceraldehyde-3-phosphate. If these are added, the pathway is initiated and proceeds to transduce energy and manufacture carbohydrate. The chance formation of three molecules of glyceraldehyde-3-phosphate from nine molecules each of CO_2 and H_2O, as well as three molecules of phosphate, either from ATP or its degradation products, is not forbidden thermodynamically. It is much less probable than the requirement of the preceding example in which two molecules of ATP formed spontaneously from two molecules each of ADP and P. The occurrence of this in the carbon cycle system is not more likely than the spontaneous ignition of a candle wick.

These examples illustrate that in some cases the mere presence of an energy source generates a cyclic process, whereas in other cases special and even severe requirements may exist for the initiation of the cycles, even though all other essential ingredients are present and the process is latent in them. It is these latter cases that are indicative of the problem faced when the origin of energy metabolism is contemplated.

The Origin of Energy Metabolism

The view of energy metabolism given in Part II of this book corresponds with what is known about contemporary microorganisms like *E. coli*. The collection of anabolic and catabolic energy pathways described in Chapter 8 involves at least 50 distinct proteins as catalysts, apoenzymes, or cytochromes. (For this number, overcounting to account for multisubunit enzyme complexes, such as pyruvate dehydrogenase with its 72 subunits, has been avoided. Such subunit counts greatly exceed 50.) All of these proteins are used to make ATP by transducing sunlight energy into phosphate bond energy. The manufacture of these proteins is the primary function of the cell. To do this, the cell uses its protein biosynthesis machinery, which includes all of the genetic apparatus. The biosynthesis machinery for polynucleotides and proteins involves at least another 100 distinct proteins and at least one large DNA genome. The ribosome alone accounts for 50 proteins. The DNA genome must contain genes for each protein and additional genes for tRNAs, rRNAs, and operon control components. As was seen in Chapter 10, the biosynthesis of these polymers requires abundant amounts of ATP for the activation of monomers. This feedback relationship between ATP and polymers was diagrammed in (10.54). In the present context, it can be viewed as an ATP cycle driven by sunlight, in close analogy with the water cycle discussed above.

Surely, this ATP cycle did not spring forth full blown in the primitive oceans just because sunlight was abundant. All of the proteins and polynucleotides would have had to form spontaneously and simultaneously from fluctuations in the molecular broth of the primitive ocean. This is an incredibly remote possibility because the free energy requirement for that many polymers is very large, and according to (10.14) this implies a probability proportional to the exponential of a large negative number; moveover, there is a requirement for specific sequences of monomers in these polymers in order to confer on them their remarkable specificities.

This conundrum is solved if energy metabolism could have arisen before the stage of evolution exhibited by the gene-directed protein biosynthesis machinery already described in this book. Lipmann [2] has proposed that primitive energy metabolism preceded life as it is known today. He suggests that the coupling of redox energy to phosphorylation energy "might have been the first event on the way to life." In the rest of this chapter, the case will be made that energy metabolism could have already become competent enough in early cells to be able to generate pyrophosphate or ATP, even though these "urcells" had not yet developed a gene-directed protein biosynthesis capability. This takes us back in time to before the evolution of true anaerobic prokaryotes, which do possess gene-directed protein biosynthesis of the kind described in Chapter 10. Two earlier stages are perceived in the ensuing discussion: (1) the earliest stage, in which energy metabolism evolves before any gene-directed protein biosynthesis; and (2) the second stage, in which a primitive gene-directed protein biosynthesis, which is the mechanistic ancestor of the contemporary mechanism, evolves and develops. The feedback structure of the machinery for primitive gene-directed protein biosynthesis will be referred to as the "primitive uroboros" to distinguish it from the "contemporary uroboros" depicted in (10.54) and (10.55). It will be discussed in Chapter 14 and shown in (14.4).

A search for the origin of a primitive uroboros on the primitive earth is an extraordinarily difficult quest. The fossil record was only very recently extended backward, from about 570 million years to about 3.5 billion years ago. Schopf [3] recently summarized the fossil record as shown in (13.1). In his article "The Evolution of the Earliest Cells," from which (13.1) is borrowed, he argues con-

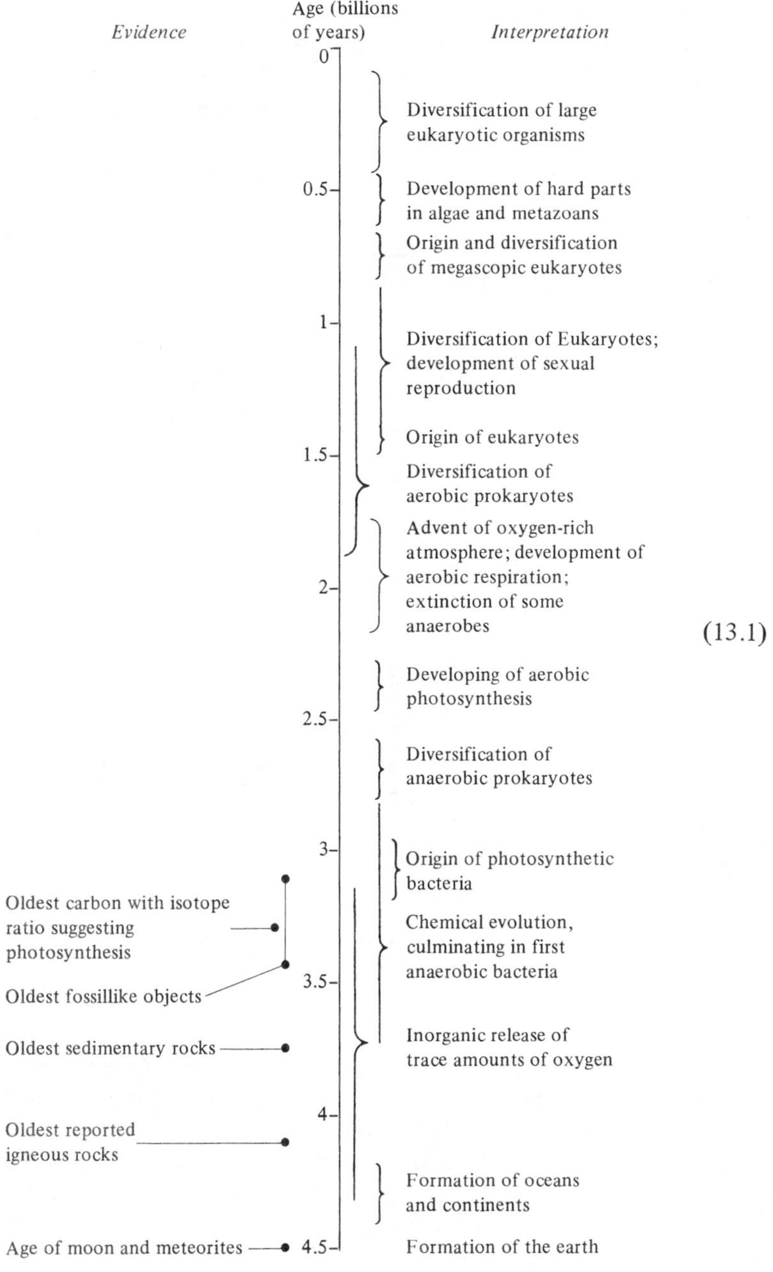

Evidence
Age (billions of years)
Interpretation
0
0.5
1
1.5
2
2.5
3
3.5
4
4.5
Diversification of large eukaryotic organisms
Development of hard parts in algae and metazoans
Origin and diversification of megascopic eukaryotes
Diversification of Eukaryotes; development of sexual reproduction
Origin of eukaryotes
Diversification of aerobic prokaryotes
Advent of oxygen-rich atmosphere; development of aerobic respiration; extinction of some anaerobes
Developing of aerobic photosynthesis
Diversification of anaerobic prokaryotes
Origin of photosynthetic bacteria
Chemical evolution, culminating in first anaerobic bacteria
Inorganic release of trace amounts of oxygen
Formation of oceans and continents
Formation of the earth
Oldest carbon with isotope ratio suggesting photosynthesis
Oldest fossillike objects
Oldest sedimentary rocks
Oldest reported igneous rocks
Age of moon and meteorites

(13.1)

Redrawn from Ref. 3

vincingly for the interpretations listed to the right of the time axis. Some of the earliest events are uncertain temporally, especially regarding earliest limits. New findings will always have the potential for pushing the starting dates still farther back. I believe that the search for a primitive uroboros should correspond with a time of about 4.3 billion years ago and that the lowest portion of (13.1) should look like (13.2). This belief is justified in the paragraphs that follow.

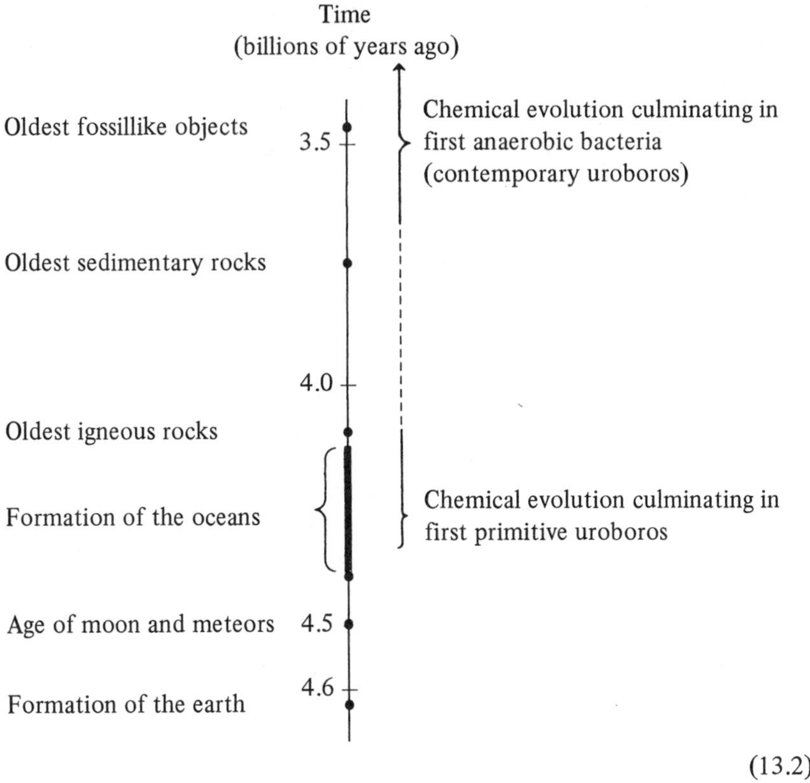

(13.2)

The first anaerobic prokaryotes appear to have possessed gene-directed protein biosynthesis of the contemporary variety. The most primitive known contemporary microbes are the methanobacteria [4]. Nevertheless, they possess ribosomes, tRNAs, synthetases, and polymerases of only slightly more primitive character than those found in other anaerobes. They are still representative of a later stage of evolution than the primitive uroboros. To ensure evolutionary continuity, the primitive uroboros must be richly enough endowed functionally to be capable of further evolution into an organism with a contemporary machinery for gene-directed protein biosynthesis.

Many researchers [5-7] have shown that, under simulated primitive earth conditions, the abiogenesis of many types of biologically significant small molecules and monomers is readily achieved. Sugars, amino acids, purines, and pyrimidines, as well as many other species of molecule, have been generated in this way. The primitive earth's atmosphere probably had no oxygen in it. The oxygen-containing atmosphere is believed to have emerged about 2 billion years ago, after the evolution of aerobic photosynthesis in "cyanobacteria." The earlier atmosphere contained mainly N_2 and some H_2, H_2O, and H_2S, which gave it a reducing character. With mixtures of some or all of the simple molecular species (H_2O, CO_2, NH_3, CH_2O, CH_4, H_3PO_4, H_2SO_4, etc.) and with energy inputs of a variety of types, including electrical discharges, emissions from radioactive decay, ultraviolet light, and heat, a diverse population of molecules is obtained under reducing conditions. Simple heating above the boiling point of H_2O, or in hygroscopic media at lower temperatures, has been shown to be an especially versatile mechanism of abiogenesis. Molecular species of all the important biological classes are generated in this way. Such results strongly suggest that the environment of the early earth, when its oceans and continental masses were first emerging about 4.3 billion years ago, was ideally suited for extended abiogenesis in numerous geophysical niches. Vigorous thermal activity and extensive interfaces between aqueous zones and hot-dry zones would have emerged with the expanding littoral zone between the oceans and continental masses.

Placing chemical abiogenesis as far back in time as the origin of the oceans solves another conundrum often encountered during contemplation of the origin of life, when it is assumed that life arose only after the oceans had fully developed. The problem with this more commonly assumed scenario is that life must emerge in a *dilute* organic soup, rather than under the far more felicitous conditions attending the origin of the oceans. Since no compelling evidence one way or the other is presently available, these assumptions must be viewed as "working hypotheses" and not statements of facts. The ideas to be presented later in this chapter are most compatible with the hypothesis that energy metabolism began to evolve shortly after the beginning of the emergence of the earth's oceans and continental masses.

When amino acids, which could have been produced by abiotic synthesis on the primitive earth, are subjected to hot-dry conditions just above the boiling point of H_2O, or to hygroscopic conditions at temperatures as low as 60°C, they spontaneously polymerize. The polypeptide product, called a "proteinoid" by their discoverer, S. W. Fox, represents an "urprotein" that is not the product of gene-directed protein biosynthesis [6, 8]. When washed with H_2O, the polymeric proteinoids spontaneously self-assemble into proteinoid microspheres. These units have a radius on the order of 1 μm and strongly resemble self-assembled lipid vesicles. The proteinoid polymers contain many hydrophobic amino acid residues, which confer on them their lipidlike properties, especially their ability to self-assemble into microspheres. These microspheres are usually uniform in size, unlike unfractionated lipid vesicle preparations, and they are stable enough for electron microscopy, which is difficult for lipid vesicles. The earth's early littoral zone between the evolving oceans and land masses would have provided numerous niches in which the production of proteinoids under hot-dry, or hygroscopic, conditions could have occurred. After tidal washing with water, the polymeric proteinoid is washed into the aqueous zone, where the microspheres self-assemble. In the lower strata of the aqueous zones, these microspheres would provide an enclosed environment suitable for chemical evolution and protected from ultraviolet radiation by the upper strata of the aqueous phase in which various molecular species would absorb this light.

That the environment defined by these urcells could support a primitive energy metabolism before the advent in evolution of gene-directed protein biosynthesis is the contention of the remainder of this chapter. It is argued below that even ATP-generating energy metabolism is possible, *without* the intervention of the protein catalysts that are so characteristic of the ATP-generating machinery of contemporary microbes. In addition, it will be argued that an energy transduction, coupling redox energy to phosphorylation energy, evolved very early in this scenario. The details of this coupling provide an explicit mechanism for Lipmann's hypothesis regarding the early emergence of pyrophosphate energy "on the way to life." The crucial feature of the scheme to be discussed is that it involves a primitive cellular membrane and utilizes this membrane chemiosmotically, albeit by a primitive chemiosmotic mechanism, to provide the coupling between redox energy and phosphorylation energy.

The environment provided by the earth's primitive littoral zones is ideally suited for the spontaneous emergence of oligomeric combinations of small and monomeric molecules. Zones of dry heat that promote the dehydration reactions leading to proteinoids would also lead to myriads of oligomeric combinations of amino acids, sugars, "vitamins," and inorganic substances like H_3PO_4 and H_2SO_4. Many of the important oligomers are depicted in Chapter 1. Two likely products of such abiogenesis on the early earth are NADH and AMP because they are merely tetrameric and trimeric, respectively, requiring only three or two dehydrations each, respectively. This number of such bonds is thermodynamically inhibited to a great extent, especially in aqueous solution. Formulas (10.14) and (10.15) imply that for bonds costing only 3 kcal per mole, the equilibrium concentration of oligomers, $[P_3]$ or $[P_4]$, is given by

$$\begin{aligned}[P_3] &= \exp[-2 \times 5]\,[M]^3\,[H_2O]^{-2} \\ &= 4.5 \times 10^{-5}\,[M]^3\,[55.55 \text{ molar}]^{-2} \qquad (13.3)\end{aligned}$$

or

$$[P_4] = \exp[-3 \times 5]\,[M]^4\,[H_2O]^{-3}$$
$$= 3.0 \times 10^{-7}\,[M]^4\,[55.55 \text{ molar}]^{-3}$$

where [M] is the concentration of monomer. Even if [M] were as large as 1 molar, at least 10^3 times too big on realistic grounds, the values for $[P_3]$ and $[P_4]$ in (13.3) are bounded by

$$[P_3] < 1.5 \times 10^{-10} \text{ molar} \tag{13.4}$$

and

$$[P_4] < 1.7 \times 10^{-14} \text{ molar}$$

The volume of a microsphere 1 μm in radius is only 4×10^{-15} liter, which means that the equilibrium concentrations, $[P_3]$ and $[P_4]$, in (13.4) correspond to 0.36 and 0.000042 molecule per microsphere, respectively. Nevertheless, heat will serve as the primitive energy source to drive the production of higher concentrations of these oligomers, just as it does for proteinoid synthesis. Some proteinoids contain as many as 100 dehydration linkages, which attests to the ability of this energy source to drive polymer synthesis *as well as oligomer synthesis*. Oligomers are subject to hydrolysis and, on the primitive earth, to ultraviolet radiation. A continual cycle of synthesis and degradation would have occurred. The interior of microspheres, in lower aqueous strata, would protect oligomers from the ultraviolet. Oligomers other than AMP and NADH are to be expected, including FMN and FAD and many other coenzymes. Indeed, the structure of all "holoenzymes," with their "apoenzyme" proteins and their nonprotein "coenzymes" or prosthetic groups, strongly suggests a primitive system in which only the coenzymes exist because no gene-directed protein biosynthesis yet exists. The coenzymes are oligomers and provide the holoenzymes with their catalytic function. The apoenzyme confers specificity for substrate, regulation, and usually an increase of rate. The coenzyme, nonetheless, catalyzes reactions at its own nonnegligible rate. Thus, an urcell containing thermally synthesized oligomers would possess primitive catalytic functions. Proteinoids containing catalytically active amino acid residues possess catalytic functions as well. This scenario also includes primitive heme molecules based on the protoporphyrin structure, which can be generated in experiments under simulated early earth conditions. It is essentially a tetrapyrrole and could give rise to a primitive chlorophyll in addition to heme. The heme could function as a primitive cytochrome without the protein component, which would arise only after the advent of gene-directed protein biosynthesis.

The emphasis on dry-heat or hygroscopic conditions reflects the simplicity of mechanism by which heat functions in promoting polymerizations. These dehydration sequences will become spontaneous if the H_2O molecules released during synthesis are physically removed by either vapor-generating heat or hygroscopic absorption. This may be viewed in terms of the discussion in Chapter 10 of the activation of monomers as a prerequisite of polymerization. The dehydration step, and thereby the H_2O, is removed in (10.19) and (10.32) to a remote process of activator (ATP) generation, that is, energy metabolism. With heat, the H_2O formed in the dehydration step is *physically*, rather than *chemically*, removed, and polymerization proceeds. Because proteinoids arise thermally, their amino acid sequences do not reflect gene-directed protein biosynthesis in which specific sequences are coded by the DNA; they reflect intrinsic chemical biases between amino acid residues. With oligomer synthesis, only a few sequences are possible, and presumably the coenzymes NADH and FAD, and other oligomers possess sequences determined by natural chemical biases between their monomeric constituents. This should be thoroughly checked experimentally.

Lipmann has argued that pyrophosphate (PP^{3-}) may well have been the primitive activator for early polymer synthesis. Adenine and ribose are formed in abiogenesis experiments, and AMP is merely a trimeric combination of these two molecules with phosphate. Therefore, AMP could have arisen very early as a carrier of pyrophosphate, that is, as ATP. When (10.19) and (10.32) are studied, it is seen that activation in each case involves phosphate derivatives of the monomers, with the adenine and ribose playing an incidental role. Thus, pyrophosphate may have played the role of ATP. In the earliest stages of urcell evolution, pyrophosphate might well have had a purely thermal origin in the earth's littoral zones. However, at pH 7, pyrophosphate is PP^{3-}, which is not soluble in lipoid microsphere membranes because of their hydrophobic quality, which excludes charged ions. The urcell needs PP^{3-} on its inside, and in reasonable concentrations. At a lower pH, the pyrophosphate has a decreased charge, PP^{2-}, but it is still not soluble in the urcell membrane, and no transport protein could have evolved yet. One hypothetical solution to this problem involves coupling redox energy to phosphate transport.

The basis for primitive redox coupling is provided by *quinone*. Quinones are universally found in chloroplasts, mitochondria, and bacteria, where they are essential members of electron transport chains. They carry out their function by freely diffusing across the membrane, as shown in (11.17), (11.20), and (11.22). No protein component is required for their function in these contemporary membranes. Quinones are produced in abiogenesis experiments. In (13.5) the generation of a transmembrane pH differential is shown using quinone, Q, as the membrane-soluble carrier. A double-layered membrane is schematized in (13.5)

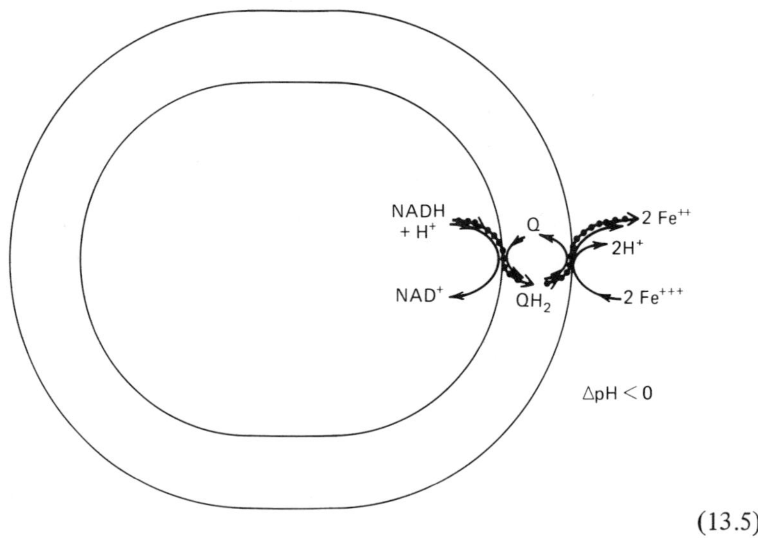

(13.5)

because electron microscopy shows that proteinoid microspheres possess double-layered membranes in keeping with their lipidlike properties. The oligomer NADH is shown as the reductant that is produced in the interior via the chemical processes of the internal environment; it is present as a result of abiogenesis.

The electron acceptor in (13.5) is ferric iron. This ion could be associated with other molecules rather than being free as drawn. Probably a great deal of ferrous iron, Fe^{2+}, was dissolved in the early oceans, and ultraviolet light could have converted it to ferric iron in combination with organic substances like ferricyanide, which would have become available as electron acceptors at lower aqueous strata.

Because quinones carry whole hydrogen atoms, the redox sequence in (13.5) produces a transmembrane proton potential that is purely $-2.3(RT/F)\ \Delta pH$. Therefore, a primitive chemiosmotic mechanism could have arisen before protein biosynthesis. This ΔpH would support the electroneutral transport of the substances that are neutralized by combination with protons, phosphate transport being a prime example. In (13.6) P^- is neutralized by H^+ and becomes freely soluble in the urcell membrane. No symporter protein is involved as in the contemporary scheme shown in (11.18). Once inside, P^- converts to P^{2-} because the pH is higher inside as a result of the redox sequence activity. This makes it insoluble in the membrane, thereby trapping the phosphate inside the urcell at the expense of redox energy. The concentration of phosphate inside the urcell is not great enough to lead to pyrophosphate formation as a simple consequence of mass action. Equilibrium concentrations of pyrophosphate yielding more than

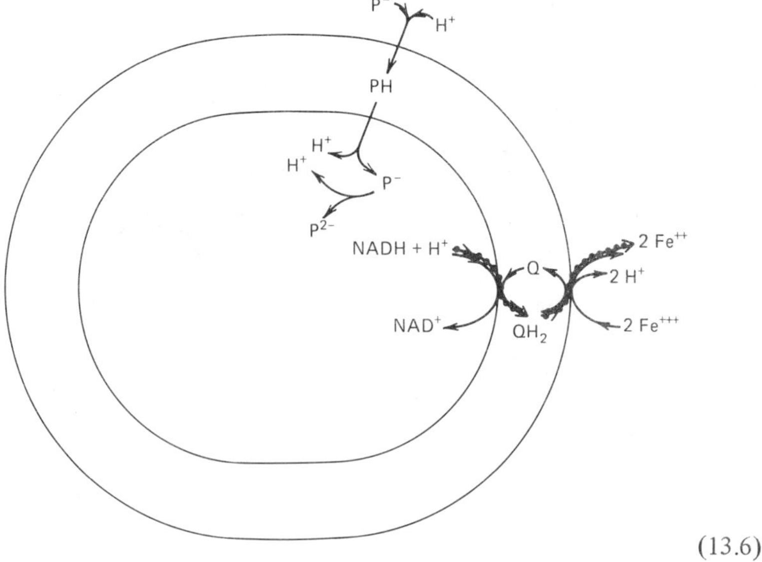

(13.6)

one molecule of per urcell would require internal concentrations of phosphate in excess of 10 molar, which cannot be expected from the mechanism in (13.6) or any other mechanism. A scheme for phosphorylation is, however, possible through the introduction of just one more molecular component. This additional component also serves to explain how NADH is regenerated once it has reduced the quinone and become NAD^+.

Abiogenesis experiments starting with enough ingredients, such as CH_2O, CO_2, and H_3PO_4, should yield substances like glyceraldehyde-3-phosphate, a three-carbon molecule that plays a significant role both in the carbon cycle and in glycolysis. The chemical coupling mechanism of oxidative phosphorylation during glycolysis was depicted in (5.12). The oxidation step is mediated by NAD^+ and phosphate is incorporated, yielding 1,3-diphosphoglycerate. The incorporated inorganic phosphate found in the 1-position of 1,3-diphosphoglycerate is an "activated" phosphate and is used to phophorylate ADP to ATP in the next step of glycolysis. The mechanism of the contemporary glyceraldehyde-3-phosphate dehydrogenase is very simple and suggests a primitive precursor mechanism that could have emerged before the evolution of gene-directed protein biosynthesis. The mechanism utilizes the sulfur atom of a cysteine residue to coordinate the reaction. The contemporary enzyme is denoted by E–SH in (13.7).

A primitive sulfur-containing proteinoid could serve as the site of the necessary sulfur atom for this sequence of NAD^+ oxidation and the "phosphorolysis"

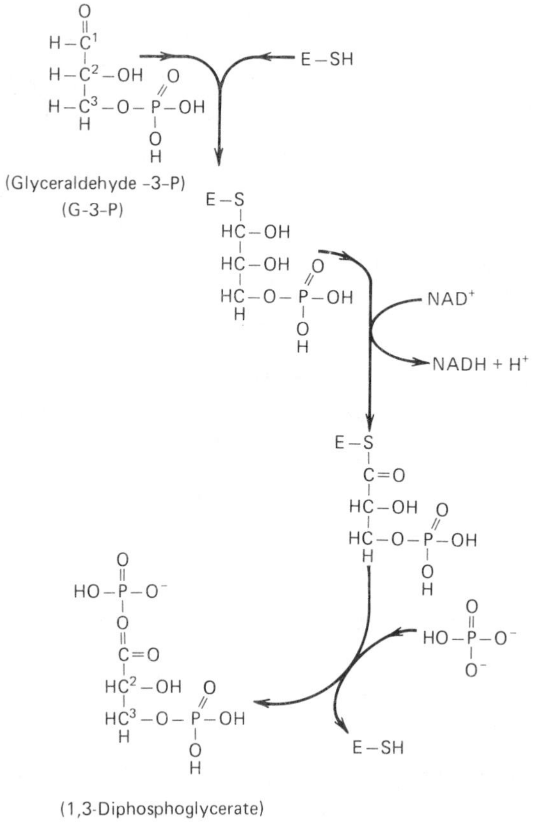

O
‖
H–C¹
|
H–C²–OH O
| //
H–C³–O–P–OH
H |
   O
   H

(Glyceraldehyde -3-P)
(G-3-P)

→ E–SH

E–S
|
HC–OH
|
HC–OH O
| //
HC–O–P–OH
H |
    O
    H

NAD^+ → $NADH + H^+$

E–S
|
C=O
|
HC–OH O
| //
HC–O–P–OH
H |
    O
    H

O
‖
HO–P–O⁻
|
O⁻

→ E–SH

O
‖
HO–P–O⁻
|
O
‖
C=O
|
HC²–OH O
| //
HC³–O–P–OH
H |
    O
    H

(1,3-Diphosphoglycerate)

(13.7)

by phosphate of the "thiohemiacetal" intermediate, glyceraldehyde-S-E complex. The product of this sequence, 1,3-diphosphoglycerate, provides active phosphate for the phosphorylation of phosphate, to pyrophosphate, of AMP to ADP, or of ADP to ATP. These possibilities are shown in (13.8), which uses the following abbreviations: G-3-P, glyceraldehyde-3-phosphate; 1,3-diPG, 1,3-diphosphoglycerate; 3-PG, 3-phosphoglycerate; P, phosphate; PP, pyrophosphate; PH, neutral phosphate H_3PO_4; (Q/QH_2), quinone redox pair; (Fe^{3+}/Fe^{2+}), any iron-organic complex; and ($NAD^+/NADH$), the redox pair of nicotinamide-adenine dinucleotide. The primitive catalyst, shown by the large dot, is simply a sulfur-containing proteinoid. In this scheme, many of the essential ingredients of modern energy metabolism are combined.

Pyrophosphate, rather than ATP, may have been the original carrier of phosphate potential for polymer synthesis, as well as for other energy require-

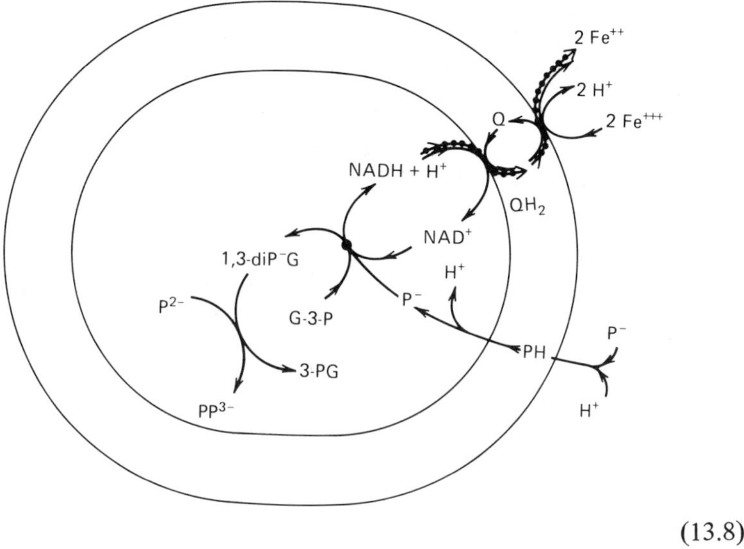

(13.8)

ments. Several reactions are suggestive, such as Harland Wood's reaction

$$\text{oxaloacetate} + \text{PP} \rightleftharpoons \text{phosphoenolpyruvate} + CO_2 + \text{P} \qquad (13.9)$$

which mimicks a cellular reaction that uses GTP in place of pyrophosphate. This reaction exhibits a potential early mechanism of CO_2 fixation and coupling between the intermediates of glycolysis and the citric acid cycle. In photosynthetic bacteria, pyrophosphate has been shown to serve as an energy donor that supports membrane functions in the dark. It will even drive the formation of ATP in the dark in these organisms.

The Evolution of Energy Metabolism

There are many uncertainties in determining the sequence of emergence of more and more sophisticated energy metabolism mechanisms and structures. Schopf's chronology in (13.1) places all forms of photosynthesis well after the emergence of the first anaerobes, which already possess gene-directed protein biosynthesis. There are several reasons to believe that photosynthetic energy metabolism did not emerge earlier. Unlike the glycolytic sequence in anaerobic prokaryotes, which contain a chemical coupling mechanism for ATP synthesis, the simplest photosynthetic bacteria extant possess a cyclic electron transport mechanism that generates a transmembrane proton electrochemical potential that in turn drives ATP synthesis via a membrane-associated ATPase complex. The requirement for an ATPase complex strongly suggests a prior requirement for gene-directed protein biosynthesis.

On the other hand, the idea that photosynthetic energy transduction preceded true organisms has been entertained by several researchers. The abiogenesis of protoporphyrins has already been mentioned. It is therefore possible that primitive precursors of chlorophyll and cytochromes existed very early on. In conjunction with iron–sulfur proteinoids, which could have been the forerunners of contemporary iron–sulfur proteins and ferridoxins, a scheme for light-driven cyclic electron transport in urcells before gene-directed protein biosynthesis is conceivable [see (13.10)].

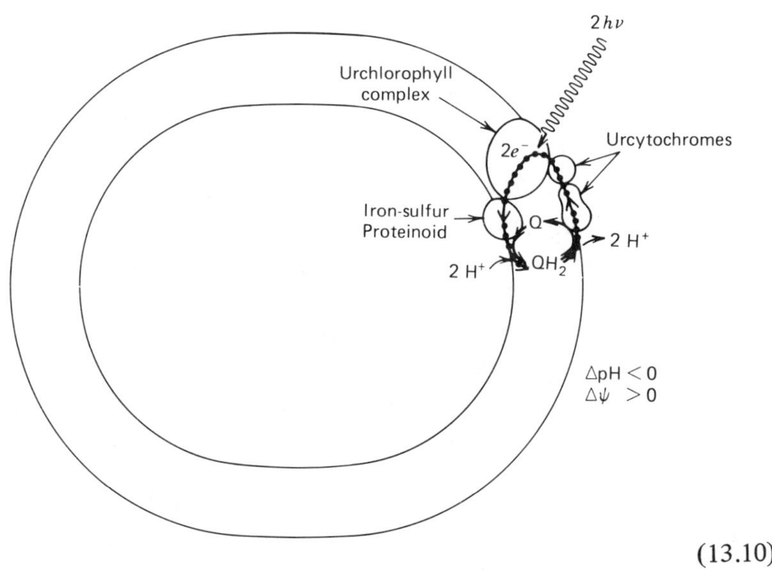

(13.10)

Scheme (13.10) is very similar to the arrangement of iron–sulfur proteins, cytochromes, and quinones in the *chlorosomes* of contemporary photosynthetic bacteria [9] like *Rhodopseudomonas*. The iron atoms on the outside of the urcell in (13.8) are now finding their way into the membrane of the urcell in (13.10). This is, of course, a characteristic of the advanced structures of energy organelles, as depicted in (11.20) and (11.22). Moreover, these additional components permit the generation of $\Delta\psi$ as well as ΔpH. However, this scheme only shows how chemiosmotic mechansims could have developed and does not help to explain an early photochemical origin of ATP or pyrophosphate synthesis. Therefore, it does not help in solving the problem of the development of polymer biosynthesis and the underlying energy metabolism that yields ATP or pyrophosphate.

It is the view of the author that, once energy metabolism got to the stage schematized in (13.8), the next development was the emergence of polymer biosynthesis, and especially of a primitive gene-directed protein biosynthesis. After that development, a more sophisticated energy metabolism, with its marked dependence on proteins, could have developed. Only much later would eukaryotes evolve and with them, the organelles of energy metabolism, the mitochondria and the chloroplasts. Much has been written about the evolution of these organelles, but in this book only the earlier stages of evolution will be considered in any detail. The reader is referred to the book by Broda [10] for an introduction to the literature on the evolution of energy organelles.

The specific questions to be answered with regard to the evolution of the earliest stages of anaerobic energy metabolism, starting with the mechanism of (13.8) include the following:

1 When did the FAD, iron–sulfur protein loop depicted in (11.17) evolve between NADH and quinone?
2 Did ATPases evolve before aerobic energy metabolism?
3 Did ATPases evolve before cytochromes? [11]
4 Did ATPases evolve before or after primitive anaerobic photosynthesis such as depicted in (13.10)?
5 When did the glycolysis pathway begin to be extended in both directions, toward glucose and toward pyruvate?
6 How early on did the citric acid cycle get going?
7 When did the formation of acetyl-CoA occur?

There are many other such questions that research must confront.

In this chapter, a rudimentary energy metabolism has been described for very early urcells. The structure of this metabolism is not uroboric, however, because it is not yet a self-begetting mechanism. In the next chapter, a primitive gene-directed protein biosynthesis will be described. In conjunction with the rudimentary energy metabolism described in this chapter, this biosynthesis provides the basis for an uroboric process.

References

1 Michael Faraday, *A Course of Six Lectures on the Chemical History of a Candle*, edited by W. Crookes (Harper, New York, 1861).

This book is a series of Christmas lectures delivered by Faraday to children. Adults will find it most enjoyable, too.

2 Fritz Lipmann, "Projecting Backward from the Present Stage of Evolution of Biosynthesis," in *The Origins of Prebiological Systems*, edited by S. W. Fox (Academic Press, New York, 1965); *Wanderings of a Biochemist*, (Wiley-Interscience, New York, 1971).

By 1940 Lipmann had postulated the central role of ATP in the energy transfer cycle, by 1950 he had isolated and characterized coenzyme A, by 1960 he had provided fundamental insights regarding the bioenergetics of carbamyl phosphate and "active" sulfate, and by 1970 he had made major contributions to the story of protein biosynthesis on ribosomes.

3 J. William Schopf, "The Evolution of the Earliest Cells," *Scientific American*, **239**, 110–138 (1978).

4 G. E. Fox, L. S. Magrum, W. E. Balch, R. S. Wolfe, and C. R. Woese, "Classification of Methanogenic Bacteria by 16S Ribosomal RNA Characterization," *Proc. U.S. Natl. Acad. Sci.*, **74**, 4537–4541 (1977).

5 M. Calvin, *Chemical Evolution* (Oxford, New York, 1969).

6 S. W. Fox and K. Dose, *Molecular Evolution and the Origin of Life* (W. H. Freeman, San Francisco, 1972).

7 S. L. Miller and L. E. Orgel, *The Origins of Life on the Earth* (Prentice-Hall, Englewood Cliffs, N.J., 1974).

8 R. E. Dickerson, "Chemical Evolution and the Origin of Life," *Scientific American*, **239**, 70–86 (1978).

9 R. E. Blankenship and W. W. Parson, "The Photochemical Electron Transfer Reactions of Photosynthetic Bacteria and Plants," *Ann. Rev. Biochem.*, **47**, 635–653 (1978).

10 E. Broda, *The Evolution of the Bioenergetic Process* (Pergamon, New York, 1975).

11 R. E. Dickerson, R. Timkovich, and R. J. Almassy, "The Cytochrome Fold and the Evolution of Bacterial Energy Metabolism," *J. Mol. Biol.*, **100**, 473–491 (1976).

CHAPTER

14

Gene-Directed Protein Biosynthesis

Polymer biosynthesis requires energy for the activation of monomers. This fact was discussed in detail in Chapter 10. The mechanism by which early urcells could have manufactured pyrophosphate for this purpose was described in the preceding chapter. Pyrophosphate may have served as a precursor to ATP for the activation of nucleotides and amino acids, yielding nucleotide triphosphates via two phosphate transfers in the first case and amino acyl carboxyl phosphates in the second case. The origin of gene-directed protein biosynthesis raises additional questions regarding the origin of the genetic code and the associated machinery for protein biosynthesis.

In Chapter 10 the section titled "The Genetic Code and the Mystery of the Second Code" discussed the missing component needed for understanding fully the mechanism of gene-directed protein biosynthesis in a contemporary uroboros like *E. coli*. The gap in understanding involves the ability of the tRNA synthetase enzymes to recognize both a specific amino acid and its cognate tRNA. All other aspects of protein biosynthesis–including gene replication, transcription, and translation–were shown to depend on the base-pairing rules. Only the synthetase-tRNA recognition step does not use base pairing, the mechanism of recognition being still unknown. It is probable that an understanding of this particular problem, the mystery of the second code, requires an investigation of the *origin and evolution* of the genetic code and the associated machinery for protein biosynthesis.

The origin of the genetic code has been addressed by many researchers. A popular line of reasoning has posed the question in the traditional "chicken-egg" form: Which came first, DNA or proteins? It is argued that the structure of DNA is ideally suited for carrying the genetic heritage and for replication, so that descendants will have a copy, whereas proteins are the catalytic agents that perform all of the metabolic functions. Usually, this line of reasoning reaches a point where it is appreciated that some type of coevolution of polynucleotides and proteins must have occurred. This is the case for the treatments by Woese [1], Orgel [2], Crick [3], and Eigen [4]. In Eigen's paper a clear statement of the mystery of the second code is enunciated.

Only Eigen [4] proposed a concrete, hypothetical macromolecular solution to the problems raised above. Chapter VI of his paper is devoted to these issues. In Section VI.1 he restates the problem of the origin of the genetic code and the associated translation machinery. In Section VI.2 he presents his "hypercycle" model, which purports to be an answer. In fact, the hypercycle model explicitly assumes a "functional translation system," as stated on page 504 of Eigen's paper: "The presence of a translation system ensures sufficient precise translation from I_i to E_i." Unfortunately, this aspect of the model begs the question. In Section VI.3 of his paper Eigen returns to the original question of the origin of the code, which he states clearly and well, but for which he has no answers in molecular, mechanistic terms. More recently, other researchers, particularly Crick, have expressed the view that these questions are too difficult to answer in terms of the hypothetically available scenarios for the primitive earth. They entertain a modern version of Panspermia, which has life coming to earth from outer space instead of arising, de novo, here. It is the purpose of this chapter to postulate a scheme, plausible for the primitive earth, for the origin and evolution of the genetic code and the associated machinery for protein biosynthesis, the primitive uroboros. This is done in part to demonstrate that the conceptual problems are really not insurmountable.

The primitive-uroboros model presented below exhibits a variety of requirements proposed by Crick [3] and Eigen [4] for the evolution of the genetic code. Specifically, *continuity of the code* and its mechanism of translation seem essential as evolution proceeds from the simple primitive uroboros toward the contemporary uroboros. This means inter alia, that codon assignments for amino acids, codon size, and protein-polynucleotide colinearity must be preserved as the molecular mechanisms evolve, as opposed to a sequence of discrete and dramatic changes in mechanism as evolutionary complexity accrues. Such properties for the evolution of the primitive uroboros described below will be emphasized as they arise.

The Origin of the Genetic Code and the Associated Machinery for Protein Biosynthesis

The objective of this section is to postulate a primitive mechanism for gene-directed protein bioysnthesis that could have been the precursor of the contemporary mechanism, which requires ribosomes, tRNAs, synthetases, and polymerases [5]. The preconditions for the emergence of this primitive mechanism will be discussed in greater detail in the next chapter. The starting point here is an urcell in which primitive energy transductions generate pyrophosphate (perhaps ATP). The pyrophosphate energy is used to activate nucleotides from their monophosphate forms into their triphosphate forms, a process that requires two phosphorylations using two molecules of pyrophosphate (or two molecules of ATP). The nucleotide monophosphates are trimeric oligomers that could have arisen spontaneously in the thermal zones of the primitive earth, as described in the last chapter. Consequently, it is assumed that some polynucleotides are generated through the activation of the monophosphates and their subsequent polymerization. This means that a primitive urcell with primitive energy metabolism could have produced some RNA molecules inside itself as a natural consequence of its energy transduction mechanisms. It is not necessary that these RNA molecules be as large as they are found to be in contemporary organisms, but they must have been at least several tens of nucleotides long if they were to have functioned in the manner described below. A bias in favor of formation of $5'$-$3'$ phosphodiester linkages over the $5'$-$2'$ linkage may have required the catalytic function of a proteinoid.

The key behind the model of primitive gene-directed protein biosynthesis is that an RNA molecule served simultaneously as a primitive gene—that is, as an "urgene"—and as a primitive messenger RNA. No DNA was involved at this early stage, and neither were ribosomes, tRNAs, or synthetases. All of these components evolved later.

The mechanism by which a protein can be synthesized in accord with the sequence of bases in an RNA molecule is hypothetical, but it does not involve either reaction chemistry or stereochemistry not already known to occur with real RNA molecules. The mechanism is the following:

1 Amino acids are activated by pyrophosphate, forming amino acyl carboxyl phosphates (or activated by ATP, forming amino acyl adenylates) [see (10.27)].
2 The amino acyl carboxyl phosphates react with the $2'$-hydroxyl groups of the RNA molecule, forming amino acyl RNA esters [see (14.1)].
3 These amino acyl RNA esters are "energy rich," just like the very similar esters at the $3'$ acceptor arm end of contemporary tRNAs. Unlike the tRNA case, in this case esters form throughout the RNA molecule at any of its $2'$-hydroxyl groups [see (10.36)].
4 The amino acyl esterified RNA molecule undergoes a *conformational change* once it has enough amino acids in close proximity bonded to it. The conformation is a right-handed helix of three nucleotides per turn, with the phosphates on the inside and the bases on the outside. The esterified amino acids are also on the outside. In (14.2) amino acids that are esterified to every *third* ribose of the RNA are found to be adjacent to each other. (See Appendix 14A for three-dimensional views.) The positively charged amino groups of the amino acids partially neutralize the negatively charged phosphate

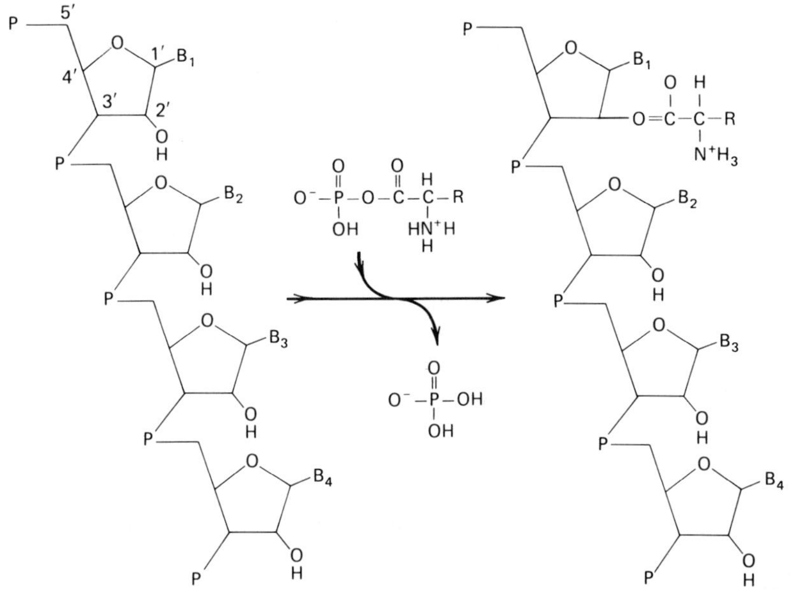

(14.1)

groups of the RNA, thereby stabilizing this otherwise unfavorable conformation. In addition, bases B_2, B_5, and B_8, as well as the other sets of "adjacent" bases (B_1, B_4, B_7 and B_3, B_6, B_9) of this helical conformation engage in "base-stacking" interactions, which tend to stabilize the conformation also. These statements are based on CPK models.

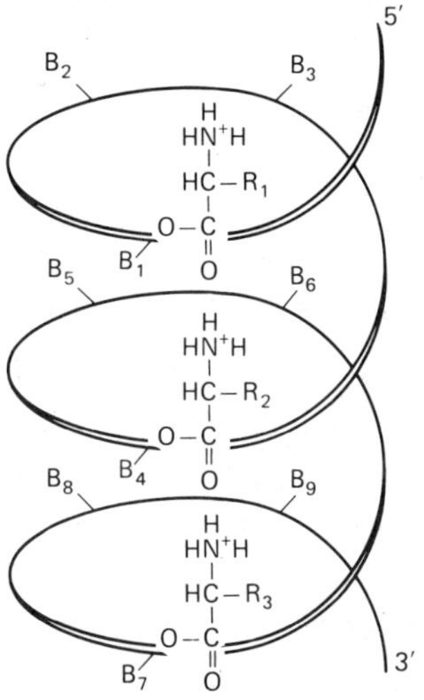

(14.2)

5 The adjacent amino groups and carboxyl esters react to form peptide bonds. The finished polypeptides leave the RNA, which reassumes its "random coil" conformation, ready for more activated amino acids to form 2′-hydroxyl esters. If the helical conformation does not form, then amino acyl RNA esters cannot come close enough together to make peptide bonds. Two amino acyl esters at two adjacent ribose moieties of the random coil conformation are 7 to 8 Å apart, so that no peptide bond formation is possible. Only amino acyl esters at *every third ribose* become "adjacent" in the helical conformation, and then they are touching, as shown in (14.2).

6 The CPK models of this mechanism indicate that, if D-ribose is used, stereochemical requirements prefer the use of L-amino acids. Furthermore, the amino acids form esters that are oriented by charged interaction with phosphate groups such that their amino groups point in the 5′ direction of the RNA. This means that the polarity of the polypeptide from the N-terminus to the C-terminus corresponds with the polarity of the RNA from 5′ to 3′. This is identical with the relationship between proteins and their messenger RNAs in contemporary mechanisms for protein biosynthesis; it implies a *direct* coupling between codons and amino acids.

7 The three-base spacing of amino acyl esters for peptide bond formation in the helical RNA conformation implies that the amino acid sequence corresponds to a sequence of triplets of bases, just as in the contemporary situation.

8 Which amino acid is esterified between bases B_1 and B_2 in (14.2) is hypothesized to depend on the residue R_1 and the two bases that are closest to it on esterification. These are the bases B_1 and B_2 in the CPK model. Base B_2 is closest as the ester is formed, but moves away some when the subsequent helical conformation forms. Base B_3 is remote from R_1 throughout the entire process. This means that, even though there is a *three-base spacing*, there is only a *two-base code*, initially. The selective interaction of the residue of the incoming amino acid with the bases closest to it as an amino acyl RNA ester forms has not been demonstrated experimentally. Should it be verified, it will imply that the origin of the code lies in stereochemical requirements attending the formation of esters between activated amino acids and RNA 2′-hydroxyl groups. This is in sharp contrast with the idea that the genetic code assignments are entirely "accidental," having no physicochemical basis.

In this primitive mechanism proteins are translated directly on RNA genes, without the intervention of ribosomes, tRNAs, or synthetases. The mystery of the second code for synthetase-tRNA recognition is *circumvented* at this stage because neither tRNAs nor synthetases are involved. Later, the evolution of these components will be considered.

The idea of a two-base code is strongly suggested by several lines of reasoning. The "wobble" phenomenon discussed in Chapter 10 indicates that the third base is less important in determining a cognate amino acid than are the first two bases. In Table 10.1 it is seen that the third base is irrelevant for leucine, valine, serine, proline, threonine, alanine, arginine, and glycine. Similarly, the only amino acids with acidic residues, aspartic acid and glutamic acid, have identical first and second bases, GA, thereby making the third base irrelevant as far as determining an amino acid that is *acidic*. One possible doublet code that provides for hydrophobic, basic, acidic, sulfhydryl, and polar residues, as well as for "punctuation" is the code in Table 14.1. Note that these codons, like GA above, correspond to the first two bases of the code given in Table 10.1, which is defined for codons on *mRNA*. These are complementary to the triplets on DNA in the contemporary system. This distinction will be examined further below. The five amino acids not coded for here—asparagine, glutamine, methionine, tyrosine, and tryptophan—are viewed as evolutionarily more recent. Each of them requires third-base discrimination in the contemporary code. This is also true for the additional "end" codon, UGA, which must be distinguished from the cysteine and tryptophan codons.

In the primitive model described above, the RNA serves as both the urgene and mRNA. The amino acids interact directly with the *mRNA*. This means that the codon assignments have significance as *physicochemical groups* when mRNA

Table 14.1

Doublet Codon (in RNA)	Amino Acid
5′ AA 3′	Lysine
AC	Threonine
AG	Serine or arginine
AU	Isoleucine
CA	Histidine
CC	Proline
CG	Arginine
CU	Leucine
GA	Aspartate or glutamate
GC	Alanine
GG	Glycine
GU	Valine
UA	End (punctuation)
UC	Serine
UG	Cysteine
UU	Phenylalanine

codons are compared with their cognate amino acids. Later, it will be argued that the emergence of DNA and of tRNAs, with their anticodon triplets, does not alter this basic relationship. This is just one aspect of the "continuity of the code" property [3].

For this stage of evolution, the primitive mechanism for gene-directed protein biosynthesis is schematized in (14.3), in which the following abbreviations

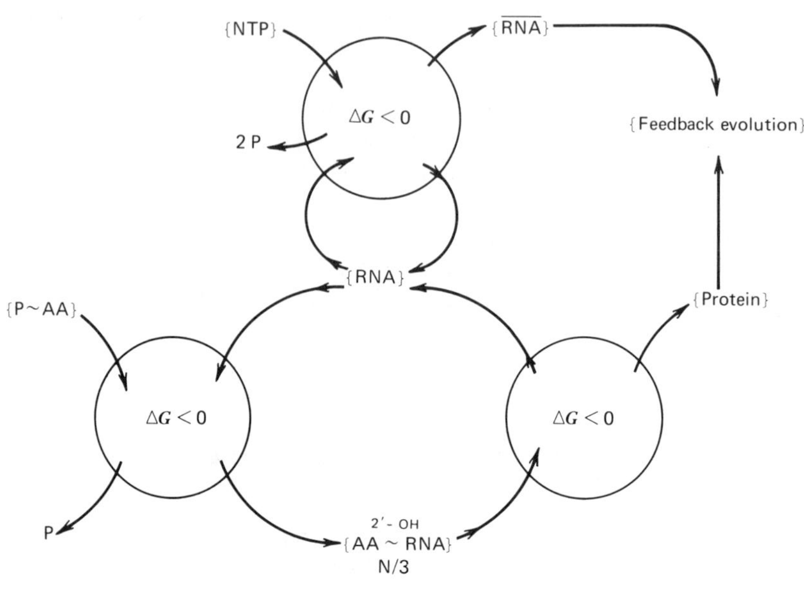

Primitive uroboros

(14.3)

are used: AA, amino acid; $\overline{\text{RNA}}$, RNA complement according to the base-pairing rules; P ~ AA, amino acid carboxyl phosphates (or amino acyl adenylate); AA $\overset{2'\text{-OH}}{\underset{N/3}{\sim}}$ RNA, RNA with amino acid esters on the 2′-hydroxyl groups of the RNA at every third ($N/3$) ribose.

One of the earliest urgenes to confer evolutionary advantage would have been an urgene for a primitive RNA polymerase. An RNA polymerase would have enabled the urcell to make more RNA and, by virtue of base pairing, to make replicas of an RNA through second-order complementarity: RNA → $\overline{\text{RNA}}$ → $\overline{\overline{\text{RNA}}}$ ≡ RNA. A fully modern polymerase is not required at first; only a primitive, relatively small RNA polymerase would have been needed. An urgene for the sulfur protein implicated in (13.8) would help to secure energy metabolism and would perhaps have been an early acquisition as well. Mutations by base changes are already possible in the primitive scheme in (14.3) because of $\overline{\text{RNA}}$ complementation errors.

By combining the energy metabolism scheme in (13.8) with the urgene-directed protein biosynthesis scheme in (14.3), a primitive uroboros is produced. With just the two urgenes mentioned in the preceding paragraph, this urcell has a relatively stable energy metabolism and polymer biosynthesis capability. If it can be shown that the evolution of this system into the contemporary uroboros can proceed step by step, one protein at a time, then the problem of initiating an uroboros is basically solved. Indeed, the thrust of the arguments presented up to this point is that, given the conditions of the very early earth, urcellular "life" is inevitable, as opposed to being the result of very improbable simultaneous fluctuations.

The Evolution of the Genetic Code and the Associated Machinery for Protein Biosynthesis [5]

The mechanism responsible for the selective esterification of activated amino acids to 2′-hydroxyl groups of RNA ribose moieties is the basis of the genetic code in the primitive uroboros. It is likely that this selectivity is not very specific. Consequently, during the course of early evolution, the development of better and better fidelity of coding would confer selective advantages on urcells possessing this improved fidelity. This is surely the evolutionary impetus behind the emergence of ribosomes, tRNAs, and synthetases.

Preservation of the genetic message is also advantageous to urcells in their competitive milieu. Preservation is enhanced if the functions of urgene and messenger can be separated, because the function as messenger involves the RNA in chemical reactions during which its degradation is a possiblility. The presence of an urpolymerase produces multiple copies of the RNA gene. This helps, but a greater advantage accrues if, through a mutation in this RNA polymerase, there evolves a DNA polymerase that reads RNA and makes a DNA complement. Such a polymerase would function as an analog of the contemporary RNA-directed DNA polymerase. The advantage in copying RNA into DNA complements is that *DNA does not have the 2′-hydroxyl groups* on its ribose moieties, because, by definition, they are *deoxy*ribose moieties [see (1.8)]. Therefore, DNA cannot serve as a template for direct translation into proteins via the mechanism of 2′-OH amino acyl ester intermediates. Along with the emergence of the RNA-directed DNA polymerase, there would also need to evolve a DNA replicase and a DNA-directed RNA polymerase. The replicase would copy the DNA directly, and the DNA-directed RNA polymerase would be the first transcriptase. How all these polymerases would arise can only be guessed at, but within the context of the present model the significance of deoxyribose seems clear enough.

A system that is engaged in energy metabolism and RNA synthesis could achieve a stage in which it produces excesses of RNA. Similarly, it could be producing excesses of proteins as well. Consequently, a variety of polymer complexes and fragments will accumulate in the urcells, and some of them may have conferred evolutionary advantages. Precursors of tRNAs may have arisen in this way, as a class of excess RNAs of a particular length. Similarly, the hydrolysis products of RNA molecules, including a class of triplets or other short RNA oligomers, could have accumulated as well.

Suppose that an urcell starts to make a protein on an RNA gene which has the property that it binds nucleotide triplets. This could be achieved by having two arginine residues located in such a way that they bind the phosphates of the nucleotide triplets through electrostatic bonds. This binding would not be triplet specific in terms of the *sequence* of bases on the triplet; it would be specific only for the *length* of the triplet [see (14.4)]. As a result of the evo-

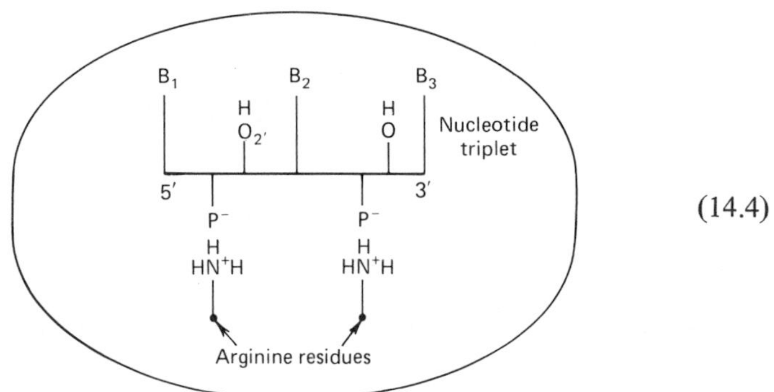

(14.4)

lution of this one urgene for one protein, an entire class of protein-triplet complexes is produced, one complex for each triplet sequence. This class of complexes will serve as the ursynthetases.

The ursynthetases must recognize and bind both an activated amino acid and its cognate ur-tRNA. The ur-tRNAs were mentioned two paragraphs ago. It is assumed that they are merely short RNA loops that provide an anticodon triplet that through base pairing with the codon triplet is recognized and bound to the ursynthetase. As shown in (14.5), the ur-tRNA may possess complemen-

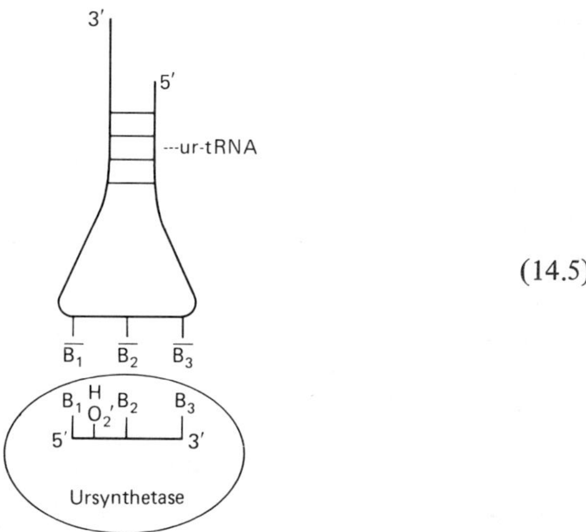

(14.5)

tary bases along its "stem" and therefore have a "hair-pin" duplex structure. The anticodon loop contains the complementary bases $\overline{B_1}$, $\overline{B_2}$, and $\overline{B_3}$. Recognition of the activated amino acids is assumed to occur in a manner paralleling the selective interaction postulated (14.1) for the direct reading of an RNA, except that in the case of the ursynthetase the activated amino acid becomes esterified with the 2′-OH residue shown in (14.5) between bases B_1 and B_2 of the ursynthetase-bound nucleotide triplet. The fidelity of this selectivity is assumed to be the same as in the case of direct reading of RNA. Thus, the recognition and binding of *both* the activated amino acid and the ur-tRNA by the ursynthetase are explained. This explanation was raised for contemporary synthetases back in Chapter 10, but it was pointed out there that contemporary synthetases are pure proteins containing no bound nucleotides.

At this stage, the amino acid must get transferred to the 3′ end of the ur-tRNA and then get carried to an urribosome on which peptide synthesis is catalyzed. The transfer of the amino acid is assumed to occur in a manner suggested by the U-conformation model of tRNA amino acylation, which was described in Chapter 10 in (10.49). That is, once the amino acid is in place on the bound triplet of the ursynthetase, the ur-tRNA folds over so that its 3′ end will accept the amino acid by a "transesterification" from the triplet bound on the ursynthetase to the 3′ terminal ribose of the ur-tRNA. This event triggers the release of the amino acylated ur-tRNA and leaves the ursynthetase free for another round of ur-tRNA activation. The folding over of the ur-tRNA would be facilitated if near its 3′ end it possessed a base that would pair with the B_2 base of the bound triplet. This would both coordinate the transesterification and displace the anticodon already bound there, facilitating the release of the activated ur-tRNA. This mechanism may explain the origin of "discriminator" bases, alluded to in Chapter 10. It suggests that the discriminator is correlated with the middle *anticodon* base.

The function of the urribosome would be to bind the RNA urgene, which is now serving as a true messenger, and to coordinate binding, by the base pairing of activated ur-tRNAs. It would also *catalyze* the peptide bond formation. As already mentioned, the fidelity of coding is no greater with the mechanism given above for the ursynthetase recognition of activated amino acids. The catalysis of peptide bond formation by the urribosome, and the coordination of amino acid placement by the ur-tRNAs and the RNA messenger attached to the urribosome could have been the evolutionary advantage of the emergence of these components. The helical conformation mechanism of (14.3) was probably distinctly less suitable for peptide bond formation.

The subsequent evolution of ursynthetases of greater selective fidelity for amino acids can also be imagined to have evolved, *one protein at a time*. It is this step-by-step emergence of improvements that renders the originally stated problem of the initiation of the contemporary uroboros less formidable. Imagine that the evolution of urcells, with primitive energy metabolism, has reached the stage in which ur-tRNAs, urribosomes, and ursynthetases of the type just described are operating. These urcells could add to their repertoire of urgenes new urgenes that code for improved ursynthetases whose function it is to recognize and bind their cognate activated amino acids as enzymes bind substrates. They could also be improved by binding their ur-tRNAs by means of a variety of recognition sites that depend on protein–tRNA interactions. In short, the ursynthetases with bound codon triplets could be replaced, *one at a time*, by pure protein synthetases that perform these recognition functions more like contemporary proteins. In an urcell with mostly ursynthetases, a new, pure protein synthetase would probably confer only a slight enhancement of fidelity as it functioned in concert with the unimproved ursynthetases. Eventually, however, all ursynthetases would be replaced by pure proteins and overall fidelity would be measurably enhanced. The key feature is that this replacement process proceeds *one protein at a time*. Each new synthetase may be improved with respect to amino acid binding and then again with respect to tRNA binding. Thus in the end the contemporary uroboros could have ended up with a variety of different types of synthetases that perform their recognition functions by a variety of different methods. This could be the evolutionary explanation of the difficulty presently encountered by researchers in their quest for the mechanism of the second code. There is no reason not to suppose that modern ribosomes evolved piece by piece as well, rather than all at once.

Many aspects of the function of the contemporary uroboros, as described in Chapter 10, have not been explained by the scheme outlined above. The principal objective above was to indicate that even the problem of the origin of the synthetases and their functions is conceptually within the grasp of present molecular biological thinking. Rather than requiring miraculous coincidences among highly improbable fluctuations, the uroboros evolution appears instead to be driven inevitably from early, simple, primitive stages into the complex contemporary stage by the relentless flow of energy, mediated by a primitive energy metabolism that is intrinsic to urcells. The model presented purports to explain the following.

1 The origin of a three-base code.
2 The third base redundancy.
3 The N-terminus → C-terminus and 5′ → 3′ colinearity of proteins and their cognate mRNAs.
4 The complementarity of D-ribose and L-amino acids.
5 The origin of RNA translation into protein.
6 The significance of deoxyribose and DNA as the genetic material.
7 The evolution of tRNAs and tRNA synthetases.
8 The origin of polymerases for RNA and DNA.

Many of these features are stereochemical properties of the molecules involved, as can be demonstrated with CPK space-filling models. No unusual chemistry is required. The *only hypothetical mechanism* is the *selective* esterification of 2′-hydroxyl residues on RNA ribose moieties by activated amino acids. This could be tested experimentally in model systems, and the helical conformation required for primitive peptide synthesis could be tested for stability. Such experiments could be performed by currently available techniques.

Appendix 14A

Three-Dimensional Perspective of AA $\underset{N/3}{\overset{2'\mathrm{OH}}{\sim}}$ RNA

The three-dimensional perspective of AA $\underset{N/3}{\overset{2'\mathrm{OH}}{\sim}}$ [see (14.2)] is achieved by viewing the stereogram while *crossing* your eyes. No special glasses or other visual aid is required. If you do use glasses, or view these stereograms by "wall-eyeing" them you must first interchange the left and right images.

The black atoms in this CPK model are carbons and the white atoms are hydrogens. Of the grays, the triangular atoms are nitrogens, which often also have a small amide cap. The other grays, the hemispherical, and the triply slotted atoms, are oxygens. An adenine is the black and gray heterocyclic group in the lower front portion of the picture. The prominent horizontal line of black and white groups sticking out just above the middle of the picture is a line of amino

 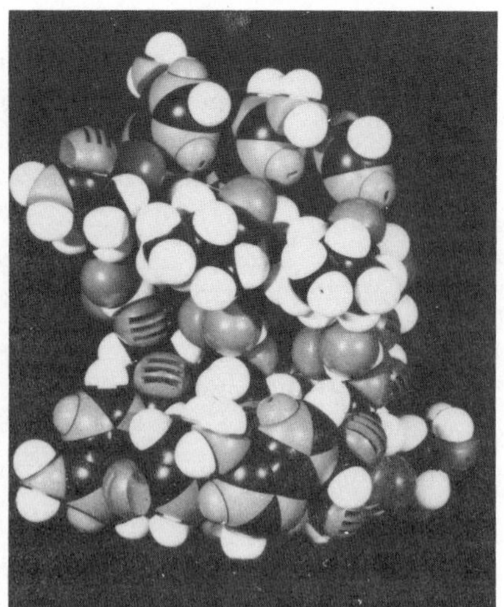

acid residues. The helical axis of the right-handed threefold RNA helix is horizontal in the picture, running from its 3′ end at the right to its 5′ end at the left. The right-most amino acid residue, for valine, is attached through a 2′-hydroxyl carboxyl-ribose ester to the RNA chain. This ester is not visible because it is directly below the hydrocarbon valine residue in this perspective. Just below this residue and at its left-hand edge are the hydrogens of its amino group, which is otherwise also obscured by the residue in this view. The amino group is pointing in the 5′ direction and is adjacent to the carboxyl-ribose ester oxygen of the amino acyl ester pictured in the middle of the line of residues. Just below the line of residues is a line of phosphates that are in close to the axis. The phos-

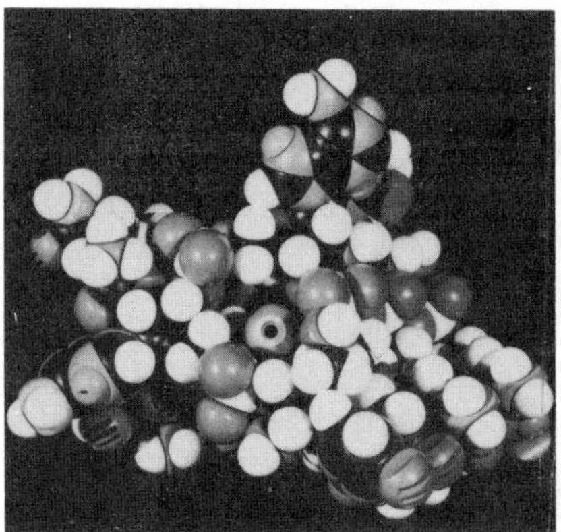 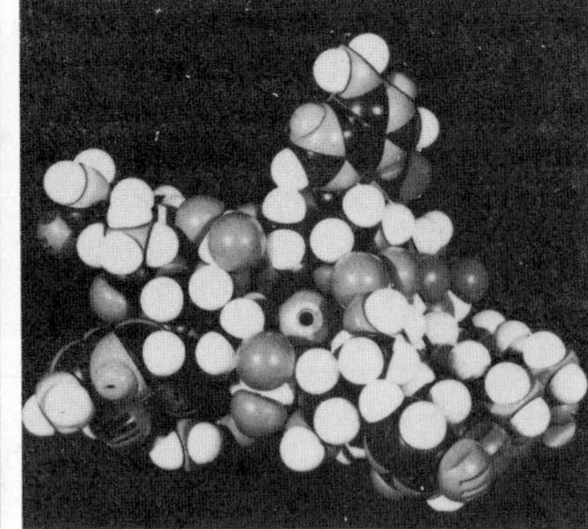

phorus atoms are whitish triangular shapes, and these are covered with oxygens. At both the top and bottom of the stereogram, the stacked bases are seen. The third row of stacked bases is behind the molecule in this perspective.

This stereogram is end-on from the 5′ end. At the lower right, at the front end of four stacked bases, is plainly seen the 5′ carbon of the 5′ end ribose. At both the 1 and 8 o'clock positions, the other two rows of stacked bases can be seen. At 10 o'clock, the amino acid line is seen, with the amino group of the end-most amino acid plainly visible. In the center, a threefold symmetric tube of phosphate groups is seen surrounding a magnesium ion. The central hole is just large enough for such ions and for H_2O molecules.

References

1 C. Woese, *The Genetic Code* (Harper & Row, New York, 1967).

2 L. E. Orgel, "Evolution of the Genetic Apparatus," *J. Mol. Biol.*, **38**, 381–393 (1968).

3 F. H. C. Crick, "The Origin of the Genetic Code," *J. Mol. Biol.*, **38**, 367–379 (1968).

4 M. Eigen, "Self-Organization of Matter and the Evolution of Biological Macromolecules," *Naturwissenschaften*, **58**, 465–523 (1971).

5 R. F. Fox, "A Model of Primitive Molecular Genetics" (1973).

This paper was originally a talk delivered at the Rockefeller University on March 28, 1973, and was later presented at a symposium held in memory of Charles Yegian by the Department of Molecular, Cellular, and Development Biology at the University of Colorado on February 8, 1974.

CHAPTER

15

Cells

This last chapter presents a summary of the content of the entire book. This goal is achieved in a natural way by discussing the origin and evolution of the cell.

According to the discussion at the end of Chapter 7, the cell is the simplest "integron" deserving to be called a living organism [1]. Even the DNA, with its quintessential property of replicability, is not yet entirely *self*-replicating. In Chapter 10 it was shown that the replication of DNA involves at least six different replication proteins. This DNA protein-dependent replication requires the simultaneous presence of a functioning machinery for protein biosynthesis. The cell, on the other hand, is truly *self*-begetting. It does more than replicate DNA—it also performs all the functions required for protein biosynthesis, including, especially, all of energy metabolism. The cell is the simplest integron deserving to be called uroboric.

The urcells, described in Chapter 13, arose spontaneously, through the self-assembly of lipidlike protenoids into microspheres. The protenoids arose in the thermal littoral zones of the nascent earth when its oceans and continental land masses began to take form. Neither energy metabolism nor polymer biosynthesis took place inside the earliest urcells. These functions, however, evolved inside the urcells until both primitive energy metabolism and primitive gene-directed protein biosynthesis were established to the degree described in Chapters 13 and 14. Such an urcell was the simplest true uroboros. Over the course of approximately 1 billion years (at least several hundred million) the primitive uroboros developed a sufficiently sophisticated gene-directed protein biosynthesis machinery to be called a contemporary uroboros, as was discussed in Chapter 10. During this evolution, the protein biosynthesis machinery was enlarged to include tRNAs, ribosomes, amino acyl tRNA synthetases, and polymerase complexes. Therefore, it is the transition from urcell to uroboros, about 4 billion years ago, that corresponds with the origin of life. At this time, the first true uroboric integron arose on the primitive earth.

The nascent earth began to develop oceans and continental land masses 4.3 billion years ago. It was a time in which the earth cooled down from a glowing orb, enough for liquid H_2O to accumulate on its surface. As the amount of accumulated H_2O grew, an extensive littoral zone emerged throughout the earth's surface. The abiogenesis of organic and biological small molecules and monomers was at its zenith during this period. Ultraviolet sunlight bathed the earth because the present ultraviolet-filtering, ozone-layered atmosphere would not have evolved for nearly 2 billion more years, along with the evolution of aerobic microbes. The atmosphere was a "reducing" atmosphere, rather than "oxidizing," as it is today because it did not yet contain any oxygen. Volcanoes erupted, and associated with these eruptions, as well as with other types of storms, there were abundant lightning discharges. The expanding littoral zone provided innumerable niches in which hot-dry domains and aqueous domains were juxtaposed. Several types of energy flux were manifest, and many of the small molecules and monomers described in Chapter 1 became abundant. As these substances became concentrated in niches, oligomeric and polymeric molecules were abiogenically produced by dehydration condensation, especially in hot-dry littoral domains washed by ocean tides and rains. Among the products were true coenzymes and proteinoids. Lipidlike protenoids, which found their way into aqueous domains, spontaneously *self-assembled* into microspherical membranes of micrometer radius. The mechanism of self-assembly in proteins and polynucleotides was described in Chapter 6.

The microspheres contained aqueous mixtures of small molecules, monomers, and oligomers. According to the discussion in Chapter 13, these constituents provide the possibility of primitive energy metabolism. With an urquinone precursor to ubiquinone present, the microspheres can couple internal and external redox reactions. Internal NADH can couple to external ferric compounds, thereby *transducing* redox energy into transmembrane, chemiosmotic, proton energy. The redox coupling, energy transductions, and chemiosmotic mechanisms were described in Chapters 4, 5, and 12, respectively. In Chapter 13, it was shown how these processes are coupled together in urcells; it was also shown how the urcells would then be able to use their transmembrane proton potential $[-2.3(RT/F)\ \Delta\mathrm{pH}]$ to transport molecules, especially phosphate, from outside to inside. Once inside, the phosphate would be used to generate polyphosphate energy, mostly as pyrophosphate. This transduction is achieved by coupling the phosphorylation step to an oxidation step that generates NADH. This coupling regenerated the NADH used to drive the transmembrane proton potential. As already pointed out in Chapter 13, this is the original "oxidative phosphorylation."

The mechanism for this transduction of redox energy into pyrophosphate energy requires a sulfur proteinoid. Because cysteine is a sulfur-containing amino acid, proteinoids would have possessed sulfur residues. The sulfur acts as the site for the molecular events concomitant with energy transduction. More sophisticated, modern uses of sulfur by organisms were described in Chapters 5, 6, and

11. The coupling of redox energy to phosphorylation also requires a carbohydrate carrier, glyceraldehyde-3-phosphate. The mechanism of phosphate phosphorylation into pyrophosphate uses an active phosphate carrier, 1,3-diphosphosglycerate, a carboxyl phosphate. Glyceraldehyde-3-phosphate is abiogenically generated. The presence of ferric iron compounds as the electron acceptors outside the urcells foretells the eventual, evolutionary, incorporation of iron into the electron transport chains of contemporary mitochondria and chloroplasts.

The urcells could occupy strata of the environment in which they and their constituents are protected from violent energy fluxes, such as ultraviolet light and lightning. Many of their constituents are rich in free energy relative to full equilibrium at the ambient temperature and pressure of the stratum occupied. A monotonic decrease in Gibbs free energy will ensue, according to Chapter 3. This tendency is what gives rise to the redox reactions and their coupling, by quinone, across the urcell membrane. In turn, this energy flux drives phosphorylation, which provides novel opportunities for energy transductions. Amino acyl carboxyl phosphates and nucleotide triphosphates are generated by phosphate transfer from pyrophosphate. Thus, the redox energy flux is ultimately transduced into the production of activated monomers, poised for polymerization. Inside the urcells, polynucleotides arise as a natural consequence of phosphorylation of nucleoside monomers.

Eventually, urcellular energy transduction will lead to urcells containing significant quantities of RNA molecules. Amino acyl carboxyl phosphates will begin to esterify the 2′-hydroxyl groups of RNA ribose. The mechanism of primitive (RNA) gene-directed protein biosynthesis, described in Chapter 14, will begin functioning. Through catalytic feedback effects, the proteins produced will confer either selective advantage or disadvantage on the RNA urgenes coding for the proteins. The urcell will have become an uroboros when it begins to accumulate efficacious proteins catalyzing steps in energy metabolism and steps in polymer biosynthesis. These steps began probably about 4 billion years ago. By 3.5 billion years ago, fossils were beginning to form from what are believed to have been primitive anaerobic bacteria, replete with a contemporary gene-directed protein biosynthesis machinery containing tRNAs, ribosomes, amino acyl-tRNA synthetases, and polymerase systems for RNA and DNA. The primitive uroboros only required a single polynucleotide, the RNA genome, which served simultaneously as urgene, messenger, and synthetase. There is no way known by which these molecular ingredients could have been preserved in fossils. The evolution from the primitive uroboros [see (14.3)] to the contemporary uroboros [see (10.55)] may have taken several hundred million years. Evolution has accelerated since then, and the evolution of all megascopic eukaryotes occurred within only the last billion years of the earth's history. Air-breathing, land-living vertebrates, such as *Eryops* and *Seymouria*, first evolved only about 300 million years ago.

Uroboric urcells became so plentiful that natural sources of abiogenically produced molecular constituents and energy-rich compounds became scarce. About 3.3 billion years ago sunlight-transducing organisms evolved and thereby eliminated the dependence of life on earth-bound energy sources. In addition, organisms had by then accumulated enough genes to support most of the energy metabolism pathways discussed in Chapter 8, as well as associated pathways for monomer synthesis from simpler precursors. These achievements reduced the dependence of organisms on ready-made molecules in the environment.

This is really the whole point of uroboric function: from certain very basic molecular building blocks, which are externally supplied, an external source of energy, such as sunlight, flows through the system and drives *the formation of uroboric macromolecular assemblies from the building blocks*. The impetus for polymer synthesis is the energy flux, and the uroboric character is a consequence of intrinsic, emergent properties of self-assembled aggregates of polymers. The catalytic function of proteins endows these aggregated assemblies with the feedback function that *closes the loop structure* of the energy flow pattern, as depicted in (10.54).

Once energy transduction has commenced, a stage is reached in which the consequences of this energy transduction are dictated by the second law of thermodynamics, as expressed in terms of the Gibbs free energy back in Chapter 3. For example, when urcells have begun to produce pyrophosphate through the transduction of redox energy available from the environment, this pyrophosphate stimulates a variety of reaction sequences, including particularly polymer biosynthesis, each of which satisfies $\Delta G < 0$. The direction of the reactions involved is governed by the second law in this way. Nevertheless, the competing processes that arise as pyrophosphate is utilized possess different rates so that the primitive uroboros can bias the outcome through the catalytic feedback effects of its protein products. Evolution proceeds as a selective enhancement of one outcome for pyrophosphate energy utilization over another outcome by virtue of selective rate enhancement. The impetus for this evolution is, nevertheless, the existence of pyrophosphate energy. The proteins are the agents by which the decrease in the Gibbs free enrgy of pyrophosphate is manifested by the uroboros. Their properties as catalysts and their ability to self-assemble may be viewed as extraordinary properties, but they are properties that obey the dictates of both physics and chemistry, and they are in harmony with the second law of thermodynamics. These properties emerge in proteins from the intrinsic properties of amino acids, which in turn arise from the intrinsic properties of the elements H, C, N, O, and S.

In summary, it has been observed that the difficulty envisaged in Chapter 13 of initiating a primitive uroboros is removed when the evolution of urcellular energy transductions is studied. The natural flow and transduction of energy by self-assembled microspheres appear to inevitably yield polymer synthesis and finally primitive gene-directed protein biosynthesis. The second law, appropriately modified for application to the Gibbs free energy, drives the free-energy-rich configurations toward equilibrium. During the transition, feedback intermediates may be produced, enabling the system to couple more strongly to available energy sources. Proteins and polynucleotides are examples of feedback intermediates that possess this property. As molecules, polynucleotides and proteins possess intrinsic properties, not possessed by their monomeric constituents, that *emerge* as these polymers are synthesized. All of this is somehow inherent in the elements, H, C, N, O, P, S, and Fe, but does not manifest itself until the macromolecular structures arise by virtue of appropriate energy flow and transductions. Can physics *explain* these *emergent* properties?

It is hoped that the preceding analysis would stimulate experimentally inclined researchers to explore energy-driven systems for their emergent properties, as well as stimulate theoretically inclined researchers to explore the formal description of energy-driven systems that are maintained in contact with temperature and pressure reservoirs. This perspective was suggested in Chapter 14, in which it was indicated that the genetic code may have had its origin in esterification reactions on RNA by *activated* amino acids. Earlier experimental approaches have often only involved tests for the *binding* of *unactivated* amino acids to RNA. In Chapter 3 the rudiments of a mathematical approach were presented in (3.11). No new laws of physics were required. Instead, it is the study of novel emergent properties for energy-driven systems that is required. For both experimental and theoretical approaches it is necessary to take a more holistic point of view, as was emphasized in the discussion of chemiosmosis in Chapter 12. In short, much remains to be learned about the dynamics of biological energy transductions and their evolution.

Reference

1 F. Jacob, "Evolution and Tinkering," *Science*, **196**, 1161–1166 (1977).

This paper provides an excellent example of the conundrums faced if *energy flow ordering* is ignored. For example, Jacob states:

> But the appearance of life on the earth was not the necessary consequence of the presence of certain molecular structures in prebiotic times. In fact, there is absolutely no way of estimating what was the probability for life appearing on earth. It may very well have appeared only once.

He also says:

> During chemical evolution in prebiotic times and at the beginning of biological evolution, all those molecules of which every living being is built had to appear.

It is instructive to study Jacob's paper alongside the preceding three chapters of this book.

PART 4

Epilogue

Are there any lessons applicable to the human predicament to be learned from the study of the origin and evolution of biological energy transductions? By "predicament," I am not referring to some malaise in man's psyche, but instead to the nature of human society. We are all aware that there is considerable world-wide malnutrition and starvation in the poorer countries, and serious shortages of energy and raw materials in the industrialized nations. Both of these situations involve, at their fundamental level, *energy transductions*. I believe that the study of biological energy transductions can provide a fresh perspective from which to view society's needs.

In the study of energy metabolism and its control processes, presented in Chapters 8 and 9, it was found that many specific proteins are required as catalysts and allosteric control subunits in enzyme complexes. All of energy metabolism appears to be the *result* of the functional properties of proteins. Consequently, in Chapter 10, we studied the biosynthesis of proteins and the associated genetic mechanism. However, it then became clear that the biosynthesis of both proteins and polynucleotides requires a considerable amount of polyphosphate energy, carried by ATP. If energy metabolism could be said to *result* from the functional properties of proteins, then it could also be said that proteins and polynucleotides *resulted* from the energy latent in ATP. This, of course, gives rise to the "chicken-egg" question: Which came first, polymers or ATP? In Part III of this book, a hypothetical, but also plausible, scenario was described for the evolution of polymer biosynthesis and energy metabolism. The outcome was that primitive energy transductions arise spontaneously, merely as a consequence of the presence of an *energy source* and the second law. Within spontaneously self-assembled microspheres, this primitive energy transduction ultimately gave rise to pyrophosphate energy, which *drove* the synthesis of polymers. It is true that all subsequent evolution is predicted on the functional properties of the polymers produced, but their existence is a *result* of the presence of the energy source, pyrophosphate. If there had been no phosphorus on the primitive earth, then the energy flow would have had to flow another way, and most probably neither proteins nor polynucleotides would have been byproducts.

The analogy between this description and a description of energy in human societies has, I feel, much validity. The key to this analogy is to view individual human beings as the analogs of proteins. The intent of this identification is to emphasize the nature of the functional role played by human beings in managing energy for societal needs. The functional role is very clearly "catalytic," like a protein's. Human beings are not as restrictively functional as proteins, at least not always. A protein performs one specific catalytic function. A human being is flexible and can do many. A protein may have a *coenzyme* or *prosthetic group* that possesses the true catalytic function. Likewise, a human being will use *tools*. A protein catalyst is regenerated by the reaction it catalyzes, although, during the reaction, modified enzyme intermediates may be involved. While working, a factotum will manipulate tools and will make direct contact with the object of his labor, but in the end, after a material change has been achieved, he will be the same person who began, and no part of him will have ended up as part of the worked upon objects. Persons perform the catalytic and control functions of society, and in this sense they may be construed as the analog of proteins *vis a vis* biological energy metabolism.

Just as the study of biological energy metabolism *at first* implied that energy metabolism *resulted* from the functional capabilities of proteins, so it is that many persons believe that the energy flows of contemporary societies are a *result* of human functions. Electricity, gasoline, and nuclear power each requires countless human acts in order for them to exist. This is indisputable. Nevertheless, mankind has tended insufficiently to appreciate the extent to which the available energy sources have *determined* the evolution and structure of our societies. The lesson from the study of the origin and evolution of biological energy metabolism is that energy itself is the fundamental impetus for all subsequent structural evolution. The emergence of pyrophosphate energy as a result of primitive urcellular energy transduction made polymers possible. The structure of the energy source (pyrophosphate) determines the mechanism of the energy transduction (monomer activation by phosphorylation) and subsequently fixes the structure of the product (polypeptide and polynucleotide polymers). The catalytic and regulatory functions of the product are determined by its structure. These ideas are the basis of another "chicken–egg" question: Which is more fundamental, structure or function? Looking only at the *products* of energy transduction, which in the biological case are the polymers, the usually stated view, especially in biochemistry texts, is that structure determines function. From the evolutionary viewpoint, however, the structure is a consequence of the nature of the energy source that is required to drive the synthesis of the structures. Applied to societies, these ideas support the view that electricity, when it became available, transformed society in accord with the variety of ways in which it is suitably used. Electric lighting, the telephone, and radio greatly transformed modes of communication. Petroleum products, including gasoline, asphalt, and synthetic rubber, revolutionized transportation. Transportation advances, in turn, revolutionized agricultural efficiency and foodstuff distribution. A half century earlier, steam engine power made similar changes on society. The purpose here is not to document the validity of this viewpoint, but to enunciate it. It is also easily seen that the nature of the energy source plays a determining role in the structure of the transducer. For example, imagine trying to use gasoline to achieve the successes in television tube function that electricity achieves. On the other hand, try to manufacture plastics from electricity. Each energy type and material is suited for a particular purpose.

One requirement for polymer synthesis is a supply of monomers. Without them, pyrophosphate (or ATP) is of no utility. In primitive times, these raw materials were abiogenically generated and were abundant during the evolution of the primitive uroboros. Similarly, the dawn of human societies was accompanied by the availability of food. If people are the analogs of proteins, then food is the analog of amino acids. Food appears to have been most abundant in the great river valleys in which early societies arose. These river valleys were well suited for this eventuality because they provided adequate moisture for plant life, and irrigation appears to have arisen quite early in these societies. The earliest energy source of significance appears to have been the heat of fire. With the advent of metallurgy, also rather early in the history of societies, *heat* transformed agriculture and warfare. Just as the emergence of iron weapons changed the fate of warriors, many hundreds of years later gunpowder did likewise, causing a major redistribution of populations. Catching wind energy with sails transformed world commerce. You would not use wind to power your TV set, yet it is infinitely superior to electricity when it comes to propelling sailing ships. Periodically, the course of history was altered more by the availability of food (the potato famine) than by energy. Nevertheless, repeatedly the course of events may be viewed, in the long run, to have resulted more from the nature of a newly available energy source and its intrinsic utilizabilities than from the conscious, prior intervention of man. Mankind reacts to and greatly diversifies the utilization of energy sources as they arise, but this is predicated on the availability and nature of the energy type.

The analogy between the cellular metabolic activity and human society, argued for above, has included several analogs: persons-proteins, foodstuff-amino acids, energy-polyphosphate (PP or ATP), and control-allostery. For the sake of the analogy, it may be added that the printed word is the analog of the polynucleotide gene. The gene directs the synthesis of protein and provides a mechanism for maintaining a transmittable genetic heritage. The printed word enables mankind to provide instructions for tool making and tool using, and it also may be transcribed for preservation for posterity. It is as mutable as DNA, indeed more so. Thus, the printed message can be modified and improved.

It would be most desirable to be able to write that the analogy runs even deeper, including more advanced regulatory processes in society to match the regulatory and safeguard capabilities of cells, and the ability to free ourselves from finitely available earth-bound energy sources through the utilization of solar energy. Organisms conquered both these problems *billions* of years ago. Mankind has used coal, which is, in a special sense, of solar energy origin, but this is not a direct use of solar energy. Moreover, some of the worst episodes in societal malfunction in the realm of controls and safeguards have been engendered by coal, at least before 1940. A more direct use of solar energy has been the burning of wood. However, England once burned up *all of its trees*. It is only in this century that mankind has learned (?) to use trees in a managed, *renewable* way. Another almost direct form of solar energy is hydroelectricity, which is coupled to solar energy by the water cycle described in Chapter 13. Dams, however, cause serious damage to rivers, and the number required by modern society to support even only 20% of its energy needs would choke all the rivers suitable for building dynamos. We are on the brink of finally solving the solar energy problem the way organisms did long ago, that is, by making solar voltaic converters. The cellular solar converters described in Chapters 8, 11, and 13 convert sunlight into cyclic electron flow or a transmembrane potential for protons. Photovoltaic devices, such as silicon devices, achieve a similar kind of function.

History has come to the point in time when society must confront its energy management and generation requirements in a deadly serious way. The evolution of cellular life confronted similar problems and overcame them. Perhaps this foretells a parallel success for mankind. At the very least, it will not do any harm if a more vigorous study of biological energy transduction and its evolution is pursued in the meantime.

Index